Beyond the Tragedy in Global Fisheries

Politics, Science, and the Environment
Peter M. Haas and Sheila Jasanoff, series editors

A complete list of the series appears at the back of the book.

Beyond the Tragedy in Global Fisheries

D. G. Webster

The MIT Press
Cambridge, Massachusetts
London, England

First MIT Press paperback edition, 2017

Set in Stone by the MIT Press.

Library of Congress Cataloging-in-Publication Data

Webster, D. G., 1975-
Beyond the tragedy in global fisheries / D. G. Webster.
pages cm.—(Politics, science, and the environment)
Includes bibliographical references and index.
ISBN 978-0-262-02955-1 (hc. : alk. paper), 978-0-262-53473-4 (pb.)
1. Fishery policy—History. 2. Sustainable fisheries. 3. Fisheries—Economic aspects. 4. Fishery management, International. I. Title.
SH328.W433 2015
333.95′6—dc23

2015011434

For Granny Lu

Contents

Series Foreword

As our understanding of environmental threats deepens and broadens, it is increasingly clear that many environmental issues cannot be simply understood, analyzed, or acted upon. The multifaceted relationships between human beings, social and political institutions, and the physical environment in which they are situated extend across disciplinary as well as geopolitical confines, and cannot be analyzed or resolved in isolation.

The purpose of this series is to address the increasingly complex questions of how societies come to understand, confront, and cope with both the sources and the manifestations of present and potential environmental threats. Works in the series may focus on matters political, scientific, technical, social, or economic. What they share is attention to the intertwined roles of politics, science, and technology in the recognition, framing, analysis, and management of environmentally related contemporary issues, and a manifest relevance to the increasingly difficult problems of identifying and forging environmentally sound public policy.

Peter M. Haas
Sheila Jasanoff

Series Foreword

Acknowledgments

Many people contributed to the successful completion of this book. First, I would like to thank Elizabeth DeSombre and three anonymous reviewers for their feedback, which helped me to articulate the complexities of the AC/SC framework much more clearly. Second, I would like to thank Dale Squires for introducing me to his work on skipper skill and other factors that reduce costs of production. His ideas and those of his co-authors are foundational to the profit disconnect. Third, I would like to thank Peter Haas and Sheila Jasanoff for including this book in their excellent series. Fourth, I would like to thank Bill Bailiff for helping me find historical information about the Inter-American Tropical Tuna commission, and Jennifer Baily, Isao Sakaguchi, Patrice Guillotreau, Ramon Jimenez Toribio, and Laure Marcellesi for their help locating and translating information in non-English sources and for reviewing specific sections of the manuscript dealing with their regions of expertise. I am also grateful to Jonathan Chipman, Director of the Citrin Family GIS/Applied Spatial Analysis Laboratory at Dartmouth College, who generated the excellent maps and cartograms found in chapter 1, and Jay Satterfield, Dartmouth's Special Collections librarian, who helped to locate many of the historic images found throughout the text. I greatly appreciate the editorial assistance and insights of Alan Berolzheimer and Beth Clevenger. I would also like to thank my indefatigable research assistant Marissa Grecco for her assistance locating and keeping track of the hundreds of sources that I used in the research for this book and for help with copyediting. Finally, many thanks to my family, friends, and colleagues who supported and encouraged me through the years of research and writing that went into this book.

Research on the Inter-American Tropical Tuna Commission and related economic factors was funded in part by an NSF Coupled Natural and Human Systems grant called Fishscape: Complex Dynamics of the Eastern Pacific Tuna Fishery (CNH-1010280). Additional financial support was provided by the Walter and Constance Burke Research Initiation Awards for Junior Faculty at Dartmouth College.

Acknowledgments

1 Introduction

The oceans are heavily overfished. There is some argument among scientists, observers, and stakeholders over the degree of overfishing, but all agree that current practices are unsustainable (Ludwig, Hilborn, and Waters 1993). Some barriers to sustainability are technical. Marine ecosystems are complex and changes in fish population dynamics are hard to predict, so it is difficult to know exactly how management measures will affect fish stocks. However, the greatest challenges in fisheries governance are economic and political (Barkin and DeSombre 2013).

This book describes how the political economy of fisheries evolved from early times to today and then highlights key patterns and leverage points that are linked to sustainable transitions in specific fisheries. The analysis is grounded in the concept of *responsive governance*. Governance is responsive when actors and decision makers respond to environmental problems rather than trying to prevent them proactively. Furthermore, because fisheries are complex social-ecological systems, synthesis across fields and disciplines is important, and the analysis extends well beyond the infamous "tragedy of the commons" to the wealth of research on fisheries governance and on social-ecological systems more broadly. This interdisciplinary lens reveals that, over the long run, fisheries governance cycles between periods of ineffective and effective governance. I call this pattern the *management treadmill*. The historiography presented here demonstrates how endogenous and exogenous forces can either slow down or speed up the cycles of economic expansion and political response that shape the management treadmill for specific fisheries.

Throughout the literature, unregulated marine fisheries are consistently treated as tragedies of the commons. G. Hardin (1968) introduced this term in the context of pasturage on common land in medieval England. He proposed that as long as access to common land was available to all, none would have an incentive to conserve the resource because any pasturage not used by one individual could be taken by another. People would therefore place too many cattle on the common pasture, and overgrazing would result. As the grass on the commons disappeared, the number of animals that could be pastured there would decline substantially. Thus, "tragically," the number of animals in the village would be reduced well below the level that could be maintained if the common land was managed collectively or even turned over to private owners. This story struck a chord with researchers studying the use of natural resources. Fisheries experts in particular recognized that Hardin's tragedy of the commons fit well with what Graham (1943, 14) described as "the general tendency of free fishing to become unprofitable" and with Gordon's (1954) work showing that the level of effort in a commons fishery would be much greater than the economic or biological optimum. This realization quickly made Hardin's tragedy the primary focus of social science research on fisheries, and it is still central to academic inquiry and policy making today.

Although most authors accept Hardin's underlying logic, many argue that his conclusions are skewed because his analysis is apolitical. The need to consider governance is well established in the extensive literature on collective action epitomized by Ostrom (1990). This critique can be illustrated by comparing Hardin's pasturage metaphor with the reality of commons management in medieval England. As Buck ([1985] 2010) established through extensive historical research, there were in fact many rules and norms that governed access to the commons in the Middle Ages. Furthermore, observed overgrazing in the period was actually driven by economic development, technological change, and the resultant appropriation of the commons by powerful outsider elites who had little interest in maintaining the resource, rather than by competition among local users. In fact, the enclosure of common lands in England and the related Highland clearance in Scotland ushered in the greatest period of environmental degradation in Britain's history—a tragedy not of the commons but rather of power disparities, myopic decision making, and the externalization of environmental costs (Fraser 2007). These problems also affect fisheries. Indeed, the pattern described for cattle in medieval England will sound familiar to anyone who has studied collective action in a fisheries context. In case after case, longstanding, sustainable institutions disintegrate in the face of new markets, new technologies, and powerful new entrants (Munro and Bjørndal 1999).

Given this, the tragedy of the commons is important but insufficient as an explanation for fisheries overexploitation globally. The nature of the tragedy—indeed, the potential to avert or escalate overexploitation—depends on many other factors besides competition under open access.

Hardin himself recognized the importance of the size of the human population relative to the resource base as a determinant of the scale of resource exploitation. In fact, his 1968 paper is actually an argument for curtailing population growth in the face of global environmental limits; calls to restrict access to specific commons resources were ancillary. Similarly, economic growth, globalization, and technological change can worsen the tragedy of the commons by increasing individual-level incentives to use the common resource (Young, Lambin, et al. 2006). On the other hand, the tragedy can be amplified politically when structures of power favor resource appropriation by a few individuals rather than collective management by communities that are dependent on the resource. This was true in the specific case of the medieval commons described above; it is also well established as a persistent problem in both private and common property regimes. Many authors also point out that negative externalities, defined as costs not included in the price of a good or a service, can increase overexploitation regardless of ownership structure (Daly 1987, 1996; Berkes et al. 2006).

On a more positive note, although Hardin advocated coercive approaches in his 1968 essay, there are more collaborative governance structures that can mitigate the tragedy of the commons. Cooperation occurs in many commons systems and is often spurred by the social-psychological characteristics of individuals such as innate cooperativeness and deeply held beliefs regarding fairness and equity (Schindler 2012). As Acheson (1997) established for lobster fisheries, through experience resource users can internalize a "conservation ethic" that ensures their adherence to conservation-based rules and norms that are perceived to be beneficial. There are many other examples of minimally coercive solutions to the tragedy of the commons in the literature on social practice. Some even argue that social practice has an effect on international management of the global commons by sovereign states (Young 2001). Moreover, research on the legitimacy of environmental governance shows that coercive methods of solving the tragedy of the commons can be counterproductive, destroying social norms and psychological attributes that support cooperative approaches (Hawkshaw, Hawkshaw, and Sumaila 2012).

The analysis presented in this book combines the insights described above, including the tragedy of the commons and its many extensions and exceptions, with the concept of responsive governance. Webster (2009) defines responsive governance as a trial-and-error process in which decision makers and other actors respond to problem signals by first applying the most expedient measures and then gradually ratcheting up their response if problem signals persist or intensify. Thus, rather than expecting proactive governance that ensures sustainable resource use, we can expect that action will occur only in response to signals of overuse such as biological depletion or economic recession. Although this runs counter to the common assumption of perfect rationality in economics and in public choice theory, it is well supported in several other literatures, including a wide array of research in political science, policy and organizations studies, and economic psychology.[1]

Figure 1.1 shows how responsive governance might work in a tragedy-of-the-commons situation. As long as management measures are weak and ineffective, the resource becomes less plentiful, which generates economic losses because of increasing marginal costs of production and other dynamics described below. This, in turn, increases political concern that amplifies pressure on decision makers to implement the most expedient management measures. If these measures are insufficient, the cycle continues, political will grows, and decision makers will try more and more costly measures, which may also be more effective. This cycling continues until there is a switch to an effective cycle of biological rebuilding and economic rebound, as depicted on the right-hand side of the figure. It is also possible that switching does not occur in time to prevent system collapse (the "too little, too late" scenario). When switching occurs early enough, management dampens exploitation, which allows the stock to rebuild, reducing costs and increasing profits. Ultimately, economic rebound reduces the political will to maintain effective management.

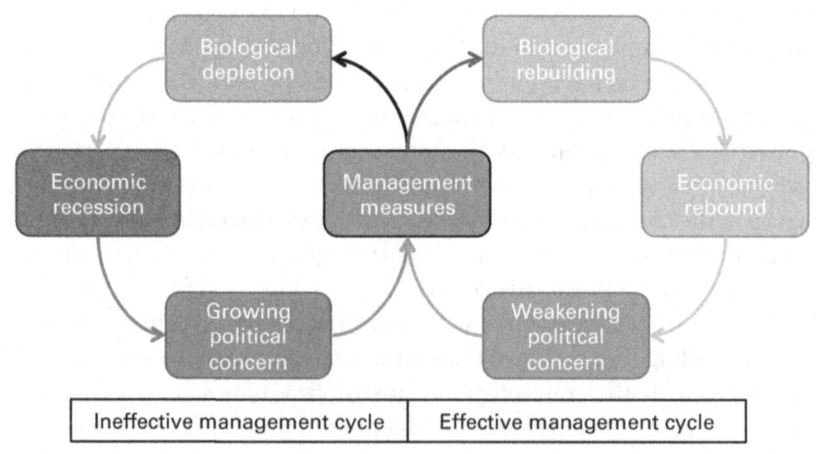

Figure 1.1
Ineffective and Effective Cycles in Responsive Governance.

Webster tested the theory of responsive governance using a fairly simple vulnerability response framework to understand variations in the international management of tuna and similar species in the Atlantic Ocean. Although Webster found much support for the concept generally, the process of responsive governance proved to be more complicated than the simple feedback loops depicted in figure 1.1. Indeed, though it was generally predictive, the vulnerability response framework did not work well in cases where endogenous or exogenous factors altered management cycles, or where non-commercial interest groups wielded significant influence. The vulnerability response framework also was highly specialized to international fisheries management and did not allow for heterogeneity of fishing interests within states. The Action Cycle/Structural Context (AC/SC) framework presented in section 1.2 below allows for a more holistic approach to understanding responsive governance. It has the tragedy of the commons at its core, but it incorporates multiple actors, a wide array of private and public responses, and both endogenous and exogenous signal disrupters.

Exogenous factors are particularly important when one is considering the current overexploited state of many of the world's fisheries. As will be described in greater detail in section 1.3, fundamental theories in resource economics such as those described by Clark (2005) and Conrad (2010) posit that marginal costs should increase with stock decline, reducing net revenues[2] and thereby limiting entry into a fishery at the point where total revenue equals total cost under open access. However, a growing body of theoretical and empirical work shows that bioeconomic factors, including the schooling behavior of fish (Bjørndal and Conrad 1987) and technical innovation by fishers (Hannesson, Salvanes, and Squires 2010) can negate the increasing costs associated with declining stock size, dampening economic signals of biological problems. When a fishery is large relative to a market, all else equal, an increase in supply associated with overfishing can also be expected to result in a decline in price, which should reduce normal profits, limiting total effort and related overexploitation. This price effect is often omitted from bioeconomic models, but empirical work shows that investment in cost-cutting technologies and in market expansion through advertising can be used to counter the profit effects of declining prices associated with increasing supply in a tragedy-of-the-commons situation (see, e.g., Campling [2012] on the tuna commodity frontier).

Profit disconnects like those described above occur whenever bioeconomic signals drive resource exploitation above sustainable levels.[3] This is different from the tragedy of the commons, in which resource users ignore economic signals because of collective action problems. However, a wider profit disconnect does result in a higher "equilibrium" level of effort and so can exacerbate the CPR dynamic. Even without the tragedy of the commons, a profit disconnect can occur when there is a gap between the privately efficient level of production and the sustainable level of production because of market failures such as negative externalities.[4] Another profit disconnect is called temporal myopia, or the more extreme hyperbolic discounting, in which decisions are made based on short-run profits and future revenue flows are heavily discounted. Although the term "profit disconnect" is not used, all of these problems are already well known in economics but are mainly considered under equilibrium conditions where the *ceterus paribus* or all else equal assumption holds. However, there are many factors that constantly alter economic conditions in fisheries, so it is necessary to consider changes in the signals received by fishers when *ceterus paribus* fails and equilibrium shifts are common and can be continuous.

As an overarching concept, the profit disconnect allows for this type of dynamic analysis. The wider the profit disconnect, the longer users can continue to profit while overexploiting the resource. When the profit disconnect is increasing—that is, when the economic "equilibrium" level of production is being pushed further away from the sustainable level of production—the biological depletion depicted on the left-hand side of figure 1.1 is occurring, but it is not matched by corresponding economic recession, because profits do not diminish (and may increase) in spite of declining biomass and expanding supply. Without the economic costs of overexploitation, political will to engage in effective management remains weak and switching from ineffective to effective cycles is not expected. In contrast, if the profit disconnect is stable or is shrinking, then economic recession will occur sooner, strengthening political will to improve management more quickly.

The above analysis of the profit disconnect assumes homogeneity among fishers, but in reality actors are seldom equal. Thus, the effects of the profit disconnect usually are filtered through political systems and can be skewed by political factors. Indeed, a *power disconnect* occurs in responsive governance when the individuals or groups that experience the costs of overexploitation are politically marginalized. When power rests with a group of actors that is also insulated from problem signals by the profit disconnect, it will further delay switching from ineffective to effective governance, because the political signals received by decision makers will favor continued high levels of fishing effort. In such cases, the politically powerful segment of the industry continues to profit as a result of the lack of effective regulation, while others languish.[5] Because of the power disconnect, it is important to understand both the economic and the political structure of a fishery. Industry groups are economically heterogeneous and receive different problem signals depending on factors like the size of their operations, the availability of alternative sources of revenue, competitive pressures from other producers, consumer demand, and marketing structures, as well as management policies and governance institutions. Noncommercial interests usually respond to biological rather than economic problem signals and so may develop political will earlier than commercial interest groups. Which groups ultimately influence decision makers depends on a wide array of factors, but political power and the ability to form coalitions are important in almost any governance system.

Though signal disconnects are important, in responsive governance even weak signals may evoke successful and relatively early response when effective solutions are politically, economically, and technically expedient. However, solutions to environmental problems are usually politically difficult, economically costly, and technically challenging (Young 1999). This is particularly true for marine systems, which are highly volatile and complex. The tragedy of the commons does play a role, as it is typically more difficult to manage open-access resources than private property. However, scientific knowledge, management technologies, civil society institutions, and the general efficacy of government bodies are also major determinants of response options. The distribution of power also affects the set of politically feasible solutions. Effective solutions are easier to implement when the costs are borne by marginalized populations. This creates practical and normative tensions. Political empowerment is an important element of environmental justice and can increase the strength of problem signals; however, empowerment restricts the set of solutions available to decision makers.

Over the long run, responsive governance can produce several different outcomes, but the management treadmill is most common. When relatively effective response occurs before biological collapse or ecological shifts, renewable resources may rebuild to sustainable levels. As the problem signal dissipates, the political pressure to engage in management declines, and a gradual return to overexploitation can be expected unless institutions are in place to prevent it. The historical record shows that this oscillation between effective and ineffective management occurs in many of the world's fisheries. However, this *crisis rebound effect* is not the only driver of the management treadmill. All the factors that dampen problem signals, particularly those associated with the profit disconnect or the power disconnect, can also "reset" the management treadmill by changing the political economy of response. In a small percentage of cases, a reset has a positive influence on resources, as when changes in environmental conditions increase biological productivity. However, in the vast majority of cases, management becomes more difficult as endogenous and exogenous factors widen the profit disconnect, weakening political resolve and increasing incentives to ignore regulations. This is a fundamental problem that cannot be explained by the tragedy of the commons alone—but it can be understood through the responsive governance lens.

This book traces the evolution of the governance of marine fisheries from early times to the present, showing how responsive governance works—or fails to work—in settings ranging from small-scale coastal fishing communities to international fisheries that span entire oceans. This introductory chapter provides a general overview of the state of the world's fisheries before describing the AC/SC framework and methods used in the historical analysis. The chapters in the first part of the book examine forces like the profit disconnect that disrupt economic problem signals, document the expansion of fishing effort in scope and scale, and show how the industrialization of fishing created hierarchies within the industry as those with access to capital invested in larger and larger fleets while those without such access struggled to compete in smaller niches. The chapters in Part II explore how governance institutions coevolved with the economics of fisheries expansion. Specifically, they show how the power disconnect increased through history as larger commercial operations with greater economic and political power eclipsed small fishing communities. These chapters also explain how problem signals are processed by decision makers in many different regions and how the set of actors and management solutions changed, ultimately altering the process of responsive governance. Chapter 9 concludes the book with an evaluation of the results presented in the earlier chapters, identifying leverage points that can generate earlier, more effective responses, and calls for greater attention to exogenous forces that increase the speed of the management treadmill.

1.1 An Overview of Global Fisheries

The world's fisheries are important social-ecological systems that provide food, jobs, and economic development to many people. At the same time, fishing substantially alters inland and marine ecosystems in ways that can be difficult to understand. Overfishing can deplete specific species, disrupt ecosystems, and cause related socioeconomic losses as fishers struggle to adapt to declining harvests. By-catch—the incidental catch of non-targeted organisms—can also severely affect marine systems. Both targeted and non-targeted fishing mortality can alter ecosystems, sometimes causing irreversible shifts in the species composition of a particular area. This section provides an overview of the current biological, economic, and legal context for fisheries globally. It also defines various technical terms that will be used throughout the book and serves as a brief primer on fisheries science and management.

1.1.1 Production

Commercial fishing is a vast, lucrative, and highly differentiated industry. In 2010, total reported production of fish and other living marine and aquatic resources reached 148 million tonnes worldwide. The value of this harvest was estimated at US$217.5 billion (FAO 2012d, 3). Fish are also an important source of foreign exchange in developing countries, surpassing many traditional cash crops such as coffee, cocoa, and bananas. About 86.5% of the 2010 harvest was destined for direct human consumption, amounting to 6.5% of the world's consumption of protein from both animal and plant sources.[6] The rest of the harvest was used for consumer products such as cosmetics and pharmaceuticals as well as for industrial and agricultural uses such as animal feed, silage, fertilizer, and landfill. Approximately 54.8 million people earned a living as fishers or fish farmers in 2010; ancillary jobs in areas such as fish processing and distribution, production or maintenance of boats, gear, and other supplies, and research and development, supported between 660 million and 820 million people (10).

Overall production can be divided between capture fisheries and aquaculture and also between inland and marine systems. Figure 1.2 shows the growth in both capture production and aquaculture production in marine and inland waters from 1950 to 2010. For reasons that will be described in detail in Part I, capture production grew rapidly until 1990. Since then, capture production has leveled off while the rate of growth of aquaculture production has increased. Currently, capture production is about 59.7% of total production and aquaculture is 40.3%. Much of the increase in aquaculture occurred in inland waters (69.7% of the 2010 total), but marine aquaculture or mariculture has also grown slowly (30.3% of the 2010 total). The pattern is reversed for capture fisheries, with the majority occurring in marine waters (87.4%) and some slight growth in production from inland waters (12.6% of the 2010 total, as calculated from table 1 on page 5 of FAO 2012d). In the rest of this section and most of the rest of the book, the focus is on marine capture production.

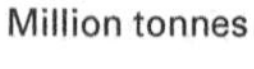

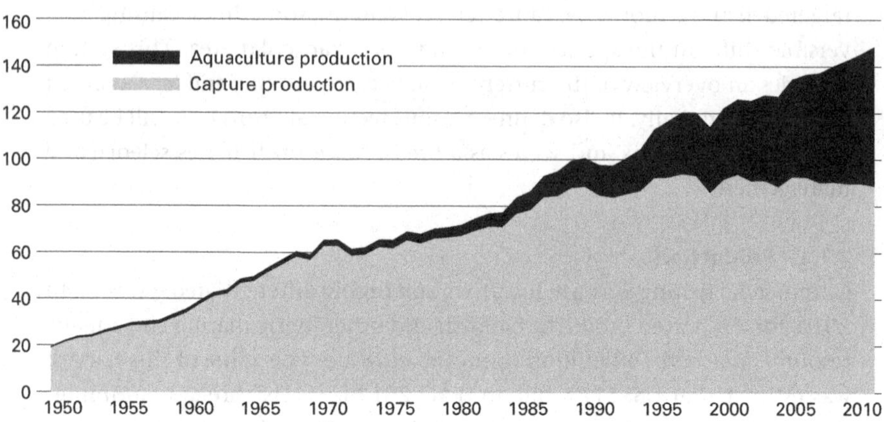

Figure 1.2
Global Fisheries Production 1950–2010. Source: FAO 2012d.

Capture production is the harvesting of wild species of aquatic organisms, including many species of fish but also aquatic plants, mollusks, crustaceans, and mammals. Throughout the rest of this book, I follow the example of the Food and Agriculture Organization of the United Nations (hereafter referred to as the FAO), simplifying the text by using the generic term *fish* to refer to all aquatic organisms harvested in fisheries. The majority of total capture production (84%) is made up of the primary group Pisces, or fish, mainly either cartilaginous fishes (Chordata Chondrichthyes) or bony fishes (Chordata Osteichthyes). Mollusks (phylum Mollusca) and crustaceans (Arthropoda Crustascea) are the second largest component of global landings at 7.4% and 6.8%, respectively. Plants (PlantaeAquaticae) and invertebrates (Invertebrata Aquatica) are 1% and 0.6% of capture production. Mammals (Mammalia), like the combined group of Amphibians (Amphibia) and reptiles (Reptilia), are a miniscule proportion of the total landings (0.0006% and 0.004%, respectively). Although absolute landings of each major group changed historically, these relative percentages are fairly stable over the period for which data are available (1950–2010; FAO 2011).

In contrast, at the species level the composition of landings has changed considerably over the last 60 years. In 1950, Atlantic cod and Atlantic herring were the most important species. At over 41 million tonnes, these two species made up about 22% of total landings. Their dominance waned in the 1970s, however, partly because landings of these stocks declined and partly because landings of other species increased. Now the top three species are anchoveta (Peruvian anchovy), Alaska pollock, and skipjack tuna. Together those species made up only 10% of total landings in 2010. Atlantic herring is now the fourth-largest component of landings and Atlantic cod comes in tenth. In fact, the top ten species in 2010, which included chub mackerel, European pilchard, Japanese anchovy, yellowfin tuna, and the aptly named largehead hairtail, made up only 21.5% of the total harvest, even though landings of these species are about five times as large as the combined landings of Atlantic herring and cod in 1950. Annual harvests of most of the top ten species exceed 1 million tonnes (FAO 2012a).

Thousands of other species are harvested in smaller amounts. Examples range from Atlantic mackerel (887,314 tonnes, 0.99% of global production in 2010) to a deep-water species called snaggletooth (1 tonne, 0.00000011% of global production in 2010). The FAO currently has landings data on 1,480 known species of fish, of which only ten accounted for more than 1% of global production in 2010, 96 for between 0.1 and 1%, and 1,374 for less than 0.1%. The FAO also keeps track of 296 "not elsewhere included" ("nei") groups consisting of fish that cannot be identified at the species level; examples include "marine crabs nei," "various squids nei," and "sardinellas nei." Two categories contain even less information on the type of fish landed: "marine fishes nei," (about 12% of total landings in 2010) and "freshwater fishes nei" (about 7%). In total, about 40% of the world's fish harvest consists of fish not identified at the species level (FAO 2012a).

It is important to note that the FAO dataset used here provides measures of "landings," "production," or "harvests," all of which indicate the proportion of the catch that is actually brought to land and (usually) sold at market.[7] In capture fisheries, catches can be larger than landings because of by-catch. Species by-catch occurs when a non-targeted species is caught incidentally with targeted species because the gear used is not selective. Size-group by-catch occurs when fishers catch fish of the target species that are too big or too small to sell. Size can matter either for legal reasons or because of processor or consumer preferences. Less valuable by-catch is often discarded at sea in a process known as high grading. When discarded, by-catch is often either dead or too weak to survive, and so incidental catches can still contribute to fishing mortality—or the amount of fish killed by fishing activities, which is usually measured in biomass rather than number of fish. Davies et al. (2009) estimate that the volume of global by-catch is equal to 40% of total landings. The composition of this catch is even more varied than that of targeted species, and the resulting fishing mortality can drive non-targeted populations to very low levels.

Even if corrected for by-catch, the FAO data would still underestimate total production because they record only reported landings, and therefore do not include harvests by fleets from countries with limited data-collection capacity or production by illegal, unreported, and unregulated (IUU) fleets, which fish in contravention of international and domestic regulations. At present, countries (technically, "states") have responsibility for management of living marine resources in the water column up to 200 miles off their coastlines. Defined by the United Nations Convention on the Law of the Sea (UNCLOS), this area is called the Exclusive Economic Zone (EEZ). Some states are able to enforce their domestic management programs within their EEZs; others lack the capacity to do so, and there may be considerable illegal and unreported landings in their EEZs. Furthermore, although most commercial fishing boats are privately owned, they are supposed to be regulated by national governments. Thus, according to UNCLOS, a fishing vessel should carry the flag of its home country ("flag state")—the state that is responsible for ensuring that the vessel abides by all international maritime and fisheries laws on the high seas (outside of EEZs). Most of the data on fisheries landings described here were collected by flag states from legal fishing operations and then reported to the FAO. IUU harvests are not included in these datasets. Good data on IUU landings are not available, but it is clear that these fleets can have substantial affects, particularly on high priced fishes[8] such as bluefin tuna, Patagonian toothfish, or orange roughy. Management capacity is also important, as weaker regulations or a lower likelihood of getting caught increase the benefits of fishing illegally (Sumaila, Alder, and Keith 2006).

Figure 1.3 shows the distribution of reported landings by flag state in 2009. It is a cartogram—a map that has been transformed by an algorithm so that the size of each country reflects the relative value of an associated variable. One can see that Asian countries landed the largest portion of the world's harvests, followed by Europe (including the former Soviet bloc), South America, North America (including the Caribbean), Africa, and Oceania. Asia's landings increased rapidly over the last 60 years and now account for about 55% of all capture production in the world. Over the last two decades, most of this growth occurred in China, which currently supplies 17.5% of world fisheries production. Africa also shows slow but steady growth in production over the entire period 1950–2010, but European and North American harvests peaked in the late 1980s, South American landings peaked in the early 1990s, and production by fleets registered in Oceania peaked in the early 2000s (FAO 2012a).

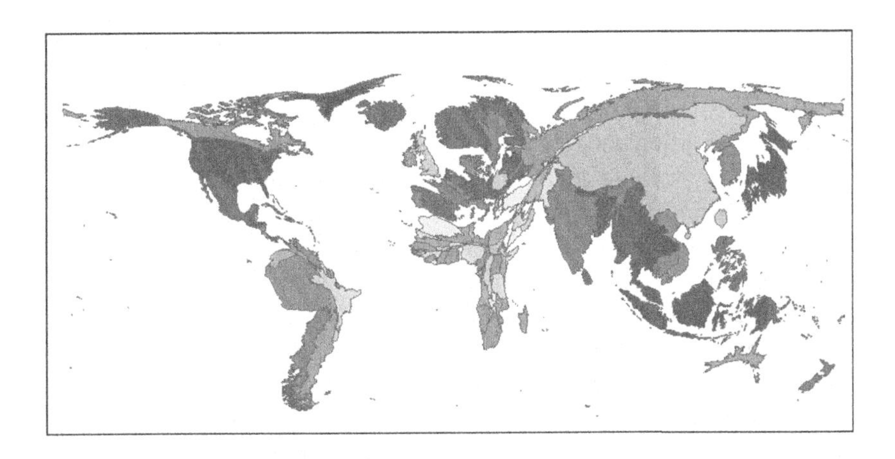

Figure 1.3
Cartogram of global capture production in 2010. Source: FAO 2012a.

At the country level, variation in capture production is substantial. The FAO has data on 240 states and other "fishing entities" (for instance, island groups that are protectorates of former colonial powers). At present, landings by the top ten fishing states account for about 60% of the world's capture production. The top 15 countries harvest about 69% of total production, and the top 25 are responsible for 80%. As has already been noted, China is the largest producer in global capture fisheries. Landings by Chinese fleets rose rapidly from about 3.1 million tonnes in 1980 to 15.5 million tonnes in 1998 and have been fairly steady since. Other top fisheries producers include the EU member states, Indonesia, India, the United States, Peru, Japan, and the Russian Federation. Some countries, including Thailand, Chile, Myanmar, Norway, and the Republic of Korea, land large harvests in some years but not others (FAO 2012a).

In general, production by developing countries increased since 1950. Production by developed countries, which were historically dominant, began to decline around 1990. Though most of the growth in production over the last three decades occurred in developing and emerging economies, even so-called least developed countries have increased production in recent years. As a result of these trends, countries like Canada, Germany, Norway, Spain, and the United Kingdom, which were in the top ten in 1950, were gradually replaced by countries like Chile, Thailand, the Republic of Korea, and Myanmar (each of which was in the top ten for at least a year in the 1990s or the 2000s). Other countries, like Japan, the United

States, and the USSR/Russian Federation managed to stay in the top ten but were eclipsed by China, India, Indonesia, and Peru.

1.1.2 Consumption and Trade

In addition to varying in their levels of capture production, countries around the world vary in their volume of trade in fish products, in their consumption of fish, and in the number of livelihoods supported by the fishing industry. Fisheries are a major source of food, providing 16.6% of the world's protein intake in 2009 (FAO 2010a). Figure 1.4 maps the global distribution of fish consumption for food (kilograms consumed per person per year). In some countries, like Portugal, Spain, Norway, Myanmar, Malaysia, the Republic of Korea, and Japan, per capita fish consumption exceeds 40 kg/person/year. In much of Africa and Central Asia, consumption is less than 10 kg/person/year. Per capita fish consumption tends to be lower in developing countries and higher in developed or emerging economies. This is somewhat surprising, because fish is a more important source of protein in developing countries (19.2% of animal protein intake) and in "least developed" countries (24.0%; FAO 2012d, 5).

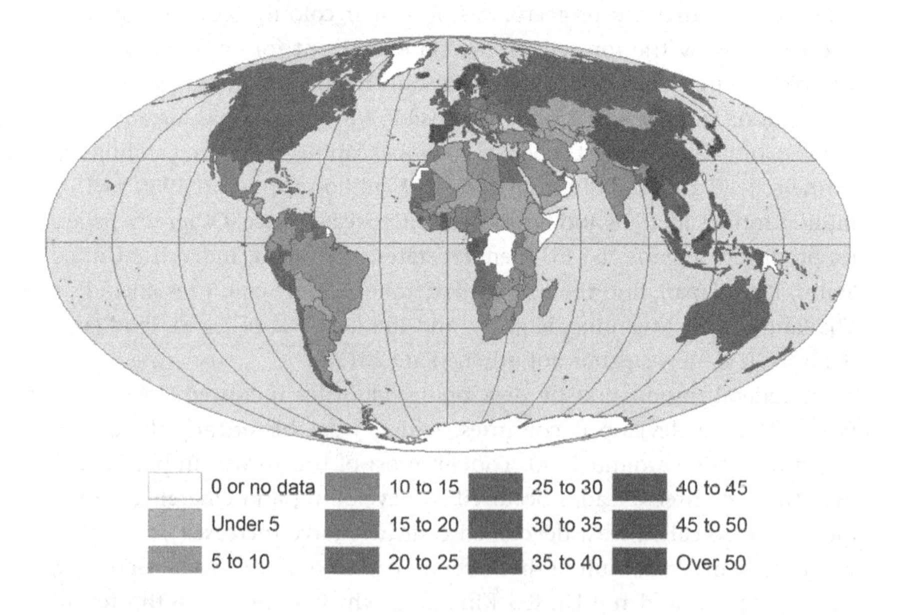

Figure 1.4
Reliance on fish as a source of protein in 2009 (kg/person/year). Source: FAO 2012b.

At the local level, fish consumption depends on geographic and cultural factors, as well as income and access to transportation. In general, people who live in coastal communities are often heavily dependent on fish as a primary source of nourishment, and people who live inland consume less fish. This is particularly true in countries where the transportation infrastructure cannot accommodate fresh fish and in countries where processed fish products are expensive imports rather than cheaply produced domestic goods. Conversely, when coastal communities have greater access to food from other parts of the country or the rest of the world, they can diversify their diets and need not depend so heavily on fish from local waters. Thus, in developed countries more people eat fish but fewer depend on fish for survival, and in many developing countries fewer people eat fish but more depend on fish as a major component of their diet.

About 40.5% of the fish obtained through capture production are consumed fresh or chilled, with little post-capture processing. The remainder is processed into food products (45.9 %)—including canned tuna, breaded fish sticks, and frozen bluefin tuna carcasses that will eventually become sushi or sashimi in high-end restaurants—and non-food products (13.6%), such as animal feed pellets, fish oils, and beauty products (FAO 2012, 13). Processing adds value to fish by either making products more desirable to consumers or allowing landings to be transported to areas where the price for the product is higher than in the region of capture. However, fish are not necessarily processed in the same part of the world in which they are captured. Fleets from the Americas produce more than twice as much raw material as European fleets, but Europe (including the former Soviet bloc) processes much more fish than North America and South America combined. In addition, Asia's output of processed fish products is slightly larger than its contribution to capture production, even though China is globally dominant in both. Africa and Oceania, whose fisheries rank low in capture production, rank even lower in output of processed fish products.

As might be expected from these levels of consumption and these variations in production, trade in fisheries products is also distributed unevenly around the world. Figure 1.5 contains two cartograms, one illustrating the value of imports of processed fish products and the other illustrating the value of exports. It is clear that Asia is the largest exporter of processed fish products by value, followed by Europe, North America, and South America. Africa, the Middle East, and Central Asia export very little processed fish by value. Western Europe imports the most processed fisheries products by value, followed by the United States and Japan. Africa, the Middle East,

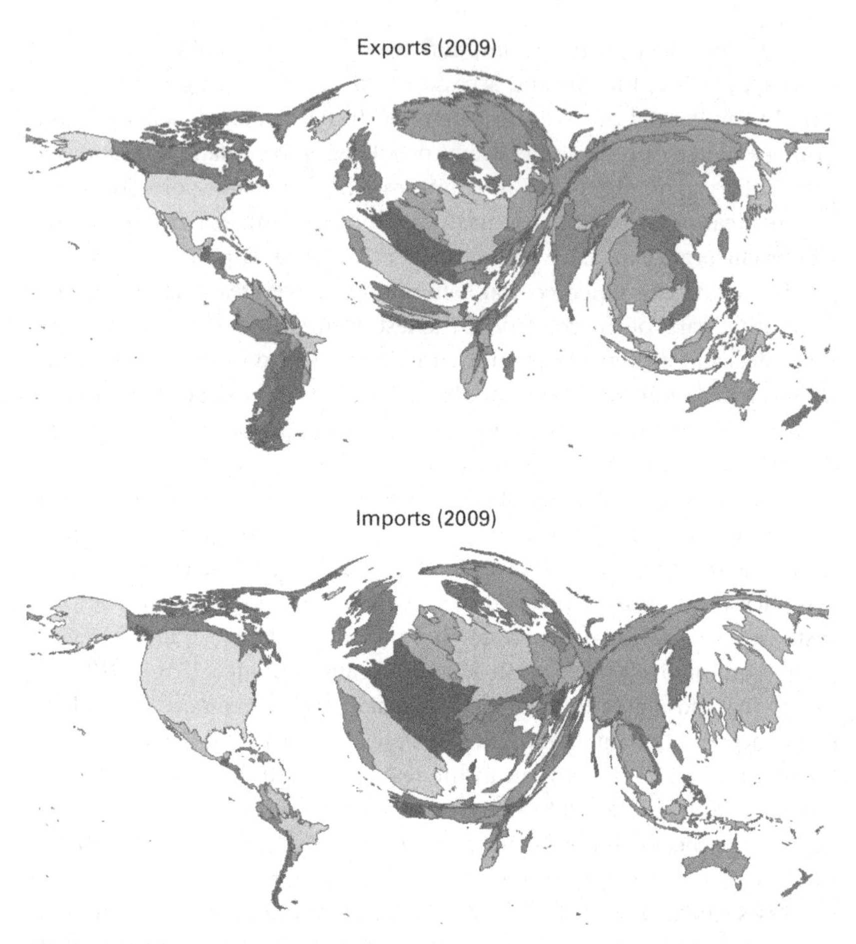

Figure 1.5
Cartograms of value of world trade in fisheries products in 2009. Source: FAO 2012a.

Central Asia, and Central and South America import very little processed fish by value.

These aggregate numbers hide considerable variation. The world's fishing fleets range from small-scale inland and coastal fleets that target a wide range of species to highly specialized distant-water fleets that target the same species throughout the oceans. Furthermore, the fishing industry has many sectors, including boat builders and gear manufacturers, the fishers themselves, fish-processing corporations, product marketers, importers, exporters, wholesalers, and retail distributers of fisheries products. All these sectors provide livelihoods for their workers. Those involved in processing, marketing, and distributing fisheries products can also add value to basic capture production. The economic benefits of fishing are not evenly distributed across these groups, and conflicts often arise among segments of the industry because each has different interests. For instance, processors prefer to keep ex-vessel prices (the amount they pay fishers for their landings) relatively low, whereas fishers would prefer to be paid more for their harvests. Alternatively, fishers using different gear but targeting the same species often clash over rights of access to the stocks. These differences in interests are complicated by variations in economic and political power, which shape responsive governance.

1.1.3 Sustainability

Concern about the sustainability of marine fish stocks and ecosystems is increasing. Perceptions regarding the current state and future potential of marine fisheries depend on beliefs regarding both human and natural systems. As will be described in chapter 8, there are multiple perspectives on what constitutes a healthy fishery or what is acceptable as effective fisheries management. In this subsection, I will review assessments of effectiveness based on three primary management standards from the fisheries literature: the single-stock maximum sustainable yield (MSY) benchmark, fisheries-centric ecosystem approaches, and ecosystem-focused approaches to management.

First, based on the MSY approach, the UN's Food and Agriculture Organization began assessing the state of the world's marine commercial fish stocks in 1974. Figure 1.6 shows changes in the level of overexploitation since that time. Though fully exploited stocks (at biomass that will support MSY) remain relatively stable at around 50% of assessed stocks, stocks not fully exploited (that is, biomass above that which supports MSY) declined from 40% to just over 10% in the period, while overexploited stocks (biomass below MSY levels) climbed from 10% to about 30%. From these estimates it appears that overexploitation of marine fisheries is increasing around the world, but at a decreasing rate—particularly for overexploited stocks.

Representing the fisheries-centric perspective, Hilborn et al. (2005) recognize that fisheries around the world are in crisis, but they view pockets of successful management, such as the individual transferable quota (ITQ) programs of Iceland and New Zealand, as signs that improvement is possible if appropriate measures are implemented. While accepting that ecological risks are present, fisheries-centric authors assert that most of the worst stock collapses are not caused by single-stock MSY-based management *per se*. Instead, they blame insufficient management for fishery failures. From

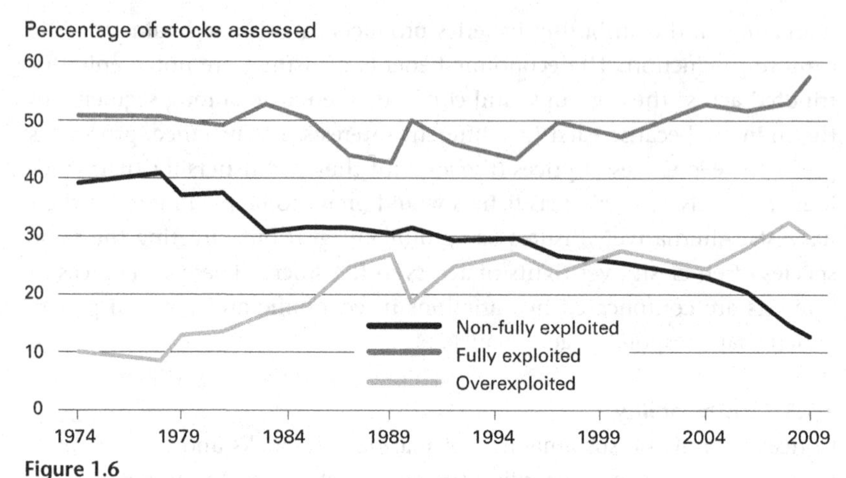

Figure 1.6
Global trends in the state of world marine fish stocks since 1974. Source: FAO 2002a.

a fisheries-centric perspective, this means that technical solutions to the management problem are already sufficient, but that political and economic factors prevent effective implementation.

In contrast, proponents of the ecosystem-focused perspective paint a much bleaker picture of the current state of fisheries resources. Indeed, some scientists and conservationists believe that world fisheries are on the brink of collapse. Like the contributors to Sumaila and Pauly (2011), some authors assert that MSY approach to management underestimates problems associated with overfishing because effects on ecosystems are ignored. They argue that drastic reductions in fishing mortality and extensive efforts to restore habitats are necessary to prevent ecosystem collapse around the world. Because they focus on the potential for catastrophic and irreversible shifts in ecosystems caused by trophic cascades and other complex feedback processes, these authors are risk averse and highly sensitive to the ecological repercussions of fishing. On the other hand, while they recognize the importance of fishing communities, ecosystem-focused researchers tend to downplay the current costs of their proposed management schemes—which are often quite high—in favor of future economic and ecological benefits. This makes sense from an ecological perspective; however, it can be counterproductive politically, since short-term costs often ignite fishers' opposition to management.

These political concerns are very important. In view of the scientific uncertainty surrounding marine fisheries, it is not likely that the debate between fisheries-centric and ecosystem-focused epistemic communities will be settled any time soon. However, since all sides agree that action needs to be taken sooner rather than later, it makes sense to focus on finding ways to remove political roadblocks to each type of management. This does not mean that fishers' voices should be excluded from political processes or that the industry should be marginalized. In fact, quite the opposite is true. Without industry acceptance, management tends to fail spectacularly. Yet there are ways to reduce political opposition to management without excluding key participants. The historical record shows that there are a number of switching mechanisms that precipitate transition from a downward spiral of depletion under ineffective management to an upward cycle of rebuilding under effective management. The AC/SC framework helps to identify both factors that delay response and factors that facilitate switching.

1.2 Responsive Governance and the AC/SC Framework

Responsive governance is a simple concept that is analytically challenging. Empirically, governments are demonstrably responsive in many issue areas, including environmental management. There is a considerable literature on responsiveness related to various types of crises, issue attention cycles, and interest-group activities (Higgs 1987; Downs 1972; Baumgartner et al. 2009; Grossman and Helpman 2001; Hilborn and Walters 1992). Research on learning and organizations also supports a trial-and-error model, although true learning goes beyond experience to the internalization of ideas and norms (Simon 1955; Newell and Simon 1972; Sprout and Sprout 1979). The literature on economic psychology further underlines the responsive nature of decision making when individuals are faced with complex problems, information is limited, transaction costs are high, and causal connections are difficult to identify. Temporal myopia is a well-known phenomenon, and heuristics (mental short-cuts) often bias action toward responsive rather than proactive behavior.[9] For instance, the availability heuristic usually skews action toward local-scale problems that are easily imagined (vivid) and perceived to be important (salient) in day-to-day life (Kuran and Sunstein 1999; Sunstein 2006; Finucane et al. 2000). Surveys also show that direct experience with environmental costs is a necessary precursor to political action, even when individuals acknowledge risks (Moser and Tribbia 2006; Moser 2007; Brody et al. 2008; Zahran et al. 2006). Thus, there is ample evidence that individuals and decision makers are responsive, at least in the area of environmental governance.

Responsiveness among fishers, processors, and other economic actors is also supported in the literature. The tragedy of the commons itself fosters short-run decision making, and this effect is amplified by high levels of uncertainty about future prices and the size of the fish stock (Clark 1973). Even when property rights are well defined by policies like individual transferable quotas, fishers often have fairly short time horizons and high discount rates, so investment calculations are heavily weighted toward current catch and profit levels (Asche 2001). In this fishers do not differ much from other business entities. *Short-termism* occurs when businesses invest only in projects with high expected payoffs in the short run (Cooper et al. 2002; Dallas 2012). Whether for the fishing industry or for other businesses, these short-term outlooks lead to behavior that is more responsive than proactive.

Although responsive behavior is ubiquitous in the issue area covered here, response for any given issue and time period is shaped by the context of decision making and by the existing institutional structures. In addition to learning, each political response to problem signals builds up new institutions, constituencies, and capacities that alter the problem structure for future decisions. Decision makers, defined as those individuals, governments, or states that have direct control over the formal rules and norms of exploitation, can establish new formal institutions that change incentive structures and power relationships (see subsection 1.3.3). Other actors can also alter the strength of problem signals, present new solutions, pressure decision makers, and change the informal rules and norms of governance (Sabatier 1987; Ellickson 1991; Wilson, Yan, and Wilson 2007). These actions may be directly spurred by the use of the resource (endogenous) or generated by external forces (exogenous). Lastly, the aggregation of these different attributes depends on the politics of power and on broader governance structures, such as the form of government (e.g., democracy vs. authoritarianism) and the scale of the problem (local, regional, national, or global).

With this complexity, it is useful to have a framework that delineates the different aspects of responsive governance and demonstrates how these factors fit together (Collier 2011). This is why I developed the Action Cycle/Structural Context (AC/SC) framework to guide the analysis presented in this book. It is inspired by Giddens' (1979) concept of *structuration,* which states that the actions of individuals and those of groups can affect institutional structure even as structure affects actions. This was Giddens' answer to the agency-structure debate in sociology. It contrasts with theories that focus only on the role of actors and with theories that view structure as the sole determinant of outcomes. The AC/SC framework also draws on work from the literature on domestic and international governance, on bioeconomics, and on social-ecological systems (SESs; also called coupled human and natural systems). It fits into a growing literature on middle-path approaches to the action-structure debate and also engages multiple disciplines, which is important when one is analyzing environmental issues (Clapp 2011).

Like most other frameworks for understanding social-ecological systems, the AC/SC framework comprises broad categories and relationships rather than the more detailed specifications of a model or a theory. It is not a replacement for existing theories and models. Instead, it provides a set of unifying concepts that encompass multiple inter-disciplinary and intra-disciplinary ideas. Although this limits specificity, it provides flexibility and allows for accommodation of variation between different contexts, which is necessary because of the interdisciplinary scope of the analysis and complexity of SESs (Anderies, Janssen, and Ostrom 2004). The primary difference between the AC/SC framework and other frameworks in this area of the literature is the theory of responsive governance and related treatments of change, including the potential for learning, institution building, and cycling between ineffective and effective periods in the management treadmill.

Much of the AC/SC framework, including the structural context and the linkages between actions and problems, is designed to synthesize multiple theoretical perspectives. In this, the AC/SC framework is similar to other frameworks in the field, including the well-known framework proposed by Ostrom (2009; see also Ostrom 2007). All the factors listed in Ostrom's framework, including resources units, resource systems, resource users, governance systems, and feedbacks can be accommodated in the AC/SC framework. The primary distinction between the two frameworks is that the framework of Ostrom assumes decisions based on expected costs and benefits, whereas the AC/SC framework is based on the paradigm of responsive governance. Another distinction is that there is little discussion of power relationships in Ostrom's (2009) interactions, which include self-organization, networking, and lobbying, as well as related outcomes in terms of various social and ecological indicators but neither power nor politics are discussed. Young et al.'s (2006) portfolio approach to analyzing complex human and natural systems can also be applied using the AC/SC framework. However, the purpose of that approach was to improve institutional design rather than explain institutional change. In contrast, explaining institutional change is the primary focus of the AC/SC framework, although evaluating effectiveness is also an essential component of the analysis. Indeed, the point is not just to define variables and processes that generate changes in fisheries governance but to understand how social-ecological systems evolve in response to those forces. Though such evolution is not based on random mutation, trial and error are crucial components of the system, and actors' responses can be both adaptive and maladaptive. In all, the analysis comes down to the linkages between the social construction of context (action), the effect of context on actor behavior (structure), and the interactions between humans and the environment that shape both action and structure.

Giddens developed the theory of structuration for sociological studies of closed communities, but the AC/SC framework is a tool for parsing the political economy of responsive governance in open systems. It is therefore quite different from Giddens' original formulation. Intellectually, it is closely related to iterative approaches in political science such as that of Cohen, March, and Olsen (1972) and that of Kingdon (2011). First, I associate agency with problem solving through an *action cycle*. Actors choose how to respond to signals they receive about some underlying problem, such as an economic recession, a terrorist threat, or the overexploitation of a stock of fish. The nature of both the individual response and the aggregate response depends on the *structural context* in which decisions take place. Over time, however, responses in the action cycle can alter the structural context and thereby affect the behavior of the system as a whole. Figure 1.7 illustrates these points by embedding the action cycle in the structural context, indicating that action is constrained by the structural context but that the context is itself created by the compounding of actions. Exogenous forces also drive, catalyze, or limit the action cycle, and they may operate at different "speeds" than the action cycle itself, as per the panarchy concepts put forth by contributors to the volume edited by Gunderson and Holling (2002) and further advanced by various authors including Liu et al. (2007), who systematically investigate cyclical and spatial characteristics of coupled human and natural systems.

First, the Action Cycle consists of a *problem* (or multiple problems) that generates *signals*, which, in turn, trigger *responses* by actors. For instance, pollution (a problem) can cause negative health effects (a signal) which then generate political action (primary response) and, where governance favors those harmed, regulatory action (secondary response). Responses may help to solve problems but can also simply dampen signals, allowing problems to become worse in later periods. As is discussed by Webster (2009), it is assumed that actors generally prefer expedient responses that have low political and economic costs. However, when a response is

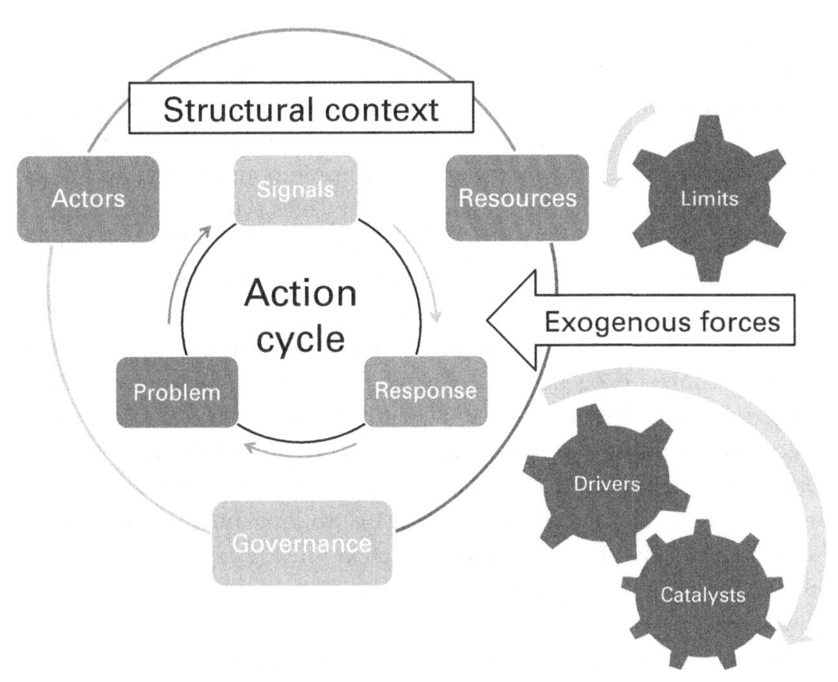

Figure 1.7
The Action Cycle/Structural Context (AC/SC) framework

institutionalized, the structural context of the system is altered and the response to a problem can accelerate. Specifically, the formation of institutions or agencies modifies the governance mechanisms that circumscribe the system. For instance, the US Environmental Protection Agency was created in part as a response to several well-publicized pollution events, but was not disbanded once those problems were solved. Rather, it remains the central agency for pollution regulation in the United States. It is important to remember that responses can be both adaptive and maladaptive, in that some increase the long-term resilience of the system and others hasten the system's collapse.

Turning to the structural context, according to Young (1994), *governance* factors include formal laws, regulations, and agencies, as well as informal rules and norms governing the behavior of actors. The AC/SC framework is designed to accommodate all levels of analysis and to highlight points of similarity or difference when used comparatively, as in the historiography presented in this book. *Actors* themselves play a part in the structural context because institutions circumscribe the roles and resources available to them. Though some actors may work to change their roles in the system, this behavior is driven by the action cycle. Furthermore, actors may be individuals, groups, governments, or states, depending on the case under consideration and the needs of the researcher. *Resources* include natural resources, capital goods and finance, technological and managerial capacity, and political power bases. As with other structural components, resources regulate the action cycle but they can also be altered as actors work to achieve solutions to signaled problems. Where resources are limited, the system as a whole is limited, unless actors can find a viable substitute.

All of the above factors are considered to be endogenous to an AC/SC system. However, other factors can change the action cycle or alter the structural context exogenously. These factors generally fall into two major categories: *limits* and *drivers*. The latter can be further broken down into direct drivers and catalysts of change. Exogenous limits are forces that prevent growth in a system. They may be related to governance, economics, or natural systems, and they may include institutional interplay, macroeconomic factors, and biophysical factors such as finite reserves of non-renewable resources and the potential for exhaustion of renewable resources. Ecosystem effects are particularly important limits when one is dealing with living resources such as fish and other marine fauna. Exogenous drivers are forces that increase the speed and intensity of the action cycle. Like limits, these drivers can be related to governance (as when governments choose to subsidize industries), to economics (as when economic development drives increasing production), or to the environment (as when changes in oceanographic conditions improve the productivity of a stock of fish). Technological innovation that is not directly associated with the action cycle can also catalyze sudden change in the cycle and/or in the structural context. Catalysts do not work independently; rather, they amplify existing drivers, as represented by the interlocking gears in figure 1.7. Sometimes these shifts mitigate the underlying problem, as with renewable energies and climate change, but more often technology increases environmental problems by reducing the costs of production; hence the association of catalysts with drivers rather than limits.

Temporal cycling is a common characteristic of complex social-ecological systems, and timing can be an important determinant of resilience (Folke et al. 2005). When dealing with time in the AC/SC framework it is important to differentiate between the speed of the action cycle itself and the speed of transition from the "ineffective" to the "effective" cycles shown in figure 1.1. Speeding up the action cycle means that problems worsen more quickly than they would otherwise, so that problem signals should escalate faster. This can generate a response that occurs earlier in time but later relative to the severity of underlying problems. Speeding up the switching response entails ensuring that "solutions" are in place when problems are relatively small and manageable. Thus, in a system in which the action cycle is slow (for instance, a small artisanal mining operation), effective response may occur after many years of exploitation but still be "early" if environmental problems remain small. Alternatively, in a system where the action cycle is fast (such as one dominated by large industrial mining corporations), switching that occurs after a few years may still be "later" than in a slower action cycle because the amount of pollution or degradation produced in the ineffective portion of the action cycle is greater, even though the period of ineffective management is much shorter.

According to Young (1999a) and many others, institutions are effective only when there are direct causal connections between rules and norms, human behavior, and related change in the resource level or environmental condition. Furthermore, these changes must be sufficient to meet predetermined institutional goals. However, tracing and testing these causal connections is very difficult in SESs, and in many instances perceived effectiveness differs from actual effectiveness (which usually is not determined until much later). For instance, if changes in environmental conditions cause the size of a fish stock to increase just after management measures are implemented, these measures are usually perceived to be highly effective; however, if environmental conditions cause the size of a stock to decline after implementation, the measures are usually perceived to be ineffective. This dynamic can either reinforce or undermine management, depending on the direction of the difference. Because of these complications, I treat the action cycle of responsive governance as "ineffective" as long as the underlying bioeconomic problem persists and as "effective" when that problem is mitigated, even if mitigation is caused by non-governance factors. This can be thought of as *relative effectiveness*, with governance evaluated on the basis of its effectiveness relative to the underlying environmental conditions. Moreover, it is also possible to evaluate response on the basis of criteria such as the level of monitoring and enforcement or the level of compliance with scientific advice. This will be discussed more in the introduction to chapter 7.

1.3 Applying the AC/SC Framework to Fisheries

Because the AC/SC framework is cyclical, the analytical process is iterative. For any given topic or case, the first step is to outline the scope of the analysis and identify a beginning or "start point" for the study. Defining the core problem helps in this process. If the focus of the analysis is a single event or case, the start point should coincide with the initial human action contributing to the core problem in that case. For instance, if one is interested in the collapse of a specific stock of fish, a logical start point is the initiation of the fishing activities that eventually led to the collapse. The next steps are to articulate the primary endogenous drivers generated by the core problem, to describe the structural context, and to show how the action cycle functions at that start point. With this foundation, the analysis moves "forward" in time as responses are observed and as effects on the action cycle and on the structural context are documented. This method is known as process tracing and may be a purely descriptive exercise or may be combined with hypothesis testing.[10] In either case, it is also necessary to anticipate and account for exogenous factors throughout the analysis. This includes description of external influences and their effects on the structural context, on the action cycle, or on both the structural context and the action cycle. In this section, I provide a generic overview of the fisheries application of the AC/SC framework with a focus on explaining the economics of expansion and the related co-evolution of the structural context through responsive governance.

1.3.1 Core Economic Problems and the CPR Driver

Three core economic problems characterize any application of the AC/SC framework to fisheries: overfishing, overcapitalization, and ecosystem disruption. *Overfishing* (also known as overexploitation) occurs when the level of catch exceeds the maximum sustainable yield (MSY)—the most that can be sustainably harvested. Furthermore, a stock is overfished (overexploited) when the size of the stock (usually measured as "biomass") is reduced below the level that will support MSY. *Overcapitalization* occurs when fishers invest more than the economically optimal amount of time and money in their fishing operations. A fishery is *overcapitalized* when the total amount of capital invested in the fishery is greater than the level that will produce the maximum economic yield (MEY)—the most profitable level of production that can be sustained (Clark 2005). Both of these problems are based on single-stock perspectives, but fishing can also cause *ecosystem disruption* through extraction of targeted species, by-catch mortality, and habitat destruction (Pikitch et al. 2004). Most marine systems are highly resilient, and can recover from substantial levels of disruption, but a number of cases of long-term fishing-induced shifts in the species composition of ecosystems are known (Bakun, Babcock, and Santora 2009; Lindegren, Diekmann, and Möllmann 2010; Edwards et al. 2013). These ecosystem shifts will be discussed further in the next two subsections. The simpler model described here covers a closed, single-stock fishery, so overfishing and overcapitalization are the only pertinent problems.

The primary driver of these core problems in most fisheries is the tragedy of the commons, or, more formally, the common pool resource (CPR) dynamic. CPRs have two specific qualities. First, they are *rival*: if one user extracts a unit of the resource, that unit is no longer available for capture by other users. For instance, if a fisher catches a fish it is no longer available to other fishers, whereas if a photographer takes a picture of the fish it will still be available to other photographers—at least until it swims away. Thus, fishing is rival and photography is non-rival. Second, CPRs are *non-excludable*, so there is no simple way to limit the number of resource users or assign property rights to a user who could then have a greater interest in maintaining the resource. In theory, the combination of rivalness and non-excludability leads to overfishing and overcapitalization as users invest large amounts of time and money to capture as many fish as they can as fast as they can, rather than leave any to be taken by others (Barkin and Shambaugh 1999).

Figure 1.8 shows how this basic CPR model of a fishery looks through the AC/SC lens. In the structural context, fishers are the only actors.[11] Here "fishers" refers to individuals in the fishery who make decisions regarding the level of fishing effort. Though labor markets certainly contribute to the effort decision, captains and vessel owners are the more pertinent actors in most analyses. The generic fishers represented in this basic CPR model are assumed to be homogeneous and independent actors. Their resources are limited to time available for fishing, existing technology, and a single fish stock. Fish population dynamics and exogenous economic factors are stable; for example, growth rates and prices are fixed. The only outside influence is the entry of new fishers, who are drawn in by the CPR dynamic; exogenous forces are omitted. In this model there are no institutions governing the behavior of fishers except the norm of non-cooperation, which is inherent in the CPR dynamic.

In the action cycle, the CPR dynamic is the core driver and overfishing and overcapitalization are the core problems. Short-run profits (total revenue minus total costs)[12] are the primary signal in this system, and the only response available to fishers is a change in the level of fishing effort. Since technology is constant, the only way to change effort is to increase or decrease the amount of time spent fishing. For any individual fisher, time available for fishing is limited by human physiology, personal preferences about leisure, and the economic constraint that total revenues must cover

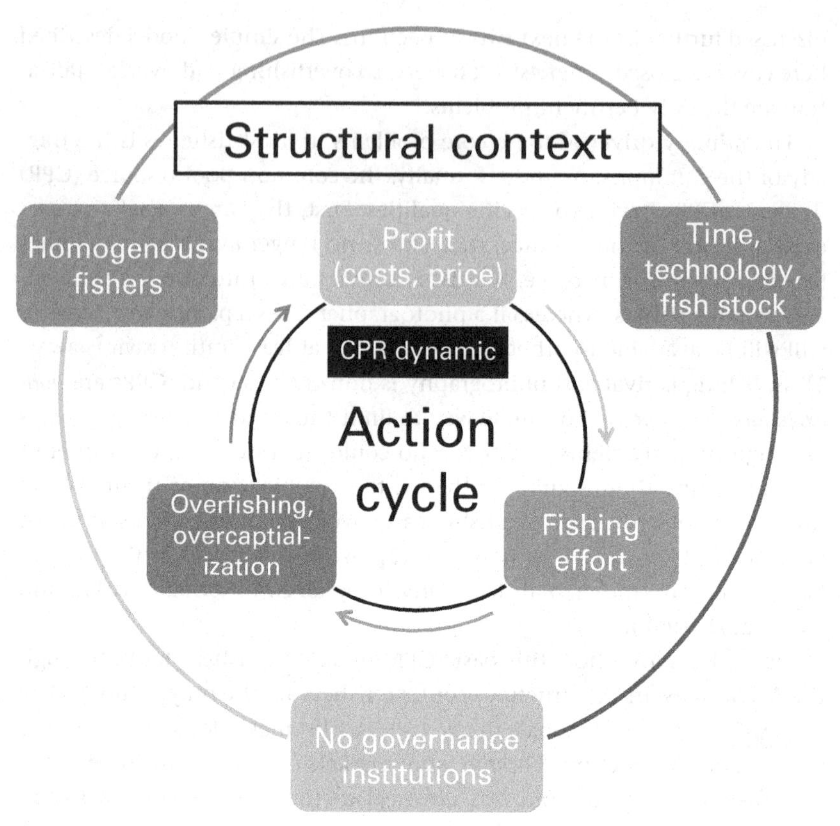

Figure 1.8
AC/SC for a basic common pool resource.

total costs. Since fishers are homogeneous, all will choose the same level of fishing effort, which is the maximum possible in a specified time period. However, new fishers may enter the system, and are expected to do so as long as total revenue per fisher is greater than total costs and profits are positive.

The start point for this model occurs just as the first fisher enters the fishery. The stock is large relative to the resources available for the fish, and there are few fishers. Working forward through the model requires understanding how the action cycle affects the core problem and the structural context. In this closed model, the only changes in the structural context occur with the entry of new fishers (who are drawn in by potential profits) and with the depletion of the stock of fish (which alters the resources available). In the most basic CPR models, profit initially increases with higher levels of effort because the fish stock is large and fishing actually reduces competition among fish over scarce resources, increasing the sustainable yield. However, with each turn of the action cycle, if profits are positive, effort increases, catch increases, and the fish stock declines further.

As fishers deplete the stock, catch per unit effort declines because the fish are harder to find. Clark (2005) refers to this as the *stock effect*. Though costs per unit of effort may be fixed in the short run, fishers must use higher levels of effort to catch fewer fish when the stock is small, and thus the marginal cost of catching fish is increasing above the point known as *maximum economic yield*. This is one component of the profit signal and can be thought of as the *cost signal*. It is directly linked to the population biology of the fish stock, which, as a resource in the structural context, acts as an internal limit on the system.[13] Similarly, costs may increase with the entry of new fishers, owing to the crowding effect noted by Wilson (1982). These two aspects of the cost signal can be divided into *appropriation externalities* and *technological externalities* (Schlager 1994). If the fishery is large relative to the market, a second economic limit arises because, all else equal, price should decrease with the increased supply of fish (Clark and Munro 1975). This *price signal* is another component of the profit signal, because total revenue is equal to the product of the price and the quantity supplied.[14] With increasing marginal costs and declining prices, profits decrease even more quickly than they would if the fishery were small relative to the market and prices were assumed to be exogenous and stable.

In either case, effort continues to increase as profits decline, but it increases at a decreasing rate. One reason for this is that fishers cannot afford to invest as much time in fishing when profits are low as they did when profits were high. Also, lower profits entice fewer new entrants into the fishery. Even though effort is increasing at a decreasing rate, continued entry will still result in a smaller stock biomass, further decreasing profitability in the fishery. However, because of the CPR driver, effort will continue to increase until total revenue is equal to total costs and profits are equal to zero. At that point, the system reaches an equilibrium level of effort and the action cycle effectively stops. In most cases, this equilibrium is greater than the bioeconomically efficient and sustainable level of effort, but this is not a profit disconnect per se, because profit signals are experienced but are largely ignored due to concerns about competition. All else equal, the downward spiral generated by the CPR driver ends in either stagnation or collapse (Conrad 2010). When costs of production are high and/or prices are low, the open-access level of production tends to be closer to the bioeconomic optimum (a narrow profit disconnect).

Economists note that open access fisheries are highly inefficient because of the potential for scarcity rent. Technically, rent can be thought of as any profit that is not dissipated by competition.[15] There are many types of rents, including monopoly rents, political rents, and resource rents. Scarcity rent exists when production depends on a scarce resource that is owned by some producers and not others. For instance, a landlord can charge more for a lakeside home than for an identical home without access to a lake because the number of lakeside homes is limited by the size of the lake. In an open-access fishery this scarcity rent is dissipated because no one owns the resource—the tragedy of the commons again. Because rents can be identified only over the long run, it is often difficult to differentiate between rents and profits,[16] particularly when the all-else-equal assumption does not hold. Furthermore, rents are a subset of profits, so I use the latter term in its inclusive sense throughout the book, referring to rents only when the distinction is analytically important.

As Clark (2005) points out, the basic model described above is deficient because it is based on normal costs rather than opportunity costs, which include the losses from the second-best opportunity forgone when a fisher chooses to work in one fishery rather than another. Economic profits account for these opportunity costs; normal profits do not. Where private ownership is in place, economic profits are usually calculated using the net present value function—that is, the discounted sum of the expected flow of net revenue over a specified period of time.[17] High levels of uncertainty regarding environmental conditions and future changes in both economic and biological factors in the profit function undermine long-term planning in fisheries. In the 1970s, economists and fisheries scientists began to model investment decisions by fishers under open access as a dynamic optimization process in which expectations of future revenues are heavily influenced by recent and present flows of net revenue. Clark and Munro (1975, 99) refer to this as a *myopic* rule, because fishers' decisions are based almost entirely on current information. These decisions may also be constrained by fishers' need to remain economically solvent despite high variability in the size of the stock and large fluctuations in prices. In response to these risks, fishers invest during times of high profit and save or diversify during times of low profit in a fishery (Lane 1988; Holland and Sutinen 1999). Thus, most fisheries economists accept a high degree of responsiveness by fishers and key profit disconnects in fisheries due to considerable uncertainty regarding future conditions.

1.3.2 Cycles of Economic Expansion and the Profit Disconnect

Even with the addition of opportunity costs and discounting, the basic CPR story is completely unrealistic. There are many stagnant and collapsed fisheries around the world, but most small, isolated communities find ways to manage the marine commons sustainably. We see stagnation and collapse much more often in open fisheries in which overexploitation is driven by powerful exogenous forces that override local institutions. Furthermore, many lucrative fishing operations manage to maintain high levels of profitability in spite of the CPR dynamic. Indeed, the fishing industry as a whole frequently employs economic strategies to circumvent the local bioeconomic limits of the traditional tragedy-of-the-commons model, something that widens the profit disconnect by allowing profits to increase even as the core fisheries problems worsen. As a result, we see cycles of expansion in the scope and the scale of fishing operations throughout history. This process ultimately transformed a relatively uniform system of small, isolated fishing communities into a highly stratified industry that spans the seas and harvests myriad varieties of fish and other marine organisms. The economic implications of a more realistic application of the AC/SC framework help to explain the cyclical process of economic expansion in fisheries.

Options for Economic Responses First, to explain the current scope and scale of the world's fisheries, it is important to recognize that fishers have a wide array of response options. As figure 1.9 shows, fishers may still respond by changing the level of fishing effort, but three other broad categories of economic response are possible: (1) explore in order to find new fishing grounds, (2) innovate and invest in more efficient and larger-scale vessels and gear, and (3) work to open new markets. The first two options are usually undertaken in response to the cost signal associated with the stock effect and with crowding, although they may also be triggered by changes in the size of the catch. Indeed, for subsistence fishers who harvest fish for their own consumption, catch is the primary problem signal of overfishing or ecosystem disruption. When prices are volatile, catch is an important problem signal for commercial fishers as well, and many view the amount of targeted fish caught as a strong indicator of the health of a stock, even if this is not the case. For instance, when the adult population is small it may still be possible to harvest large quantities of juvenile fish. In a number of cases, fishers clearly indicated that these deceptively large harvests of juveniles signaled that the adult population must still be healthy, in spite of scientific evidence to the contrary (see, e.g., Mason 2002 or Boerma and Gulland 1973). In general, catch signals do not trump profit signals, but they can increase concern if both point in the same direction.

Exploration to find new fishing grounds—either in response to a decline in catch or in response to a decline in profits—is common in the fisheries action cycle. As the opportunity cost of remaining in an overfished fishery

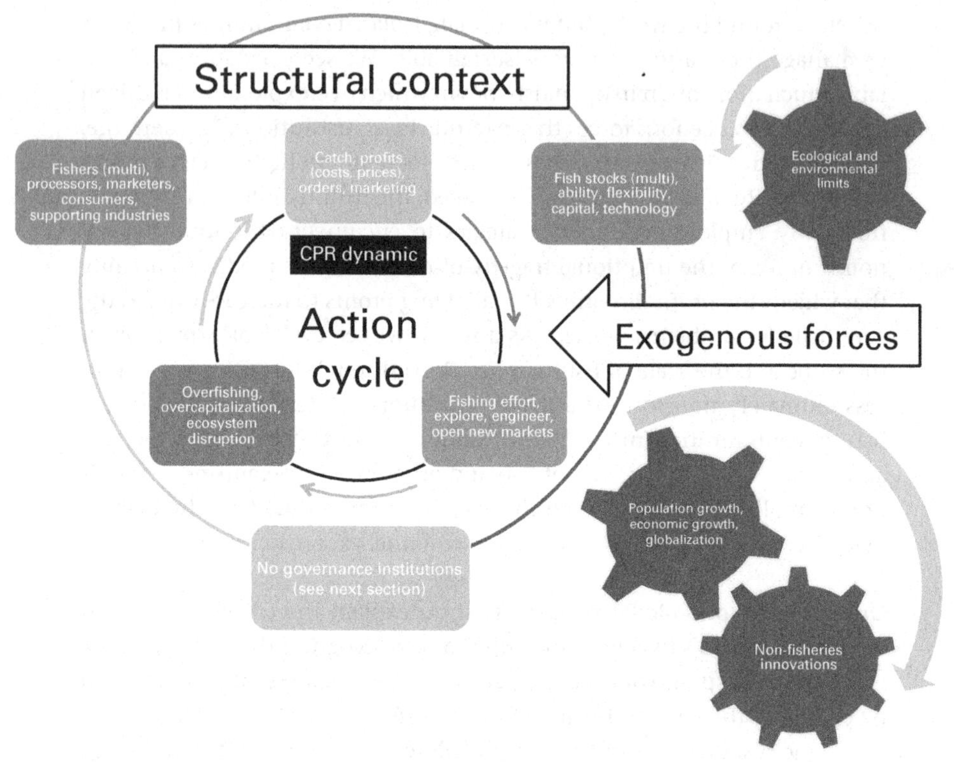

Figure 1.9
Expanded AC/SC application, economics only.

increases, fishers who have the option usually choose either to move to new fisheries or to exit the industry. Clark (2005, 75) refers to this as a *switching time* in the fishery, although when fishers are heterogeneous each will have his or her own individual point at which the opportunity cost of remaining in a fishery is negative and may therefore choose to switch to another option. Because of self-selection by fishers and the highly specific nature of fishing skills, there is considerable stickiness in the labor and capital markets for fisheries, so exit is not common (77). Transition to new fisheries or diversification of targeted species is a more frequent response to declining profits associated with overfishing. Fishers are heterogeneous in their preference for local exploration (same area, more species) over geographic diversification (more areas, same species), and this has had a profound effect on the evolution of fisheries economics and governance (Christensen and Raakjaer 2006). Finding new stocks alters the opportunity costs of fishing in all existing fisheries and can trigger rapid changes in the distribution of fishing effort.

Exploration is risky but can yield great rewards.[18] If fishers can find unexploited stocks of fish, either locally or at a distance, they effectively increase the biomass available in the structural context. Because effort is low and biomass is large on the new fishing grounds, all else equal, catch will be higher, costs of production will be lower, and profits will be greater because the stock is not yet overfished or overcapitalized. Many of the most successful captains begin to explore for new fishing grounds fairly early in the action cycle, when profits just begin to decline, and then move on to new locations as soon as catch or profits start to fall more quickly. These leaders usually strive to keep their new locations secret but are eventually followed. As long as they can prevent others from discovering the new fishing ground, leaders are able to capture entrepreneurial rents (sometimes called quasi-rents; see Schumpeter [1942] 1976). Exploration increases the resources in the structural context by expanding the set of known fishing grounds or the set of targeted species, but it does not halt the action cycle. Rather, this response simply leads to the proliferation of exploited fisheries and generates cycles of geographic expansion. This is an important prediction from the AC/SC framework. Evidence of these cycles of geographic expansion will be provided in chapter 2.

Fishers also respond to catch signals and cost signals by innovating to increase harvests and to reduce cost per unit effort in an attempt to negate the stock effect. They do this by investing in bigger and faster vessels, more efficient gear, and better fish-finding technologies (Squires 1987; Squires and Vestergaard 2013). Bigger and faster vessels enable them to exploit a larger number of stocks and reduce costs through economies of scale. For instance, large distant-water vessels cost millions to purchase and operate, but, because they harvest so much on each trip, the cost per metric ton of catch can be as low as a few hundred dollars. Efficiency improvements allow fishers to catch more fish every time they deploy or "set" their gear, thereby countering the stock effect. In this way, even small changes can add up to large savings. Lastly, better fish-finding technologies reduce the time and the amount of fuel spent in the search process. By reducing costs and increasing catch through innovation, fishers dampen the cost and catch signals, widening the profit disconnect. In formal language, innovation reduces marginal costs at a given stock size, shifting the supply curve out and increasing the distance between the open-access level of effort and the optimum sustainable level of effort. In AC/SC terms, investing in innovation intensifies the CPR dynamic and speeds up the action cycle.

Both of the above cost-reducing responses shift the supply curve out, triggering a price signal when affected fisheries are large relative to the market. In fact, there are many examples in which technological change or geographic expansion led to spikes in supply and related sudden drops in prices for specific fish products. Though these dynamics drive some fishers out of business, they drive others to respond by opening up new markets through advertising and improved preservation or transportation technologies. Advertising allows fishers to create new markets by convincing consumers to buy novel species or products. Of course, advertising can only go so far; fishers also must ensure that their products appeal to consumers. Better preservation and transportation technologies facilitate market expansion by allowing for sales of a highly perishable product at long distances from the fishing grounds. Without these technologies, inland areas would have little access to fish, and distant-water fisheries would not be feasible because fish would be inedible by the time the vessel returned to port. In fact, the scale of modern fisheries would not be possible without international markets and the ability to convince consumers to try new fish products. However, this response dampens the price signal for fishers and thereby increases the profit disconnect, further exacerbating core problems identified when using the AC/SC framework.

Table 1.1 summarizes the possible responses available to fishers and their effects on the core problem and related signals. Only one possible response—reducing effort—actually helps to counter the core problems of overexploitation and overcapitalization. This is also the least likely response, as fishers usually resist exiting the industry or switching to non-fishing alternatives. Instead, when only economic factors are considered, fishers usually choose to pursue one or more of the other responses listed in the table, all of which dampen problem signals without providing any long-term solution to the underlying problems in the action cycle. Exploration provides short-term relief from these problems by increasing the biomass available for exploitation, but this ultimately increases the scale of the core problems as more and more fisheries become overexploited and overcapitalized. Engineering and marketing can have similar effects when they bring new species into the mix but merely exacerbate the core problems in existing fisheries. These responses are palliative, in that they dampen problem signals but do not solve the problems themselves.

Consider the application of this extended economic model to a small-scale subsistence fishery. The start point for analysis occurs as the first fisher enters the system. Initially profits are high, so more fishers enter, driving up effort and driving down stock size. However, once the cost and catch signals set in, instead of reducing effort, fishers explore for new fishing grounds and engineer better technologies, thereby increasing catch while reducing the

Table 1.1
Expanded economic responses and expected effects

Possible responses	Effects
Reduce effort/exit fishery	Reduce and/or limit overexploitation
Explore for new fishing grounds	Expand biomass available for harvest, ↓ cost and catch signals, ↑ price signal (without new markets)
Engineer bigger and better capture technologies	Speed up/increase overexploitation, ↓ cost signal and catch signals, ↑ price signal (without new markets)
Open new markets	Speed up/increase overexploitation, ↓ price signal

costs of production as the geographic scope of fishing effort expands, the total biomass declines, and the threat of ecosystem disturbance increases. As long as markets remain small, the system is still limited because as supply increases, price should decrease, causing profits to decline. However, if fishers are able to open new markets, they can neutralize the price signal. All of these responses allow fishers to overcome local biological limits; they are able to increase the scope and scale of effort as much as possible given the constraints of current technologies and capital availability. Thus, with each "turn" of the action cycle—or, more generally, with the passage of time—the core problems get worse while fisher response dampens the problem signals. This is a widening profit disconnect. The *ceteris paribus* (all else equal) assumption does not hold, and fisher response consistently shifts the open-access level of effort farther and farther away from the optimum sustainable level of effort. This can have profound effects in closed systems, especially if complemented by exogenous drivers and secondary actors.

Limits, Drivers, and Catalysts Although fishers have more economic options in this expanded application of the AC/SC framework, their responses are still limited by the structural context. Specifically, resources circumscribe fisher response at any given point in time. In fisheries, resources include exploitable fish stocks, existing technologies, capital available for investment, potential new markets, and the costs of inputs. Exploration increases the number of exploitable fish stocks, but it is closely tied to innovation, since remote or deep-water stocks require fairly advanced vessels and fishing gear. Innovation itself is limited by existing technologies and availability of capital, and all three, in combination with population size and the level of affluence in the economy, regulate the potential for new markets. The costs of production are also closely tied to broader economic systems, such as markets for labor, fuel, steel, nylon, and related inputs. The state of global trade is a key determinant of both cost and price signals, as greater trade opens up new markets but also allows fishers to shift their operations away from high-cost locations to regions where costs of production are lower. Secondary actors and exogenous drivers can increase the resources available to fishers and thereby speed up the action cycle. In fact, without these factors fisheries would still be relatively small-scale, isolated operations in most of the world.

Though fishers receive the majority of attention in most fisheries models, other economic actors are also important. Secondary economic actors include all individuals and groups who make economic decisions based at least in part on signals generated by the fisheries action cycle. They are "secondary" because they do not interact directly with the fish stocks but rather receive indirect signals from fishers and related supply chains. These actors include processors, marketers, and supporting industries, which, together with fishers, make up the fishing industry as a whole. Consumers are also secondary actors, insofar as their willingness and ability to pay for fish products defines demand and therefore has a substantial effect on the price signal. However, the problem signals that consumers receive tend to be quite weak, because they are usually separated from the fish and the fishers by several layers of processors and marketers. Indeed, because of the resourcefulness of the industry as a whole, most consumers rarely receive any information about the state of the stocks that are raw inputs for the fish products they purchase. Information on the health of related ecosystems is even less accessible to consumers (Burger et al. 2004; Jacquet et al. 2009). They have no direct effect on the system, except to the extent that their willingness to accept new products—their response to the marketing signals sent by the fishing industry—limits the types of species and products that can be exploited. So far, such limits tend to be temporary, as processors and marketers refine their product lines to suit consumer tastes.

The specific interests of fishers and other industry actors are often at odds, largely because a price charged by one group is a cost to another, but all of these actors have a vested interest in the expansion of the industry as a whole (Campling, Havice, and Howard 2012). Thus, supporting industries, processors, and marketers often work with fishers, boosting their capacity for response in several ways. On the cost side for fishers, supporting industries like boatyards and gear manufacturers often cooperate with innovative fishers to produce new, lower-cost technologies. This benefits both groups, providing fishers with new technologies and supporting industries with new products to sell. Processors and marketers may also provide loans to fishers for vessel and gear improvements to keep their own costs of production low, and may even invest in new vessels, thereby encouraging growth in landings and exploitation of new stocks to maintain or increase the flow of raw materials for their businesses. For them, multiple sources of fish are necessary to stabilize production or sales in spite of the high level of variability in most fisheries (Prochaska 1984).[19] This type of diversification is common in many large-scale corporations. Processors and marketers also invest in transportation and preservation technologies and product marketing to open new markets, ensuring high wholesale or retail prices. This allows them to push out what Campling (2012) calls the *commodity frontier* in fisheries, expanding beyond the local, short-term limits in the traditional tragedy-of-the-commons story.

The impact of these secondary actors can be substantial because they have highly specialized knowledge of vessel, gear, and processing technologies, as well as marketing methods, and they build up substantial reserves of financial capital compared to the vast majority of fishers. Yet because of their size and diversification, these industry actors are distanced from the problem signals in the action cycle and so are even more susceptible to the profit disconnect. As with fishers, considerable uncertainty, high discount rates, and short time horizons reinforce responsive practices by secondary actors, who may make decisions based on a few years of profit forecasts rather than on the entire expected income stream from exploitation of a fishery.[20] Thus, when profits are high for secondary actors—usually when new fishing grounds are discovered through exploration, but also when new markets are opened—investment by secondary actors also tends to be high, and excess investment in processing or ship-building capacity is common during boom periods in fish production (Campling 2012). When processing firms are numerous, there is also a well-known coordination game that acts very much like the tragedy of the commons even though ownership is well established. Olson (1971) describes how profits in an open market with numerous firms are rapidly driven to zero because no subset of firms has sufficient incentive to curb production in order to keep prices high. As he points out, this is the inverse of Cournot's theory of oligopolistic competition.

Even with the assistance of secondary actors, expansionary fisher responses would still be limited if not for exogenous drivers (see the bottom right-hand corner of figure 1.9). These include both demand-side and supply-side factors. Population growth and economic growth are major demand-side factors. Population growth increases demand even if the consumption of fish per person remains stable. Economic growth generates higher levels of income per capita, which generally increases demand per person. Thus, these two factors combined cause rapid rightward shifts in the demand curve as more people are willing and able to pay more for larger amounts of fish products. This dynamic counteracts the price signal, as greater supply is met by greater demand. The resulting price signal trickles down from consumers to retailers to processors/marketers, and finally to fishers through the ex-vessel price. These exogenous forces, which are powerful disruptors of the price signal, drive economic expansion in fisheries.

On the supply side, exogenous drivers include globalization of trade and finance and technological innovation, which works primarily as a catalyst (Young, Berkhout et al. 2006). Globalization is a difficult concept to pin down, but here the word refers to increasing economic interdependence among national economies as the extent and the volume of trade and financial transactions increase around the world. As Ricardo ([1821] 2004) pointed out, trade allows for the reduction of costs of production because producers in different regions can specialize in areas in which they have comparative advantage. Globalization of finance facilitates the increasing availability of capital around the world, though unevenly, thereby contributing to the economic and technological stratification of fisheries. The implications of globalization are highly contested, but in fisheries it is clear that this force increases access to capital and lowers the costs of production, dampening the cost signal in the CPR action cycle. Thus, we can expect that globalization will tend to make overexploitation, overcapitalization, and ecosystem disruption occur more rapidly, while mitigating problem signals and delaying response.

Technology is an important determinant of fishing effectiveness and, therefore, of the costs of production and the profit disconnect more broadly (Dercole, Prieu, and Rinaldi 2010). In general, exogenous technological changes work as catalysts in the AC/SC framework. New technologies developed for military or commercial applications, such as petroleum engines and sonar arrays, can be rapidly appropriated by fishers who have access to sufficient capital, causing a swift transition from one technological state to another. This contrasts with purely endogenous technological change, which is usually gradual, although some big leaps have occurred almost completely within the fishing industry, such as the development of fish aggregating devices.[21] Which groups of fishers and other economic actors appropriate new technologies depends on their existing resources. Capital is normally a constraint, but some innovations, such as outboard motors and nylon netting, are cheap enough for appropriation by subsistence and artisanal fishers who generally have little access to capital. These new technologies are also readily adaptable to existing technologies among these groups. Nevertheless, appropriation of new technologies in any group usually depends on a few entrepreneurial individuals who are willing and able to accept the risk of trying something new. If they succeed, others will quickly imitate them if they have the capital and ability to do so (Orbach 1977).

For the most part, both secondary economic actors and exogenous forces contribute to the expansion of fisheries worldwide. However, global limits on the system do exist in the form of endogenous and exogenous biophysical boundaries. Endogenous limits were described above. Exogenous limits relate to the carrying capacity of the oceans more broadly and can be linked to both environmental and ecological conditions. Changes in ecosystem structure and productivity can have multiple causes, including human activities such as pollution, coastal development, and climate change. Non-anthropogenic factors such as currents and long-term cycles in oceanographic conditions can also affect productivity. Under favorable environmental conditions, stocks can grow rapidly, sometimes in spite of high levels of fishing effort. In such cases, these factors act as drivers rather than limits. However, highly favorable conditions are often short-lived, and most anthropogenic effects are detrimental to existing ecosystems; thus, for the most part, ecological and environmental factors can be viewed as limits on the productivity of targeted stocks.

In other words, fishers, the fishing industry, and consumers will eventually run out of new fisheries to exploit. Furthermore, although reducing the costs of production dampens the cost signal temporarily, it increases the core problems in this fisheries application, which means that more and greater innovations may be required to reduce costs with each turn of the action cycle. Similarly, opening new markets can dampen price signals associated with increasing supplies of fish, but this leads to higher levels of overexploitation, overcapitalization, and ecosystem disruption. If exogenous factors cause the global profit disconnect to widen further, capture production will eventually decline world-wide and fish may become a luxury good in the future, as it was in the past. Aquaculture is often seen as an answer to the problem of biophysical limits in fisheries, but to date fish farming is still highly dependent on capture fisheries for provision of feed, and also harms wild stocks through escapement, habitat destruction, and transfer of parasites and disease (Naylor and Burke 2005). In view of all this, governance is the only way to counter the economics of expansion, ensuring prosperity for fish, fishers, and secondary actors well into the future. Yet governance is no panacea; it can exacerbate these economic problems even as decision makers strive to provide solutions.

1.3.3 Co-Evolution through the Politics of Response and the Power Disconnect

Nearly every fishery has some sort of governance structure. Even small, isolated fishing communities usually have local institutions that govern use of marine resources. Modern commercial fisheries are governed by complicated sets of formal and informal institutions at multiple levels of analysis. For instance, lobstermen in the Gulf of Maine adhere to local sharing arrangements on which they agree collectively without government involvement. But state and national regulations also regulate the minimum and maximum size of lobsters that may be landed, protect egg-bearing females, and prohibit poaching, damage to other fisher's equipment, and the taking of life or property (Acheson 1997). The many different fisheries governance institutions currently in existence co-evolved with the cycles of economic expansion described above as decision makers responded to the competing claims of multiple interest groups. When governance mechanisms are included in the application, the AC/SC framework can help to explain this process and the key role played by the power disconnect. Because the economics of resource exploitation has a profound effect on the political responses of fishers and the fishing industry, the power disconnect parallels and is directly linked to the profit disconnect.

The concept of responsive governance is central in any analysis in which the AC/SC framework is used. Decision makers do not prevent problems; they respond to problem signals that they receive from fishers and other interest groups in a fishery system. Furthermore, decision makers prefer to implement those responses that are politically expedient first, and to try more costly measures only if problem signals persist or worsen. This assumption is based on Simon's 1955 work on *satisficing*—that is, selecting an option that is satisfactory rather than optimal under conditions of imperfect information—but could be modified to include more complicated models of adaptive decision making, such as those described by Payne, Bettman, and Johnson (1993). In view of the importance of expedience, it is necessary to understand the cost and benefits of potential management measures, which can be thought of as resources in the structural context, as well as the strength of the political signals received by decision makers (Kingdon 2011). In particular, the profit disconnect will induce fishers to resist attempts to reduce catch or effort and thereby postpone switching response until the economic costs of overexploitation and overcapitalization are manifest in profit signals. Furthermore, the power disconnect is much greater when the structural context favors interest groups that are less sensitive to the underlying problems of the action cycle and have vested interests in maintaining economic expansion.

Figure 1.10 is a modified illustration of the AC/SC application from subsection 1.3.2 which includes governance. The relationships between the boxes are the same as in figure 1.7, but are not represented due to space constraints. New elements associated with the governance application are indicated in boldface. Furthermore, I distinguish between attributes and actions associated with fishers and other interest groups (lighter areas in each box), those associated with decision makers (darker areas), and factors related to both groups (indicated by asterisks). The topmost set of boxes shows the expanded structural context. In addition to the economic actors described above, the model now includes non-commercial interest groups (recreational fishers, environmental groups, scientists, the media, and the public) as well as decision makers (individuals, governments, or states that control established institutions in the governance segment of the structural context). Where they self-regulate, fishers themselves fall into the category of decision makers, but more often decision makers are local elites, bureaucrats, and politicians. At the national level, "governments" may be treated as decision makers; at the international level, "states" are decision makers. However, the AC/SC framework can be used in nested applications in order to capture interactions between these levels of analysis.

New resources are also available to actors when governance is added to the analysis. Fishers and other interest groups wield political power and influence and may benefit from the formation of social capital such as education and civil society organizations. Decision makers can use similar political power sources, but they also have a legal monopoly on the use of force (although historically fishers also have used violence to protect their access rights). On the other hand, their work can be assisted or constrained by the management options, the budgetary resources, and the enforcement capacity available to them.

Governance Institutions and Political Response The box furthest to the right in the topmost row of figure 1.10 contains six categories of governance institutions that are common in fisheries. *Endowments* are essentially use rights that specify which actors are allowed to engage in fishing. *Entitlements* indicate which individuals have rights to benefit from that fishing effort (Kreuzer 1978a). The basic applications presented in previous sections of this chapter all assume that fishers own both endowments and entitlements. That is, anyone who chooses to become a fisher has a right to physically appropriate the resource and a right to retain all benefits from

Structural context		
Actors	**Resources**	**Governance institutions**
Fishers	Fish stocks	**Endowments (access rights)**
Other industry	Ability	**Entitlements (rights to benefit)**
Consumers	Flexibility	**Expansionary measures**
Scientists	Capital	**Restrictive measures**
Environmental groups	Technology	**Enforcement mechanisms**
Recreational fishers	**Political power***	**Policy/decision processes**
Media/journalism	**Social capital***	
The public	**Violence***	
Decision makers	**Management options**	
	Enforcement capacity	

Action cycle		
Problems	**Signals**	**Responses**
Overfishing	Catch	Fishing effort
Overcapitalization	Profits	Exploration
Ecosystem disruption	Costs	Innovation
Conflict over access rights	Prices	Open new markets
Harm to charismatic species	Marketing	**Demand/protest exclusion/access rights**
Threat to political power	**New entrants/cheating**	**Demand/protest expansionary management**
	Scientific reports*	**Demand/protest restrictive management**
	Media reports*	**Exit/refuse to comply**
	Voting	**Alter governance institutions**
	Lobbying	**Alter policy/decision processes**
	Campaign contributions	

Exogenous forces		
Limits	Drivers	Catalysts
Ecological	Population growth	Non-fisheries innovation
Environmental	Economic growth	Non-fisheries science
	Globalization	**Political/economic norms**
	Food security policies	**Institutional interplay**
	Employment policies	**Environmental movements**
	Development policies	

Figure 1.10
AC/SC application with governance.

that exploitation. This is a logical extension of the open-access attribute of the CPR dynamic. However, in many governance systems endowments and entitlements restrict open access (Wilson 1982; Schlager 1994). In sea-tenure systems, specific fishers or fishing groups are granted exclusive rights of access to particular areas or particular stocks of fish. Similarly, licensing programs and permit systems that restrict access are types of endowments in fisheries. On the entitlements side, norms allocating a share of the catch to a fishing community, rules about paying tribute to local elites, and fees or taxes collected by governments are all institutions that divide up the benefits of fishing between fishers and decision makers. Thus, even when fishers have the right to appropriate the resource they seldom have the right to retain all benefits from their labor.

By establishing endowments, fishers or other decision makers create a new problem in the action cycle: conflict over access rights, or, in formal terminology, *assignment problems* (Schlager 1994). This problem can be signaled by the entry of new (outsider) groups of fishers or by observed cheating on established rights of access by insiders. Unlike economic signals, these political signals can occur at any point in the action cycle, as long as existing fishers believe that they have established rights of access to the stock or area.[22] Fishers respond to these signals by escalating their attempts at exclusion, which is one of the earliest and most widespread responses observed in fisheries action cycles around the world. Where institutions are scarce, violence is a common method of excluding outsiders and establishing access rights. Where broader governance institutions exist and prohibit such violent conflicts, fishers must engage with local elites or legal authorities to safeguard their endowments. Decision makers often claim entitlements in exchange for protection of endowments, although rights to benefit may also be ingrained in wider political power structures. Public violence may be used to enforce fisheries regulations or to contest fishers' access to resources (Hart and Pitcher 2001; Platteau 1992). Chapter 5 of this book describes how exclusionary institutions rise and fall with cycles of economic expansion and related conflicts over access to fish stocks.

Whereas endowments and entitlements govern the allocation of scarce resources, management measures are explicitly designed to control the level of fishing effort. Expansionary management measures include all actions that support fishers economically without forcing equal reduction in catch or fishing effort. The most common are subsidies, or direct payments from the government or local elites to fishers (Sumaila et al. 2010a). Guaranteed loans and other assistance raising capital also fit in this category, as would community-funded or government-funded investment in the culture of targeted species to increase populations (and therefore the amount that can be harvested sustainably). As will be discussed in chapter 6, the vast majority of these measures are purely palliative. By reducing costs of production, increasing access to capital, and even ensuring fairly high prices, these measures all increase the profit disconnect and exacerbate the cycles of economic expansion predicted by the AC/SC framework. However, as the first US Fish Commissioner, Spencer Fullerton Baird, pointed out in the 1870s, expansionary measures are usually popular with the industry and with decision makers, many of whom are concerned with issues such as food security, employment, and economic development. Indeed, this is why Baird called these measures "positive" in contrast to "negative" management measures, which require some reduction in either catch or fishing effort (Allard 1978). I refer to the latter as *restrictive measures*.

Because expansionary management measures are politically expedient, we can expect that decision makers will use them fairly early in the action cycle. When profit signals are relatively strong, fishers often demand expansionary measures from government. In these cases, subsidies or related interventions can be classified as *supporting* because they keep struggling fishers afloat during hard times. Decision makers may also use *developmental* measures that are designed to build up new fisheries (Royce 1987). Developmental measures are usually implemented in response to exogenous factors such as food shortages or economic development programs. Both types of expansionary measures have also been used by governments to solidify claims of rights of access to marine resources as part of contestation and negotiation of international law regarding ocean use and management. Whereas expansionary management measures tend to be quite popular, non-commercial interest groups and even the public sometimes protest this governance approach, because it exacerbates problems in fisheries and because the burden of these measures increases with the worsening crises associated with the economics of expansion. That is, expansionary measures actually drive fishing effort higher, exacerbating the profit disconnect and the related downward spiral of overexploitation and overcapitalization such that subsidies or related measures must increase in order to maintain the same level of "profitability" in a fishery.

In contrast, restrictive management measures in fisheries are designed to limit either catch or effort. These measures include size limits, catch limits, area limits, licenses, and fees, as well as ecosystem-based management and other more advanced regulations (Caddy and Cochrane 2001). Some of these measures are observed in informal governance systems, but for the most part such limits are associated with modern practices of formal management by government agencies. Most people, like Commissioner Baird, expect that fishers and industry actors will necessarily resist the application of restrictive management measures, but this is not always the case. Certainly, when management threatens to diminish a fisher's livelihood that fisher—and all others who are so affected—will probably protest the new rule. However, there are three situations in which fishers are likely to demand restrictive management measures: (1) when such measures exclude other fishers but have little effect on the group demanding regulation; (2) when one type of management is seen as the lesser of two evils, as when fishers demand size limits as a less painful option compared to catch or effort limits; and (3) when all other response options are exhausted and the fishery is clearly in great economic and biological distress. Of course, which measures are selected depends on the broader political context.

Any given restrictive regulation usually creates both winners and losers among groups of actors as defined in the structural context. For instance, a minimum size limit for landings of a certain species will hurt fishers who target fish of that size, but will benefit fishers who catch bigger fish of the same species because more small fish will survive to grow to the larger size. Alternatively, a restriction on the type of gear that can be used in a fishery will force some fishers to exit because they cannot afford to acquire different equipment, leaving more fish for those who use the permitted gear. Fishers, who are well aware of the differentiated effects of management measures, often organize into groups and make demands on decision makers accordingly. Thus, Lasswell's ([1936] 2011) depiction of politics as contestation over resources is highly relevant to fisheries governance, even when the structural context favors norms of open access rather than systems of endowments and entitlements. Indeed, conservation is often used as a rationale for the adoption of restrictive management measures that exclude "outsider" groups of fishers or other secondary actors. Since exclusion signals usually occur much earlier than economic problem signals, demand for exclusionary regulations often precedes other political responses. Indeed, modern fisheries management has its roots in this political dynamic.

New Political Actors When groups of fishers demand restrictive management measures that are less onerous than other proposed policies, they may be defending themselves against attempts at regulatory exclusion by other groups of fishers or against calls for conservation-based restrictive regulation by recreational fishers, environmental groups, the scientific community, or other non-commercial actors. As will be explained in chapter 8, non-commercial actors and fishers respond to different problem signals. Non-commercial actors are usually much less concerned with the catch or profitability of fishing fleets. Recreational fishers, however, pay attention to their own catch, and particularly to the abundance of large trophy fish of preferred species. There is a long and fairly successful history of political action by recreational fishers to exclude commercial and even artisanal and subsistence fishers from inland waters, but until recent times recreational fishers were less active and less influential in marine fisheries (Aas 2007). Environmental groups are also recent arrivals in the actor category. These groups include preservationists, who lobby for bans on the killing of any animal in a preferred species group, and conservationists, who are mainly concerned with the sustainable management of marine species and ecosystems more broadly (Stoett 2011). Recreational fishers, conservationists, and preservationists all tend to rely on scientific assessments as the primary biological signal that species or ecosystems are overfished, and usually respond by demanding restrictive management measures on commercial fishing. Recently some non-commerical interests have started to appeal to consumers and the public at large to pressure the industry by avoiding products derived from overfished stocks, endangered species, or charismatic megafauna (e.g., dolphins, whales, and sea turtles).

Wittingly or not, scientists, or more broadly, epistemic communities of experts are also political actors in the structural context for modern fisheries (Haas 1989). Their research is frequently problem driven and has profound effects on the management options available in the "resources" component of the structural context. In fact, the institutions of fisheries science were built largely in response to cycles of growth and collapse in several key fish populations and the resultant socioeconomic costs in fishing communities and related sectors. These cycles escalated with the rapid expansion of fishing effort that began in the 1800s, causing increasing conflicts among groups of fishers and leading to growing demands for exclusionary management interventions. In response, governments increased funding for fisheries science aimed at understanding the causes of biological changes in stocks in order to settle conflicts between fishers. This response was influenced by exogenous forces associated with the upsurge in positivist philosophy at the time, and by the belief that government regulation should be based on the best scientific evidence available. Nevertheless, because of power dynamics that will be described below, even government-funded scientific advice often went unheeded.

The media and the public can also wield significant influence in the fisheries action cycle. Indeed, although other interest groups play important roles in agenda setting and in the development of policy alternatives, public pressure—often inspired by media coverage—can tip the balance in action cycles where commercial and non-commercial interests are at loggerheads. This is what Olson (1971) referred to as mobilization of a latent interest group. However, in general both the media and the public respond only to sensational or highly salient events, rather than engaging in prolonged monitoring and engagement on policy issues. Furthermore, novelty is important, and these groups loose interest quickly if crises are prolonged or problems stagnate (Baumgartner and Jones 2009). Fisheries are not of much interest in most regions, although the topic of fisheries may be closely watched by the media and the public in areas where the fishing industry is a major employer or where marine tourism is an important source of revenue. We can expect to see fairly continuous public engagement in these areas but expect only sporadic engagement triggered by well-publicized events at larger scales. Even when charismatic megafauna are involved, media attention and public outrage are not usually triggered until biological signals are very strong and stocks are near collapse or extinction. For fisheries more generally, public response often occurs after major collapses, driving retrospective responses in the action cycle. Of course, the influence of the public and that of the media depend greatly on governance in the structural context.

Power and Politics In view of the political divisions within the fishing industry and the interests of other actors, it is important to understand how the structural context shapes decision making. At any given time, selection of new management measures and institutions depends on existing (baseline) institutions that govern fisheries and underlying decision processes that shape the roles and political resources of different actors. Incremental behavior can be expected, as described by Lindblom (1977), but there is also evidence of large-scale revolutionary change in response to crises, as expected by Higgs (1987) or in response to vertical and horizontal interplay as explained by Young (2002).

Structural factors also affect the power disconnect and thereby alter responsive governance. When actors who are sensitive to problem signals have political influence, political response occurs earlier in the action cycle and is often more effective, although much depends on the management capacity available in the structural context. This helps to explain why sustainable management regimes are often observed in small, relatively isolated fishing communities in which both the profit disconnect and the power disconnect are minimal. However, the cycles of economic expansion described above disrupted most of these collective-action regimes and exacerbated the power disconnect through positive feedbacks between the economic scale of fishing operations and the political power of the fishing industry. Increasingly, the actors who receive early, strong problem signals are marginalized politically, while those who engage in economic responses that dampen problem signals become more and more powerful.

Before looking into the power disconnect further, it is important to understand how structural factors affect political influence. In combination with the political resources available to interest groups, including their political power, policy positions, and the set of management measures available, decision processes are major determinants of governance outcomes. There is considerable literature on this issue area, and a researcher using the AC/SC framework has many options. In general, public choice theory does not fit the framework because of the focus on maximization, but constructivist (Wendt 1992), behavioralist (Bendor et al. 2011), game-theoretic (Olson 1971, 2000; Ostrom, Gardner, and Walker 1994; Hardin 1995), cultural (Almond and Verba [1963] 1989), organizational (March and Simon [1958] 1993), institutionalist (Ostrom 1990; Haas, Keohane, and Levy 1995), and regime (Young 2002, 2010) perspectives all can be accommodated. Here, I describe the approach that I use throughout this book. It is a hybrid drawn from all of the aforementioned approaches, but it is most firmly based in the literature on environmental governance and fisheries management (see e.g., Wilen and Homans 1998; DeSombre 2002; Finley 2009; Webster 2009; Hilborn, Orensanz, and Parma 2005).

At the domestic level, the nexus between actors, resources, and governance structure can be broken down into three major aspects: access, alignment, and leverage. *Access* refers to the interest group's ability to make its arguments to decision makers, either directly or through an intermediary. If one group has access and others do not, that group is more likely to affect governance. On the other hand, *alignment* between decision makers' goals and interest groups' goals is also important. Many commercial fishing interests, including processors and marketers as well as fishers, have been able to influence government regulation because their interest in expanding the industry aligns well with policy makers' exogenous goals (e.g., high employment, food security, and economic growth), at least in the short run. However, interest groups with considerable *leverage*, defined as the ability to threaten a decision maker's power base, may be able to gain access to decision makers and force changes in governance goals.

Access, alignment, and leverage are all affected by the broader governance context. In more democratic societies, access to decision makers depends on lobbying as well as on the ability to mobilize votes and/or campaign contributions. Petitions, strikes, and protests can further contribute to leverage. Of course, no democracy is perfect, and access is often asymmetrical among interest groups, but decision makers must at least appear to represent the public interest. Thus, public-relations campaigns are often crucial elements in democratic decision processes, as is the relative power of regional representatives (e.g., members of Congress or Parliament from districts with large commercial fishing interests). Just as with any other issue area, more senior or more adept representatives are usually better able to influence legislation—particularly on fisheries topics—which are seldom of great interest to the wider public (Kingdon 2011). This is why political power is represented as a resource for both interest groups and decision makers in figure 1.10. In litigious systems, like the United States, interest groups can also sue the government to force the implementation of laws that are already on the books (Perkins and Neumayer 2007). Similarly, where they are accountable to multiple, competing interests, decisions makers often agree on legislation that vaguely promises to meet the demands of all groups, leaving interpretation up to bureaucratic decision makers who may be more closely aligned with some interest groups than with others.

In authoritarian systems or similar regimes, power depends on access and alignment, and leverage can be wielded only through drastic measures, such as blackmail, coup d'états, and civil war. Fisheries rarely cause such upheavals in modern times, but were regularly cause for military conflict in the past. More often, either fisheries are ignored in authoritarian systems, and fishing communities govern themselves, or else endowments and entitlements are used to reinforce existing power structures (see e.g., Muscolino 2009). Because these systems are usually more closed than democratic systems, in which open access tends to be the norm, they may produce more sustainable outcomes. On the other hand, authoritarian and centrally planned governments like the Soviet Union may view fisheries as a major source of economic development, leading them to invest heavily in large-scale industrialized fleets, which, in turn, creates vested interest in economic expansion within the government itself (Armstrong 2009). Much depends on the specific details of the hierarchical structure, particularly the role played by fishers and other industry interests. Hypothetically, an authoritarian government that is captured by non-commercial interests might also implement more sustainable management measures, but this has not yet been observed.

In both types of systems, the AC/SC framework predicts positive feedbacks between economic growth and political power that widen the power disconnect. That is, as fishers accumulate greater amounts of capital, produce larger amounts of fish, and provide more jobs or other economic benefits, their ability to affect the political process usually increases as well. In a more democratic system, this translates into more votes and more campaign contributions, whereas in all types of government a larger fishing industry means greater political rents for decision makers (Krueger 1974). Therefore, it is not surprising that cycles of economic growth coincide with political cycles of delayed response, no matter what type of government is in charge of fisheries management. The power disconnect widens the profit disconnect through expansionary and exclusionary management measures, while the profit disconnect widens the power disconnect by making politically powerful actors less sensitive to problem signals and more influential over management decisions. Furthermore, these determinants of the power disconnect are important at multiple levels of analysis.

A major benefit of the AC/SC framework is that it can be used to link domestic policy processes to outcomes at international negotiations by examining how the domestic power processes described above affect access, alignment, and leverage at the international level. Governments are expected to base their international policies on domestic concerns in conjunction with considerations regarding their influence in the structural context for negotiations (Putnam 1988). In this application, states are not unitary actors as per collective-action approaches to international relations; rather, they are strategic representatives of domestic interest groups. Nevertheless, I use the term *states* in international applications of the AC/SC framework to distinguish between the government as an arbiter in domestic political decision making and the government as a representative of national interests in negotiations with other states. In this context, national interests can be driven endogenously by the political process described above and exogenously based on the geopolitical goals of states. They may also be influenced by the logic of appropriateness, or social norms at the international level, as well as the logic of consequences, or pure interest in the costs and benefits of negotiation outcomes (Young 2001).

International sources of state influence can be divided between *soft power* (the ability to persuade others through argument or social pressure) and *hard power* (the ability to alter other states' behavior through the threat or use of military force or economic sanction; Keohane 1977). In modern international negotiations, there is a general norm against use of military power to control marine resources. Use of economic sanctions or the power of pursuasion can be enhanced by access to more powerful states or institutions, by alignments with other states that facilitate coalition formation, and by leverage, which can take the form of side payments or issue linkages. Side payments involve provision of some reward for cooperation, such as economic development aid or special trade status. Issue linkages occur when states make their position on one issue contingent on another state's action on a different issue (Barrett 2003). Because many international treaties are based on the norm of consensus, states may also use their *de facto* veto power to block measures that they do not like. The threat of blocking a measure may also be used as leverage to force changes in a proposal. Peterson (1995) provides an exhaustive analysis of the institutional strategies used by "leaders," "followers," and "laggards" in international fisheries negotiations. Webster (2009) describes responsive governance in the same context. Lastly, where there are strong social norms at the international level, states may use the logic of appropriateness to advocate their preferred international management measures.

As shown in the bottom right-hand corner of figure 1.10, both domestic and international decision processes can be affected by exogenous changes in international law and by widespread philosophical shifts related to political economy and governance (Young 2002). For instance, the rise of capitalism had profound effects on the economics and politics of fisheries governance. With the establishment of the rule of law in much of the world, use of private violence to exclude outsiders and punish non-compliance with informal institutions became illegal in many countries. Similarly, the expansion of democracy altered both the ways in which fishers could affect governance and the involvement of government in fisheries management. In addition, institutional interplay between fisheries-management institutions often occurs, as when measures developed by the International Pacific Halibut Commission proliferated rapidly during the 1930s and the 1940s because of perceived successes (horizontal interplay) or when states chose to extend their claims over marine resources first to three miles and then to 200 miles from shore (vertical interplay). Though these legal changes affected all fisheries, they were driven by complicated political processes tied to multiple issues, many of which had no links to fisheries. At the national level, decision makers often choose to invest in expansionary management measures because of broader macroeconomic goals. Shortages during the two world wars drove much of the concern regarding food security and control over resources more broadly. More recently, the upsurge in environmental concern in some countries increased the power of environmental interest groups and profoundly altered the management of marine fisheries around the world.

Details of the elements described above and of the institutions created by the fisheries action cycle will be covered throughout the rest of this book. For present purposes, it is important to point out that political responses can eventually switch an AC/SC system from cycles of depletion to cycles of rebuilding. Moreover, these responses lead to the creation of institutions that can foster more rapid response to future crises. This is how many of the most successful management regimes in existence today originated. However, as long as the endogenous and exogenous drivers of economic expansion persist, the task of fisheries management becomes more difficult. On the one hand, the rebuilding of the stock reduces costs associated with core problems and also diminishes political will to maintain restrictive management. On the other hand, with higher profitability due to declining costs of production and increasing prices, incentives for fishers to lobby against restrictive management or simply refuse to comply with regulations increase. Where stable regimes existed for centuries in isolated fisheries, the expansion of industrialized fleets and market access can rapidly undermine existing institutions. All of these factors create the management treadmill. Rather than stable, long-term governance solutions, the AC/SC application to fisheries predicts cycling between effective and ineffective management and suggests that governance is a chronic problem with many underlying causes rather than a permanent solution to the tragedy of the commons.

1.4 Methods and Organization

In view of the two primary goals of this book—to explain the current scope and scale of global fisheries and to explore how responsive governance works in a wide array of situations—the analysis presented is an exercise in synthesis rather than hypothesis testing. In the tradition of classical political economy, I use a historiographical approach, guided by the AC/SC framework. Historiography involves *process tracing*—that is, documenting how systems change over time. According to Collier (2011), process tracing allows researchers to identify and analyze causal connections between events. Collier further argues that this approach should be guided by some conceptual framework and that it may be used to establish "recurring empirical regularities" in different spatial-temporal contexts (824). The AC/SC framework provides the necessary expectations regarding prior knowledge and the primary purpose of the analysis is to identify regularities in the functioning of responsive governance given the various endogenous and exogenous forces detailed above. The research conducted is global in scope and extends from the archeological record through modern times. This is in part to determine if responsive governance is indeed ubiquitous and in part to advance our understanding of change in the profit disconnect, the power disconnect, and the management treadmill given a wide range of structural contexts and exogenous conditions.

As a method, historiography has a long and somewhat mottled history, particularly among social scientists. In his historiography of historiography,

Breisach (2007) argues that the approach evolved from a means of reinforcing existing hierarchies and cultures via constructed association with divinity to a means of critically examining the nature and origins of power. Most classical political economists used historiography. Smith ([1776] 1976) used an inductive approach; Karl Marx ([1887]1977) used the dialectical lens to describe history as a deterministic process. By the turn of the 20th century, the primary focus of historiographies in political economy was the explanation of continuity and construction of "driving forces" in history. With the crises of World War I and World War II, attention shifted from continuity to change, as scholars could not ignore the major political and economic shifts of the times. Finally, in the post-war period, a new wave of historiographies that described history as a cyclical process of stability and change united these perspectives (Breisach 2007). Knutsen's 1999 work on the rise and fall of world orders is a good example, as is the historiography presented in this book. Particularly because of the action cycle, the focus throughout this book is on cycles, whether they are cycles of economic expansion (Part I) or related cycles of governance response (Part II).

Breisach (2007) also documents shifts in historiography from highly subjective narrative approaches to the positivist focus on empiricism and the post-positivist acceptance of the subjectivity of knowledge. This pattern parallels similar trends in many fields, particularly political science and policy analysis, in which constructivist approaches were developed in response to the lack of realism in the rationalist worldview. However, Lejano (2006) and others note that both schools are in fact deficient, the rationalists focusing too much on structural elements (e.g., game-theoretic constructs) and the constructivists giving too much power to the actions of individual policy makers. The AC/SC framework fits well with what Lejano calls the *experiential approach*, which allows for both action and structure to affect decision making, but it also crosses levels of analysis and so is not focused only on policy makers. As such, the process tracing that I conducted for this book connects the experiences and actions of individuals, decision makers, and states. It is post-positivist insofar as all actors are assumed to be boundedly rational—indeed, responsive rather than proactive in their decision making—but does not abandon structure entirely. Rather, as has already been noted, the AC/SC framework is designed to provide a middle path between these extremes, so I cite empirical evidence wherever possible but I also accept the subjectivity of actors' experiences and related effects on outcomes.

The AC/SC framework serves as a guide for this exploration of the historical evolution of cycles of responsive governance in three primary ways.

First, when I researched specific fisheries in a given time period, I used the iterative process described above and summarized here: (1) identify elements of the structural context at a logical start point and (2) use process tracing to document how this base-line context changed with revolutions of the action cycle and pressures from exogenous forces. This analysis also involved identifying connections between fisheries (e.g., when overexploitation in one fishery generated exploration and eventual exploitation of a new fishery) and allowed me to sketch out larger patterns of geospatial change. During the initial phase of the analysis, I also looked for deviations from the expectations established in the framework. For instance, while conducting a broad, general search of the literature on global fisheries, I also searched specifically for cases in which governance was proactive rather than responsive, or in which the behavior of actors deviated from the AC/SC expectations. Although I did not find many exceptions, those that were identified are detailed in the text. Because of the extensive scope of the analysis, other fisheries described here are either representative of a large number of similar cases or are important turning points in the historical narrative, such as the development of a new technology or a systemic shift in global institutions.

Second, I used the AC/SC framework to compare the cases available within each period and to identify commonalities or differences. This coincides with Collier's (2011) "identifying empirical regularities". In this analysis I grouped fisheries on the basis of similarities in the structural context, including actors, resources, and governance institutions. Although I found great variation in the specifics, many of these factors could be classified fairly easily into broad typologies. For instance, although fishers in different parts of the world use very different types of gear, they can still be classified by scale, technological advancement, and capital requirements. Similarly, governance institutions can be categorized on the basis of broad decision processes, management capacity, and the types of regulation available. I also compared the speed of the action cycle and the predominance of specific exogenous factors in different eras. For example, for much of human history population growth was the primary exogenous driver of expansion of fishing effort (typology 1), with the exception of ancient kingdoms such as Sumer, where economic growth and technological advancement also played important roles (typology 2). This early historical period lasted for millennia; the action cycle was extremely slow because fisheries technologies and related resources were limited. The process of typology identification allowed me to recognize meso-scale trends and counter-trends in the historiography. It was also useful organizationally, as it helped me to identify the most representative cases for use as illustrations in the text.

Third, I used the AC/SC framework to identify broader patterns of change over time in order to develop theoretical insights into the evolution of fisheries governance. Practically, the theory-building process reveals important levers for improving fisheries management, but it also shows that management tends to cycle back and forth between downward spirals of overexploitation and positive cycles of rebuilding, as shown in figure 1.1. I call this dominant pattern in fisheries governance the *management treadmill*. Furthermore, there can be cycles within cycles. When fisheries are isolated, the treadmill displays a pattern of dynamic equilibrium, either because of the crisis rebound effect, in which the rebuilding of the stock relieves political pressures for restrictive management, or because of natural cycles in fish population dynamics. When fisheries are open to the exogenous drivers described in Part I, the treadmill follows a *growth cycle* pattern in which expansion of governance institutions is followed by an escalation in fisheries problems, reducing the effectiveness of fisheries governance and causing a return to bioeconomic depletion. In a growth cycle treadmill, waves of change propagate across fisheries, speeding up the action cycle and reducing switching response time. Drivers can be endogenous or exogenous, and can fit into political-economic and environmental categories.[23]

This book is loosely organized around key periods in the evolution of global fisheries management, with parallel segments that cover the economics of expansion (Part I) and the politics of response (Part II). In Part I, chapter 2 describes how population growth and other factors drove exploration throughout history, expanding the area subject to exploitation from small, coastal zones to almost all of the oceans. This extended the amount of exploitable biomass, increasing the sustainable level of production. Chapter 3 covers responses that widen the profit disconnect by reducing the marginal costs of production relative to stock size. These responses include investment in engineering to reduce costs of production and increase the technological efficiency of fishing fleets. It shows how this key fisher responses combined with exogenous technological changes to increase the scale and composition of production. Chapter 4 focuses on price signals and the profit disconnect. It describes the responses associated with opening up new markets, through improvements in processing and transportation technologies and advances in marketing strategies. These responses negate the price signal and further allow the open access level of effort to increase beyond the sustainable level of effort.

Chapters 2–4 are closely intertwined, and the responses discussed in them are heavily interdependent. For instance, expanding production would not have been as successful without improvements in processing and marketing, because demand would have been stagnant while supply increased, driving prices down and thereby amplifying the profit signal. Furthermore, all responses are deeply dependent on exogenous factors, such as technological innovations from other sectors, population growth, economic development, and globalization. Combined, these forces generate cycles of economic expansion that constantly widen the profit disconnect because increases in the sustainable level of production due to exploration are substantially outweighed by factors that dampen cost and price signals, causing exponential increases in the open-access equilibrium level of production.

Part II uses the AC/SC framework to show how political systems coevolved with these cycles of economic expansion through the process of responsive governance described above. Chapter 5 shows how incentives to exclude outsiders shaped management at the local and global levels. This includes collective action associated with endowments and entitlements as well as lobbying for government protection of access rights, which ultimately required the establishment of the rule of law and led to international negotiations regarding the Law of the Sea. Chapter 6 discusses the application of expansionary management measures globally, whether driven by demands from fishers or by exogenous political concerns with economic development and food security. This response widens the profit disconnect, ratcheting up fishing effort and exacerbating problems associated with the tragedy of the commons. Chapter 7 covers fishing industry lobbying on restrictive management measures that are put in place for either conservation or for exclusion. It shows how lobbying tends to delay management as long as fishing is profitable for the more powerful sectors of the fishing industry, but can also speed up management when the price signal is clear and profits are down. It links the profit disconnect to the power disconnect, showing how modern fisheries governance was shaped into a process of managed overexploitation. Exogenous changes in demand are important here, along with environmental changes, whether caused by fishers or by other forces.

Most of Part II focuses on fishers and decision makers. Until recently, few other actors were involved. However, in keeping with the AC/SC framework, chapter 8 describes the rise of new actors, including environmental activists, recreational fishing interests, and interdisciplinary groups of scientists, toward the end of the twentieth century. Environmental movements started to protect charismatic megafauna, and then expanded to encompass overfishing of commercial species in response to the collapse of major fisheries in different parts of the world. The overall effect of non-commercial interests is difficult to gauge because their efforts could incite considerable opposition from fishers, thereby delaying response. However, when fishers and non-commercial interest groups worked together, they often achieved improvements in management.

Finally, chapter 9 synthesizes the lessons learned from analyzing the economic trends covered in Part I and the growth of political institutions documented in Part II. It describes the management treadmill and its multiple variations. Insofar as climate change and other drivers of the management treadmill are largely exogenous to fisheries governance, one can conclude that good fisheries management is necessary but not sufficient for sustainability. That is, management is necessary for grappling with the endogenous factors that drive fisheries expansion, including but not limited to the CPR problem, but it is not sufficient to rein in the exogenous drivers identified in the AC/SC model. Therefore, long-term sustainability in global fisheries depends on a worldwide transition to sustainable development. In the concluding section of chapter 9, I highlight important parallels between the literature on sustainable development and the lessons learned from application of the AC/SC framework to fisheries case studies. From this perspective, the tragedy of the commons is less important than broader exogenous drivers of overexploitation and fisheries are shown to be much more closely related to other environmental issue areas than is usually acknowledged in the literature. Furthermore, some of the lessons learned from the case studies, particularly regarding the effect of the profit disconnect, the importance of legitimacy, and the need for accountability, are also key components in the search for sustainable development in the global economy.

I The Economics of Expansion

2 Exploration

The AC/SC framework predicts cycles of geographic expansion in fisheries as actors respond to local bioeconomic limits through exploration. Specifically, fishers will search for new fishing grounds in response to the profit and catch signals generated by the core problems of overexploitation, overcapitalization, and ecosystem disruption. Although some fishers might be driven to explore by desperation in the face of severe overexploitation, the more adventurous choose to set out in search of richer waters well before their home fisheries reach the point of stagnation or collapse. Fishers who are able to explore for new fishing grounds do not increase the profit disconnect at the local level, but they do generate regional and even global profit disconnects. They are often supported by processors or marketers who provide access to capital in order to maintain or increase flows of raw materials for their businesses. This process is endogenously driven by the common pool resource (CPR) dynamic but may also be exogenously driven by population growth and increasing affluence, both of which speed up the action cycle. On the other hand, this exploration is limited exogenously by the level of technological knowledge that is available at any given time. As long as fishers have sufficient access to capital, any innovation in navigation technologies or in vessel technologies can catalyze the rapid expansion of fishing effort into new locations

This chapter describes the expansion of fishing effort from the numerous small, isolated fishing communities of prehistoric times to the vast global scale of modern industrialized fishing operations. Section 2.1 covers the long period of coastal settlement, during which fishing and vessel technologies were largely stagnant, exploration occurred only along the coasts, and expansion was driven mainly by population growth. During this period, which spanned more than 11,000 years, expansion occurred sporadically and technological innovations were most often associated with the rise of great civilizations, including those of Egypt, Phoenicia, and Rome, and concomitant periods of technological innovation and growth in demand. As will be shown in section 2.2, this pattern changed in the 14th century, when innovations in vessel and navigation technologies allowed fishers to venture farther from shore, though most fishing stayed within 100 miles of some coast until the late 1800s. In a few hundred years, the area exploited by commercial fishers increased by several orders of magnitude compared to preceding millennia. Section 2.3 covers the last phase of expansion to distant waters, which was catalyzed by the introduction of steam engines in the 19th century and then the development of combustion engines in the early 1900s. Combined with numerous other technological innovations, these new engines catalyzed rapid expansion in the geographic extent of fisheries during this period. Few areas of the world are now left unexplored by fishers, and we are rapidly approaching global limits on exploration as a viable response in the fisheries action cycle.

2.1 Coastal Settlement

As their populations increased (see figure 2.1), the earliest peoples living near rivers, lakes, and oceans needed easily collected sources of protein. Most of them lacked appropriate implements with which to catch mobile aquatic creatures, but they could dig up mollusks like clams, mussels, and oysters[1]; evidence that they did so comes from the piles of mollusk shells found in the kitchen middens of archeological sites. As prehistoric peoples ventured out in search of new lands, they often followed watercourses. Thus, fishing activities first spread out along rivers and coastlines with the dispersal of human populations (Walpole 1884a). Some societies chose where to settle based on the availability of fish and other resources (Agarwal 2007).

With continued population growth, food production from the sea increased in scope and scale. Fishing was of great importance in Japan and China as early as 10,000 BCE and in India by the Mesolithic period (20,000–9,500 BCE; Kalland 1995; Yen 1910; Agarwal 2007). Ancient Egyptians fished extensively both for recreation and for sustenance (Adams 1884a). The earliest written record of commercial fishing activities was found in Sumer, a land fed by the Tigris and the Euphrates. Clay tablets dated to about 2300 BCE detailed the employment of hundreds of fishers organized in guilds with well-established access rights. As population pressures drove overexploitation of wild fish, some Sumerian guilds created artificial ponds and stocked them with valuable fish species (Royce 1987). Farther to the

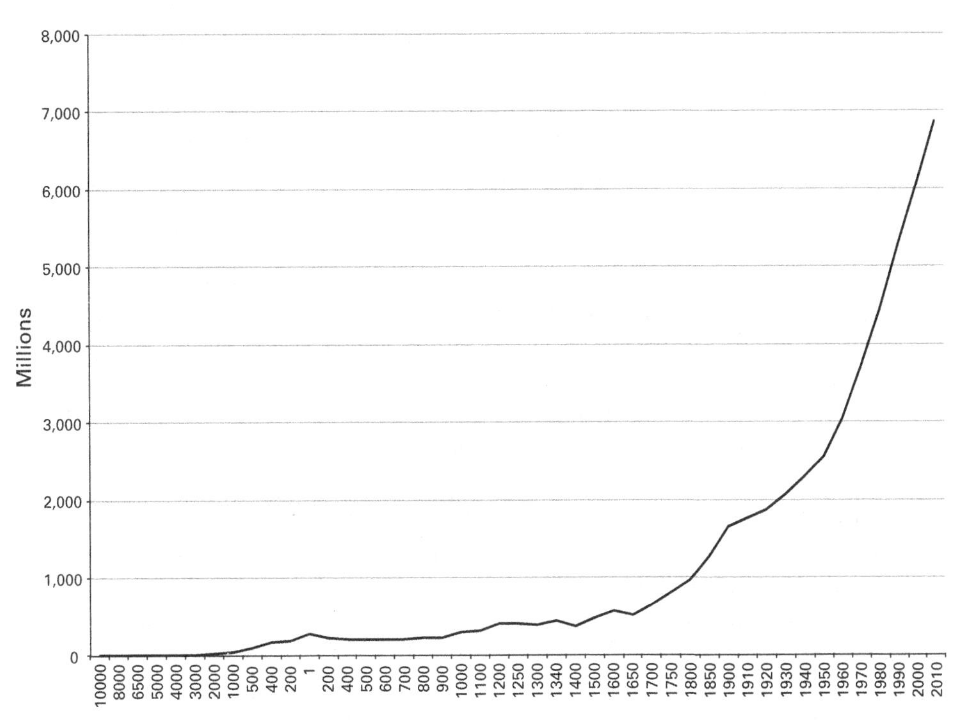

Figure 2.1
Population estimates from 10,000 BCE to 2010 CE. Sources: US Census Bureau 2013a,b.

east, aquaculture also started on inland waters, mainly in China, around 1000–2000 BCE (Rabanal 1988).

The first historical evidence of the CPR dynamic and related economic responses is observed in places like Sumer, where both populations and economies boomed. These ancient centers of commerce were also hubs of innovation that provided fishers with the vessel technologies they needed to expand their effort through exploration, thereby exogenously altering the resources available in the structural context. In Polynesia and a few other areas, there is evidence that large sea-going rafts with sails were used in prehistoric times, as these would be necessary to complete observed migrations. A birchwood paddle found in northeastern England that dates to 6000 BCE is the first direct physical evidence that human beings used boats to traverse inland waters. These were small, ephemeral vessels that would not last long at sea (Gould 2011, 92). However, commercial exploitation of fish farther from shore requires stable, decked ships with the capacity to store large quantities of fish. The earliest evidence of sea-going ships dates back to Egypt in about 2450 BCE, during the Old Kingdom, when population growth, economic development, and technological innovation converged to drive major shifts in many artisanal industries, including fishing (124). At that time, the Egyptians still navigated by using coastal landmarks, which confined their boats to near-shore waters and well established trade routes.

Driven by similar factors, other civilizations continued to advance vessel construction and navigation technologies. Early boats evolved via the Viking culture in northern Europe and on China's many rivers. During this period, the Phoenicians also worked to improve the design of Egypt's original vessels, making them larger and more stable. These innovations were central to the Phoenician dominance of trade in the region. Later, the Romans continued to improve vessel and navigation technology in order to expand both trade and territory. Population growth and economic growth associated with the rise of cities fueled these technological advances and also drove increased demand for fish products. This widened the profit disconnect, leading to increased fish production and overexploitation around ancient population centers. Technological constraints ensured that fishers remained in coastal areas, so increased demand caused higher concentrations of fishing effort in relatively small areas. Fishers could not move far from coasts until the middle of the 13th century, and even then only a few highly valuable species were commercially targeted far from land.

The lack of ocean-going vessel technology was an important limit on early geographic expansion, but other factors were also significant. On one hand, expansion was easier in some regions because fishers could spread out along uninhabited coasts and did not need to take on the risks associated with open-ocean voyages. On the other hand, fish products were highly perishable, and fishers had not yet developed many methods for preserving their catch. Drying and salting could only preserve fish for a week or two, so transport to inland areas was minimal, constraining demand outside of growing population centers (see chapter 4). Perhaps most important, fishing gear was not very efficient, and in most areas fishers had neither the ability to overexploit coastal resources nor the surplus production that would enable them to amass sufficient capital to invest in larger vessels (see chapter 3). The same was not true of inland fisheries, many of which were severely depleted even in prehistoric times (Royce 1987). Except for the great civilizations noted above, most of the early expansion of marine fisheries can be attributed to the movement of populations in search of new lands rather than to the movement of fishers in search of more fish.

2.2 Venturing Out

In the 1300s, this pattern of coastal expansion began to change as some fishers started moving farther out to sea. There were both exogenous and endogenous drivers of this transition. Increasing demand was a particularly important driver in Europe, where coastal fisheries were already highly developed. In this period, economic growth and cultural preferences amplified the effects of population growth, causing prices for fish products to increase substantially. Culturally, the rise of Catholicism generated more demand for fish, since it was the only type of meat Catholics were permitted to eat on Fridays and other holy days. These combined factors reversed the price signal and increased incentives to explore new fishing grounds and travel farther to find fish (Johnston 1965; Beaujon 1884). This also led to a widening profit disconnect in commercial marine fisheries throughout the Europe, as costs declined with increasing technological innovation, concentration of capital, and trade, while prices rose due to growing demand and improved preservation technologies.

Although the demand driver was important at this time, expansion beyond the coastal zone would not have been feasible without better technology. Here again, available resources constrained exploration while technological innovations catalyzed growth. First, the development of sails made longer trips possible. The introduction of steam-powered fishing vessels in the 1800s and vessels powered by internal combustion engines in the 1900s enabled fishers to travel even faster and farther. Second, people could also build much bigger and more stable boats than in the past, which provided additional storage capacity that made fishing trips on the high seas profitable (J. T. Jenkins 1920). Third, navigation and cartography were greatly improved as well. Most of the important fishing grounds were well known by the 16th century, and most of the world's oceans and continents were charted using longitude and latitude by the 18th century, leading to safer and more accurate navigation (R. A. Gould 2011).

Whaling was a particularly important component of European exploration and the endogenous advancement of fishing vessel technologies. Whalers were the first fishers known to systematically venture out onto the high seas. High prices for whale oil drove Europeans, Japanese, and fishers of other nationalities to target cetaceans throughout the oceans (Adams 1884a; Gibbs 1922; Beaujon 1884). In Europe, the Basques developed the first large fishing vessels, which they called "baleiniers." They used these boats to exploit local whale populations in the Bay of Biscay and, once these stocks were depleted, they explored other areas to find new whaling grounds (J. T. Jenkins 1921). Even without the aid of a compass, they captured whales off the coast of Newfoundland in 1372, more than 100 years before Columbus' famous "discovery" of the Americas (64). Their vessels were much bigger than those of any other fishers at the time. One baleinier was recorded at 400 tons whereas English, Breton, and Norman vessels were no bigger than 50 tons (66).[2] This is one instance when technology developed for a fishery transferred into other "industries," particularly the privateering and exploring that made the Basques a major power in the 12th and 13th centuries (61).

While chasing whales, the Basques discovered one of the richest fisheries in the world: the abundant cod stocks off Newfoundland, Greenland, and Iceland. They may even have harvested Georges Banks cod off the coast of what is now known as New England. Hearing of these rich fishing grounds and of the economic success of Basque fishers, British fleets started exploiting cod near Iceland in the 1400s and then moved on to other regions in the western Atlantic (J. T. Jenkins 1921). By the time that John Cabot "discovered" and claimed the Grand Banks for England in 1497, fleets from several different nations had established fisheries in the area. In the 15th century, Portuguese and French fishers (many of whom were Basques) dominated the Grand Banks fishery. The Spanish overtook them in naval and fishing power in the middle of the 16th century, largely because of growing wealth from their colonial empires and related availability of capital to invest in the large vessels needed to make such long trips profitable (Anspach 1819; J. T. Jenkins 1921).

Even though technologies continued to improve, the journey to the Grand Banks was still costly, long, and perilous for European fishers. In the late 1500s, British and French colonists began to establish permanent settlements in New England, Nova Scotia, Acadia, Quebec, Newfoundland, and other northern territories so as to better exploit the fishing resource (Anspach 1819). European fleets fished on the Grand Banks only in the summer months, because winter storms were much too dangerous for long voyages; colonists could fish all year round because the fish moved closer to the North American coast in the winter, which allowed for short trips when the weather was favorable (Goode 1884; Innis 1940). Cod was one of the important commodities that moved in the triangular trade in which slaves and guns were exchanged for sugar, tobacco, and other agricultural products. Salted cod was sent from English and French colonies in northern regions to Europe and the West Indies in exchange for molasses, guns, and other goods (J. T. Jenkins 1920).

In the 16th and 17th centuries, with the decline of the Spanish fleets, Newfoundland fisheries were again left to the French and Portuguese fishers, who continued to fish in the region. In the 18th century, the British became the dominant naval and fishing power in the North Atlantic (Innis 1940). This was largely due to the Industrial Revolution, which allowed for the technological innovation and growth in capital that was necessary for the continued geographic expansion of fishing effort.[3] By the 19th century, coastal fisheries around Europe were already overfished and even the northwest Atlantic cod fishery was highly overcapitalized. However, because of the appropriation of steam engines by fishers, British vessels were able to exploit new territories. A single vessel could fish off Morocco during the winter months, near Iceland in the spring, and off the coast of Newfoundland in the fall. Costs of production were high for first-class steam trawlers, at about £9,000 (US$1.02 million in 2010 dollars) per year, but revenues could be as great as £10,700 (US$1.21 million in 2010 dollars), so the fishery was quite profitable (J. T. Jenkins 1920, 22)[4]. The steam trawlers of the late 19th century and the early 20th century were only able to succeed because they were many times more efficient than coastal vessels, which operated at much a smaller scale. In turn, they relied on fish from many different parts of the world, mitigating risk by diversifying their fishing grounds instead of their targeted species.

Coastal expansion like that described in section 2.1 also continued from the 16th century on due to colonization. Like their European counterparts, immigrants to North America were adept at commercial exploitation of fisheries, which helped feed growing populations and generated considerable revenue. Some locations, including Newfoundland and Cape Cod, were chosen by settlers because they provided access to important fishing grounds. Other settlements, such as Jamestown, relied heavily on fish for subsistence, particularly in their early years (Goode 1884). By the late 1700s, fisheries were exploited commercially along the east coast of the United States and Canada, up through the Grand Banks, off the shores of Greenland and Iceland, and all the way to the coasts of Europe and North Africa. By the 1800s, commercial fishing was a lucrative industry all along the west coast of the Americas, from Alaska to Mexico. The British also started exploring fisheries in the South Seas during this period (Adams 1884a).

In the rest of the world, fisheries remained mainly artisanal until the early 1900s. Data on these fisheries is not widely available, but estimates from Japan—which was the most technologically advanced non-European fishing country at the time—provide some indication of the difference. Unlike the Europeans, the Japanese did not target a few high-value species far from shore; rather they employed thousands of small vessels to target many different species in their coastal zone. By the 1880s, when British fisheries employed about 118,000 fishers in approximately 37,000 vessels, more than 1.5 million fishers were plying the waters around Japan in more than 187,000 small boats called "junks" (Walpole 1884b; Okoshi 1884). In spite of their numbers, Japanese fishers produced only about 22,600 tonnes of fish, whereas British fishers produced more than 20 times that amount with a much smaller work force (Okoshi 1884; Walpole 1884a). This difference can be ascribed to variations in access to capital and vessel technologies but may also be related to fisheries governance (see chapter 5). Nevertheless, expansionary dynamics continued to drive exploration as historically dominant fishers were driven farther out to sea and fishers from other countries built up their own modern, industrialized fleets.

2.3 Distant Waters

The Industrial Revolution changed the world's fisheries radically. With steam- and gasoline-powered engines to get them farther faster, mechanical refrigeration devices to keep the catch from spoiling over long trips, and an almost virgin resource just waiting to be harvested, those fishers who were dissatisfied with the now overcrowded and depleted conditions in historically exploited fishing grounds were able to follow the profits to new territories. As in earlier centuries, whalers led the way, exploring north into the Arctic and south through the tropics. In 1891, steam-powered whaling vessels ventured into Antarctic waters. Although the first such trips were not successful, the Antarctic whaling fishery was well established by the early 1900s (Jenkins 1921). Figure 2.2 documents known whaling grounds in the middle of the 19th century. By this time, whale stocks in the Atlantic were so heavily overexploited that most whalers worked in the Pacific, Indian, and Southern Oceans. These were the first vessels in what are now referred to as *distant-water fleets,* which operate on the open ocean, hundreds and even thousands of miles from shore.

With the proliferation of improved technologies and the increasing globalization of trade, geographic expansion occurred in other fisheries as well. In the short period from 1900 to 1939, European fishers took their boats east into mid-Atlantic regions and south to the fertile fishing grounds off the west coast of Africa around the Gulf of Guinea. From New England, US vessels moved south into the Gulf of Mexico, then through the Panama Canal to the eastern Pacific Ocean. Canadians moved on the opposite tack, heading north into waters that had been too ice ridden and foreboding in

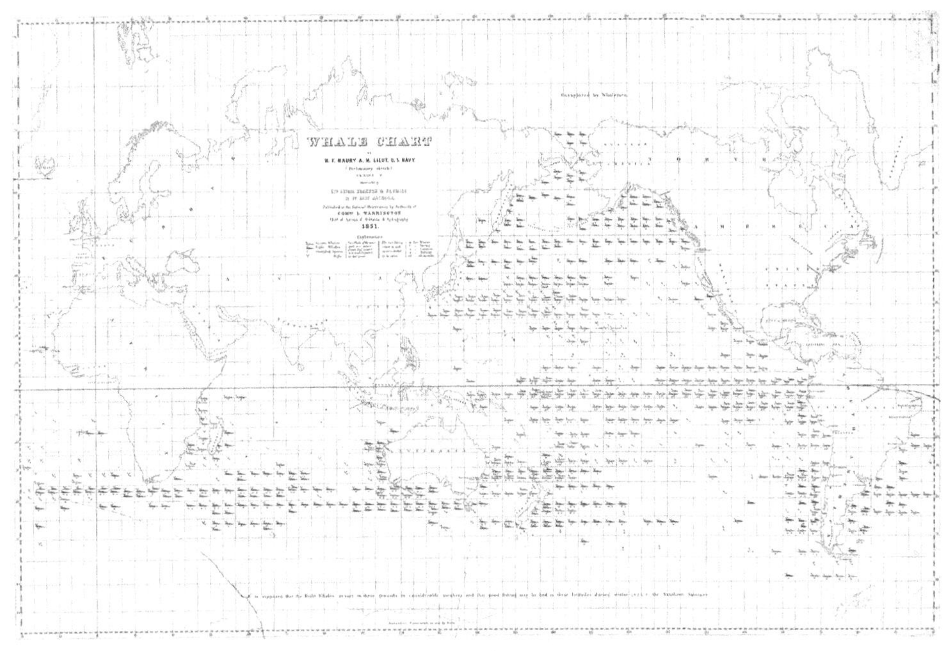

Figure 2.2
Whaling grounds, by species and season, circa 1851. Source: Maury 1851.

the days of wooden ships (see chapter 3). The Japanese built up industrial fleets and explored in several directions, first to their colonial possessions (pre–World War I), then to Taiwan and Singapore, and finally to the Indian Ocean and the South Pacific. Soviet vessels also started fishing intensively in both the Atlantic and the Pacific (Royce 1987).

World Wars I and II only temporarily dampened the geographic expansion of global fishing fleets. By 1950, Japan was fishing in the Atlantic, European vessels were fishing in the Pacific, and more than 150 states or fishing entities had commercial fleets fishing somewhere on the world's oceans. The largest of the new entrants came mainly from countries in Latin America, including Mexico, Peru, Argentina, and Ecuador, along with some countries in East Asia like Korea, Taiwan, and Indonesia (FAO 2012a). Globalization was a major driver of expansion in this period, as it generated both increasing demand and plentiful capital for investment (see chapters 3 and 4).

By the middle of the 20th century, fisheries were exploited commercially across the North Atlantic and the North Pacific but commercial fisheries were just emerging in tropical and southern temperate waters. Thus, fisheries in those areas were still coastal in 1950. However, as fishers depleted local resources, they repeated the cyclical pattern of exploitation and expansion that had already occurred in the North Atlantic. Distant-water vessels moved further out onto the high seas, and the fishing industry invested in the development of port facilities in many out-of-the-way places to harbor these mega-fleets.

This progression of fishers from coastal areas to the high seas can be seen when catch statistics are plotted against a map of the world. The UN Food and Agriculture Organization (FAO) provides a tool that can be used to explore the spatial distribution of landings of key *highly migratory species*—fish that are frequently found in the open ocean, many of which migrate exceptionally long distances (FAO 2011). Commercially exploited highly migratory species include multiple species of tunas as well as billfish, particularly swordfish. Figure 2.3 shows how quickly the fisheries for highly migratory species expanded between 1950 and 1975. There was little demand for these species before World War II, so part of the growth shown in the figure is due to market creation. The figure also shows how quickly a fleet can expand when capital and technologies needed to exploit a virgin resource are readily available.

In 1950, tuna and billfish (mostly swordfish) were captured mainly in the North Atlantic, although there were exploratory fisheries off the coast of southern Africa and in the tropical Pacific. Five years later, extensive fisheries for these species existed in the Pacific, probably owing to the rebound of Japanese fleets after World War II. Tuna was an important staple of the Japanese diet before the war, and Japanese tuna fleets quickly spread out once postwar restrictions were lifted (Suisankyoku 1915; Finley 2011). At this time the Japanese also built up a fleet of fishing vessels in Tema, Ghana, in order to start a fishery for skipjack tuna in the Gulf of Guinea. By 1960, US tuna canneries were operating in Southern California and US fishers targeted highly migratory species on the Pacific coast as well as in the tropical Atlantic. Europeans also began targeting tuna in this period, increasing exploitation in the Atlantic and the Pacific. Within ten years many of these fleets were fishing throughout the Indian Ocean and in parts of the Central Pacific (Miyake, Miyabe, and Nakano 2004).

In subsequent years, fisheries spread even farther. In some cases, fishers found new stocks of tuna and other popular species, but in other cases they began to exploit new species—for example, Patagonian and Antarctic toothfish, also known as Chilean sea bass. There are now Arctic and Antarctic fisheries for krill, toothfish, and other cold-water species. In fact, because of climate change, areas once closed to fishing because of extensive ice

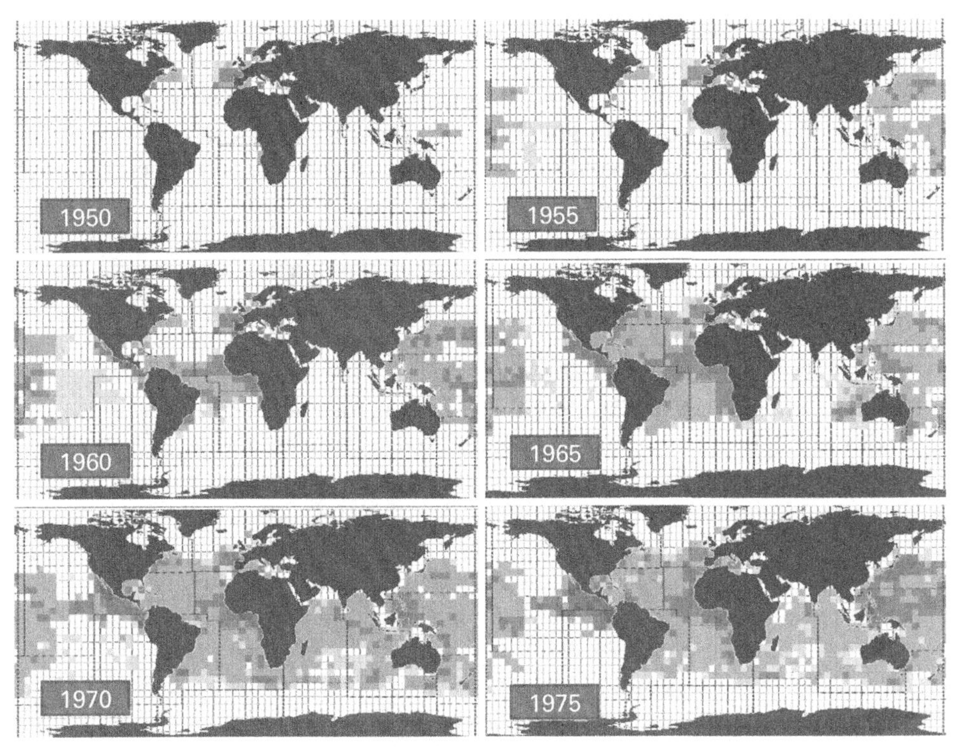

Figure 2.3
Distribution of total tuna and billfish landings by catch location. Source: FAO 2011.

cover are opening up in the Arctic and the Antarctic, and thus the total area fished may expand even more in the near future (Brander 2007).

Indeed, the only new areas available for exploitation are near the Poles where ice is retreating as a result of climate change, and in the deep sea, where marine life is sparse and very difficult to exploit. In contrast to historical exploration, entry into these new areas is already partially curtailed via regulation. In the Southern Ocean around Antarctica, most fisheries are managed by the Commission for the Conservation of Antarctic Marine Living Resources (CCAMLR), which takes a precautionary approach to fisheries management and severely restricts the entry of new vessels into the area. In addition, the fishery for southern bluefin tuna is controlled by the Commission for the Conservation of Southern Bluefin Tuna (CCSBT), which also limits entry into the fishery. As demand drivers widen the profit disconnect further, pressure for exploration of new fisheries in the Southern Ocean may overcome regulatory restrictions, leading to greater geographic expansion of illegal, unreported, and unregulated (IUU) fleets (Stokke 2009).

On the other side of the world, fishing in the Arctic Sea was limited for much of the 20th century by physical conditions, including large areas covered in ice and the risk of collision with icebergs in unfrozen areas. Even so, the Soviets experimented with fishing in the Arctic Sea from 1967 to 1971, catching just over 7,000 tonnes in the peak year (1968). Recently, Russian fishers began experimental fisheries in the area, landing small but increasing amounts—480 tonnes in 2008 and 589 tonnes in 2010 (FAO 2012a). The Arctic Ocean has no formal fisheries management organization, so fishers are free to explore the high seas there for the time being. However, states with an interest in the region are working together to move forward with a cooperative program to manage Arctic resources, and so exploration may be limited in the future. The area is rich in other resources and is also important for military and commercial shipping purposes (Molenaar 2012). Therefore, negotiations may take many years. In the meantime, the ice cover is likely to diminish further due to climate change, and so fishers from many countries will have an opportunity to establish operations in the area. Fishing in the Arctic may become even more attractive as fish stocks move north because of climate-induced changes in sea surface temperature and in ocean current systems (Cheung et al. 2010).

This chapter shows how fishers in many parts of the world were able to circumvent local bioeconomic limits through exploration. Furthermore, exploration tended to occur in cycles that varied in speed and intensity depending on endogenous and exogenous drivers, as would be expected based on the AC/SC framework. Technological innovation proved to be an important catalyst for expansion and eventually facilitated the evolution of modern global-scale fisheries. However, fishers and the fishing industry as a whole now face global limits on exploration as a response in the fisheries action cycle. Aside from the small increases in fishing areas created by climate change, which in any case are likely to be offset by ecosystem effects elsewhere, fishers have run out of space in which to spread out. This does not mean that they are out of economic options. They can invest in new technologies to increase the intensity of effort and also target and market new species, further increasing the amount of production in any given area. These two fisher responses are covered in the next two chapters.

3 Investment and Innovation

The second major fisher response predicted by the AC/SC framework is investing in innovative technologies that increase efficiency or otherwise reduce the marginal (per unit) costs of production for any given stock size. Unlike geographic expansion, innovation for efficiency does not increase the amount of fish available for exploitation. Instead, this response dampens the cost signal and widens the profit disconnect. For much of history, this effect was limited by the technologies available in the structural context, as explained in the general AC/SC application to fisheries described in chapter 1. However, the rate of innovation generally is increasing over time, which catalyzes additional innovation in fish production technologies. In fact, some economists estimate that, at its current pace, technological change could improve efficiency enough to completely negate any cost increases due to declining stock size (Hannesson, Salvanes, and Squires 2010; Squires and Vestergaard 2009). Access to capital is another limit on innovation that is determined by both resources and governance in the structural context. Since investment helps fishers and secondary industry actors to accumulate capital, it acts as an endogenous driver of innovation. Globalization of finance and trade can also magnify investment exogenously by providing greater access to capital in many parts of the world and fostering the escalation in technological advancement described above. Thus, the cost signal, which was quite strong for much of history, has been eroded over time through investment in innovation. This has important implications for our understanding of fisheries economics and the common pool resource (CPR) dynamic more broadly.

Most economists blame any and all efficiency improvements on the "race for fish" created by the CPR dynamic, but there is considerable evidence that such investments would occur even if fisheries were privately owned (Ward et al. 2004). Technological innovation is a hallmark of capitalism and occurs in all industries, including those in which private owners exploit natural resources (Grossman and Helpman 1994). In fisheries specifically, some of the earliest innovations occurred under conditions of enforced monopoly or well-established sea tenure. Of course, other important innovations occurred under conditions of open access, but only when sufficient capital and technological knowhow were available to fisheries entrepreneurs. Furthermore, the historical evidence clearly shows that some fishers acted as entrepreneurs while the majority either followed along or actively resisted changes (see chapter 5). One reason for this could be that most fishers did not have access to sufficient capital to obtain the new technologies, but it also stemmed from pure reluctance by most fishers to change their methods (Chapman 1966). In contrast, fisheries entrepreneurs such as Christian Salvensen (whose company commissioned the first factory trawler) and Harold Medina (who introduced the use of spotter planes) demonstrated persistent willingness to take large risks to improve their operations. When analyzing the growth in fishing efficiency, it is difficult to distinguish between the common pool resource dynamic and these entrepreneurial efforts. However, it is clear that the latter are important internal drivers of the profit disconnect and should not be ignored.

Heterogeneity in willingness or ability to invest in innovation generates stratification in the size of fishing operations. Strata are delineated both by the size of vessels used and by the number of vessels owned by a single individual or company. Here, an ecological metaphor is apt, since fishers and fishing companies of different sizes survive together by fitting into their own economic niches. For instance, even in the 1800s people recognized that smaller-scale vessel-gear combinations brought in higher-quality fish than the larger steam trawlers. Along with government regulation, this helped the sail-based fishery to survive and eventually evolve into a motor-based fishery with approximately the same scale and ownership structure (J. T. Jenkins 1920; Robinson 1996). Many other niches exist and can be defined based on local demand, quality-quantity tradeoffs, and even physical, social, or political limitations that inhibit economies of scale in fisheries. Nevertheless, fishers in each niche either find ways to reduce costs over time or they are forced to exit the fishery when overexploitation and overcapitalization set in. This competition—which, again, would occur even without the CPR dynamic—causes the profit disconnect to expand for every stratum of the industry as efficiency improvements shift the equilibrium level of production farther away from the bio-economic maximum.

The appropriation of new technologies is also stratified temporally, as fisheries evolve in cycles of slow growth followed by sudden technological advances. This is a dynamic process, but, unlike dynamic models such as those outlined by Clark (2005), these cycles do not converge to a stable equilibrium; rather, they move between periods of growth and recession that oscillate around a longer-term positive trend, as predicted by Schumpeter ([1942] 1976). However, the infinite growth predicted by Schumpeter is not possible, in fisheries or in other sectors, because of physical, biological, and ecological limits in the resource base. In particular, as has already been noted, the technological innovations in capture production do not expand the resource base; rather, they increase the rate of depletion in commercial fisheries due to improved effectiveness and because of increased investment in fishing capacity. This chapter shows how these cycles of innovation increased the efficiency of fishing operations over time, widening the profit disconnect and amplifying the core problems in the action cycle. Section 3.1 covers the evolution of vessel and gear technology and section 3.2 describes improvements in fish-finding technologies. Both of these sections focus on the roles of technological catalysts, globalization, and entrepreneurship as well as the CPR dynamic. Lastly, section 3.3 briefly shows how aquaculture is a separate response from other types of investment in efficiency and how, as practiced today, marine aquaculture is not a viable solution to the core fisheries problems of overexploitation, overcapitalization, and ecosystem disruption.

3.1 Vessels and Gears

Cycles of advancement in vessel and gear technologies can be divided into two categories: gradual change and sudden innovation. On one hand, fishers can make small, gradual modifications that steadily improve efficiency, either by reducing inputs (e.g., a more efficient engine) or by increasing their ability to harvest fish (e.g., gear improvements). On the other hand, stepwise improvements result when capital, technology, and entrepreneurship come together to introduce a new vessel-gear combination that is substantially more effective than existing configurations. Research on individual fisheries suggests that many small changes to existing fishing operations can be important precursors to stepwise growth (Hannesson, Salvanes, and Squires 2010; Rijnsdorp et al. 2008). This section focuses on well-documented periods of stepwise transformation in fisheries technologies; it also shows how many of the advances of the 19th and 20th centuries contributed to the profit disconnect for fishers filling multiple economic niches.

3.1.1 Manpower

There are a number of ancient methods for catching fish without tools; some survive today in indigenous cultures and as sporting events. Because these techniques are not easy to master and tend to be time consuming, early fishers had considerable incentive to invest their energies in the development of more effective technologies. Over time, people developed fishing gear of different types, including projectiles such as spears and harpoons, many kinds of nets and traps, and fishing lines, which were used with various types of poles, hooks, and bait. Artisanal fishers still use small-scale labor-intensive devices and so are much less efficient than commercial fishers (FAO 2012c). In the past, fishers were constrained both by existing technologies and by access to capital; in modern times the latter is the primary determinant of the scale and the efficiency of fishing effort. This subsection describes how various gear improvements evolved prior to the industrialization of fisheries in the 1800s.

Line Fisheries Much as with exploration, population growth and the development of civilizations drove innovation in fishing technologies for much of history. Often, communities would invest in large-scale technologies such as beach seines, fish traps and weirs, or relatively large boats in order to take advantage of the periodic appearance of schools of fish or pods of marine mammals (Kalland 1984; Royce 1987; J. T. Jenkins 1921). To increase commercial production of the popular bonito, early Japanese fishers developed a rudimentary version of the pole-and-line technique that is now used by industrial-scale vessels called "baitboats." A fisher using this method first "chums" the water with baitfish. Once the targeted fish are in a frenzy, the fisher hooks individual fish and hauls them aboard using a device that is similar to a recreational fishing pole but which has a fixed, short line. Figure 3.1 depicts a contemporary version of this ancient technique. The largest Japanese pole-and-line vessels from the early 1900s accommodated 30–40 fishers working off both sides of a vessel and could store 18–36 tonnes of fish in onboard salt-water wells. They would sometimes travel hundreds of miles from the coast to fish (Suisankyoku 1915; Joseph 2003).

Off the coasts of Europe and the Mediterranean, hand-operated longlines (also called trawl lines) were the first gear known to have been used in commercial fisheries. Early hand longlines consisted of long ropes made of hemp holding many baited hooks that would be set in the water, usually weighted down to target demersal species. Lines were set and hauled in by a single fisher who would remove the catch, then rebait and reset

Figure 3.1
A Japanese pole and line vessel Source: Suisankyoku 1915.

the line. Valuable fish were kept; "trash fish" were discarded or used as bait (Gibbs 1922). This ancient technology persisted for many centuries with incremental improvements such as preservation of the lines through tanning (in Europe) or treatment with the juice from unripe persimmons (in Japan), which helped to prevent damage due to constant soaking in salt water (Okoshi 1884). Nevertheless, there were few step-wise improvements until the Middle Ages; even then, vessel technology changed rather than gear technology.

As noted in chapter 2, the growth of trade and increases in warfare around the Mediterranean led to the production of larger and faster sailing vessels in 14th-century Europe. Fishers with sufficient access to capital started purchasing ships called schooners or hookers, depending on the region and the time. These vessels were specifically used to target cod and other demersal species in the North Atlantic, using hand longlines. In that fishery, a large sailing vessel, typically with a tonnage of 35 tons (gross), carried about eight small boats called "dories" stacked on deck. Sailors doubled as fishers and vice versa as each man, or sometimes a pair of men, would take a dory out away from the ship and set a line up to a mile long, with hundreds of hooks. By developing this *mother-ship* configuration, European fishers took advantage of larger and faster sailing vessels that were exogenously developed for the navy and the merchant marine without any alterations to their fishing gear. Indeed, this innovation multiplied the power of a single schooner tenfold, since lines were often set from the main vessel as well as from the dories (Gibbs 1922). The primary drivers of innovation were increasing demand—and related price signals—in addition to the CPR dynamic.

It is important to note that hand longlines and gear of similar scales were also used during this period in coastal fisheries in Japan, China, India, and other parts of the world, but that those fisheries did not develop the mother-ship configuration. There are several reasons for this. First, in much of the world large vessels powered by sails were not available. Even in China and Japan, where such vessels certainly existed, they were not used for fishing, primarily because fishers had neither the status nor the capital necessary to purchase them. Stringent social norms reinforced the existing class structure, in which fishers were generally categorized as laborers, and prevented entrepreneurship of the magnitude needed to transform small-scale coastal fisheries into large-scale commercial enterprises (Kalland 1995; Yen 1910; Day 1884; Adams 1884a). Governance aspects of the structural context could also limit innovation, even in this early period. For example, Chinese regulations prohibited the construction of large fishing vessels as part of an anti-piracy campaign during the Qing Dynasty (1644–1911 CE). Officials feared that fishers would collude with pirates or turn to piracy themselves, so they limited fishers to small vessels that would stay near shore and were easier to monitor for illegal activities (Muscolino 2009). Therefore, even though demand was increasing in China just as in Europe, fishers were prevented from taking advantage of higher prices by regulation and lack of access to capital—both of which are important components of the AC/SC framework.

Net Fisheries In contrast, the merchant classes of Europe in the Middle Ages saw fisheries as a major foundation of their wealth and invested heavily in innovation. As a result, Europe was the primary crucible of fishery development until the early 20th century. The first great example of coordinated investment in commercial fishing comes from the Hanseatic League, a group of German towns that dominated production of Atlantic herring in the 13th and 14th centuries. The League is often called a proto-state. Though there was no sovereign, its actions were formally coordinated by a Grand Council. However, the Hanseatic herring fleets were Danish, not German, and centered on the port of Scania, which became famous for its high-quality herring products. As merchants and governors, the Hansa provided funds to help purchase these boats, invested in processing facilities along the Danish coast, and strictly regulated the industry to minimize costs and maximize prices. They also protected the fleets from pirates, reducing the costs to fishers even further (J. T. Jenkins 1920; Zimmern 1889). Thus, even though Danish fishers were relatively poor, because of capital investments by the Hanseatic League they had access to greater resources in the structural context, and their fleet was the largest and most advanced of the period. In 1382, one observer estimated somewhat optimistically that the Danish fleet consisted of 40,000 boats and 500 large freighters (J. T. Jenkins 1920). This may seem high, but the ships were still small relative to modern fishing vessels and it is not likely that the stocks were at all overfished at the time. Thus, the Hansa profited from their investments in the fishery but did not contribute to a profit disconnect per se.

Although the Hanseatic League remained powerful in the northeastern Atlantic until the 17th century, the Dutch took over from the Danes as major producers of pickled herring in Europe in the late 1400s—partly because the fish moved away from Denmark and closer to Holland and partly because of investment in innovation by the Dutch. Like the Hansa, the Dutch fishing industry organized to raise capital for investment in bigger and better technologies. First and foremost, they developed a new technique of "pickling" the herring that resulted in a better and longer-lasting product than traditional salting techniques. Second, they purchased ships called "busses", which had enough room to process the fish on board, a key step in the pickling process (Beaujon 1884; see figure 3.2). These busses were about 80 tons, more than twice as large as other fishing vessels of the time (J. T. Jenkins 1920, 105). Third, the Dutch were the first to use large gill nets to target herring (Beaujon 1884). Gill nets hang vertically in the water column and snag fish by the gills, so no bait is required. The Hansa used smaller gill nets as well as set nets, which were similar but fixed near shore. With their larger vessels, the Dutch could haul bigger and heavier nets; the larger gill nets enabled them to bring in much greater harvests, which, in turn, filled the capacity of their vessels and allowed them to supply herring to much of Europe (J. T. Jenkins 1920).

In addition to these technological innovations, the Dutch utilized an organizational innovation to target stocks farther from shore than other fleets. Dutch fishers used tender vessels called "hospital ships" to restock the fleet and transport the harvest to port, so the fishing vessels themselves would not have to return to land so often. Today, such boats are called "transshipment vessels," but their purpose is the same. These organizational and technological advances enabled the Dutch fleets to stay out

Figure 3.2
Dutch herring busses. Source: Fulton 1911.

for months at a time, following the herring from the Baltic Sea to the North Atlantic and back (Adams 1884a). At the height of their power, the Dutch harbored as many as 2,000 busses, which could harvest an average of 40 lasts per vessel each season (Beaujon 1884, 64–65). A "last" is exactly 13,200 herring, so 40 lasts is equal to 528,000 fish (Gibbs 1922, 58). For reference, if an adult Atlantic herring weighs 700 g on average, a single last would be approximately 368 tonnes and 40 lasts would be just under 15,000 tonnes. The annual value of this harvest could be more than 30 million florins, and the annual outlay for new vessels (few could be used for more than one season), gear, and supplies was about 15 million florins (Beaujon 1884, 64–65).[1] With such high returns it is no wonder that the Dutch and foreign observers alike believed that herring was the basis of that nation's wealth. Still, given the size of herring stocks, the core problems of the CPR dynamic were prevented largely by the technological limits of the structural context in this period. That is, the open access level of production was still lower than the sustainable level simply because gear and vessel technologies remained inefficient relative to the size and productivity of the fish stocks.

Although the Dutch controlled the herring fishery both economically and militarily, they were minor players in the other major fisheries of the

17th and 18th centuries. These were the cod fishery in the northwest Atlantic and the fishery for whales in the Arctic Ocean (Beaujon 1884). There were few innovations in the fishery for cod in this period, as exploration proved more important than innovation. Schooners remained dominant in the fishery until the early 20th century, though by that time they plied the North Pacific as well as the Atlantic (Cobb 1906).

The most revolutionary technological innovation in whaling occurred in the 16th century, when a Basque named Francois Sopite invented a method of boiling whale blubber down to oil onboard ship. Much like fishing for cod and herring, whaling started as a shore fishery. Once a whale was spotted, an entire community would mobilize to kill it, tow it to shore, and process the carcass. Even when larger boats were used to target whales farther out at sea, the need for processing on shore limited the industry to local stocks. Onboard processing allowed fishers to exploit the distant whaling grounds described in chapter 2 at commercial levels. This innovation was crucial for continuation of the industry, because many coastal whale species were already heavily overexploited; it also widened the profit disconnect by reducing the cost per unit of whale oil produced (J. T. Jenkins 1921).

In the 18th century, the British took over as the major power in the North Atlantic. They also dominated fisheries for herring and cod in this period and began targeting other species like haddock and plaice at commercial levels. Some of their dominance was due to the Industrial Revolution, which reduced the cost of inputs like nets and gear while providing capital for investment to expand fishing capacity (J. T. Jenkins 1920). Naval superiority was also pivotal during this period, because fishing grounds were often contested and piracy remained widespread (see chapter 5). In addition, changes in European preferences, including an increase in demand for low-priced fish, helped the British edge out the Dutch (see chapter 4). Most of these changes were exogenous, but one endogenous innovation was necessary for Britain's ascendency as a major fish producer: the invention of the beam trawl. Trawls are nets that can be dragged across the ocean floor or through the water column to harvest fish. They are used mainly to target plaice, halibut, and other demersal species, which were valuable at the time because of changes in preservation and transportation technology. Trawling started in Britain in the late 1700s with fairly small nets and ships. Over time, fishers gradually increased the size of nets and they eventually started using larger boats called "smacks." Smacks using beam trawls reached their maximum size in the middle of the 19th century, ranging between 23 and 36 tons (Holdsworth 1874, 66). Measuring up to 50 feet, the "beams" of a beam trawl held open the mouth of the net and so limited its size (Gibbs

1922, 46). British fishers maintained their supremacy through most of the 19th century, but only by capitalizing on steam power.

3.1.2 Steam Power

In the late 1800s, transition from sail to steam power removed the technological limits that had prevented the overexploitation of high-priced species such as herring and cod and increased the profit disconnect for the most important commercially exploited species of the period. In combination with the otter trawl and other gear innovations, steam power allowed for substantial reductions in the cost per unit of effort for most species, negating the cost signal even as it increased the capacity to overexploit and overcapitalize fisheries. Because steam engines are more powerful than sails, these new vessels could move twice the maximum weight carried by even the largest sailing vessels, and most took advantage of steam-powered winches and other mechanical devices to supplement the manpower required to pull in large nets full of fish (J. T. Jenkins 1920; Holdsworth 1874). However, it took considerable time to develop fishing gears that were compatible with steam-powered locomotion. Net-based vessels were particularly affected, since they tended to tow their gear from the stern. Though not problematic for sail-powered vessels, towing gear from the stern was infeasible on steam-powered vessels because the propeller(s) would tangle in the nets.

Technological Adjustments British fishers experimented with the first steam-powered trawlers in the 1850s, but they had no success until 1876, when they developed the otter trawl, which was much larger than the beam trawl and could be deployed from one side of the boat rather than the stern (Robinson 1996, 84). The otter trawl incorporated two "otter boards" on either side of the net rather than one beam across the mouth, so the size of the net was no longer dependent on the size of a beam (J. T. Jenkins 1920; Gibbs 1922). This innovation would not have been possible without steam power, which was necessary both to propel vessels large enough to handle the otter trawl and to power the mechanical winches that were needed to haul in the massive weight of the resultant catch. Because of this symbiosis, the scale of trawling operations increased substantially. In the 1920s, trawl nets were three times as large as they had been in the 1880s, and the otter trawls of the time could catch 47% more than a beam trawl (J. T. Jenkins 1920, 29; see figure 3.3 for a typical example). The new technology also proliferated rapidly. Steam-powered vessels rigged with otter trawls were in use in Spain, France, the United States, Germany, Belgium, the Netherlands, and other major fishing countries within a few years of their first successful use by British fleets (Blake 1884; Teuteberg 2009). This increased capacity and reduced marginal costs of production for the most important commercial fleets of the time, speeding up the action cycle and widening the profit disconnect. As a result, coastal stocks were rapidly depleted, driving much of the exploration described in chapter 2.

A similar dynamic also occurred in the whaling industry. Whalers started experimenting with steam engines in the 1850s. Like trawl fishers, they were not successful at first. However, after a profitable run in 1861, steam-powered whaling vessels soon became a central part of the fishery (J. T. Jenkins 1921). Again symbiosis with whaling technology was critical for success. Specifically, since whale stocks were already overexploited in much of the North Atlantic, whalers had to travel far to find whale stocks and, in spite of high prices for whale products, the costs of coal to power steam engines for long journeys was prohibitively high until a Norwegian whaler invented the first harpoon gun. Much more effective than harpoons thrown by hand, early harpoon guns were about 4 feet long and had a range of 25–50 yards. They also sported a "bomb"—a glass vial of sulfuric acid that would break once the harpoon penetrated the whale, killing it quickly. These highly efficient weapons reduced the risks and the time involved in capturing large whales and could also be used to target smaller whales that

Figure 3.3
A British steam trawler. Note that the boat still has rigging for sails, which were used in case of engine failure. Thus, the earliest steam trawlers were hybrids rather than purely steam-powered vessels. Source: J. T. Jenkins 1920.

were too quick and agile to catch easily using hand-thrown harpoons (256; see figure 3.4). The result was the "Norwegian" method of whaling, which used smaller steam-powered vessels that were much cheaper to fuel than the larger vessels used in the "American" method, which was only economically feasible with sail technology (Suisankyoku 1915). As described above, there was already a substantial profit disconnect in these fisheries, which was widened by the transition to steam; this step-wise change in technology increased whalers' ability to maintain profitability in spite of declining whale populations and the overcapitalization of whaling fleets.

The purse seine was another type of gear that was made significantly more efficient by mechanization. US fishers pioneered the use of the giant or great purse seine for their mackerel fisheries in the 1870s. By using cotton instead of hemp, they increased the durability of their gear while reducing the weight. This innovation was beneficial to all net fisheries, since the same vessel could cover 5 times as much area with cotton nets than it could with hemp nets (Juda 1996, 18). American mackerel fishers also used steam-powered winches to pull full nets out of the water. In 1880, 468 US vessels employing 5,043 men used giant purse seines to harvest mackerel. With a total catch that year of approximately 59,874 tonnes, the catch per vessel was 127 tonnes and the catch per fisher was about 12 tonnes (Goode 1884,

Figure 3.4
An early 20th century harpoon gun on a "Norwegian style" steam-powered whaling vessel. Source: J. T. Jenkins 1921.

40). This was 15 times the average harvest per vessel (9 tonnes) or per fisher (1.21 tonnes) in the British mackerel fishery at the time, which was still sail-based and used both drift nets and traditional seines (Cornish 1884, 10–13). Lower costs of production associated with giant seines also revitalized the US fishery for menhaden in this period (Goode 1884). Menhaden provided cheap food for human consumption, but large amounts were also used as bait in line fisheries or as inputs into the manufacture of manure (38). In fact, these innovations helped the United States to pull ahead of Britain in the total amount of fish production. The US also had other advantages; stocks near its coasts were not as heavily overexploited as European fisheries and growing domestic markets for fish products generated high ex vessel prices for US fishers (Walpole 1884a; Goode 1884). Still, with increased production capacity, US fleets quickly overexploited coastal stocks, and yet again industrialization generated growing profit disconnects in many areas.

Mechanization quickly spread to other fisheries; reducing costs and increasing the open access level of production. For instance, mechanized purse seines were imported from the United States to Japan around 1880, for use in Japanese sardine and herring fisheries (Kitahara 1910). Development of mechanical gear also revitalized the demersal longline or trawl-line industry in the US and Europe, which faced tough competition from drift-nets and trawl nets at the end of the 19th century. Lines that were once only a few hundred feet long were extended to 14 miles, with 12,000–15,000 hooks. In addition to mechanized hauling, US fishers developed machines that could bait hooks rapidly, feeding sections of lines from "baskets" in a fully automated process (Goode 1884, 11). This technology was also adapted in the pelagic longline fishery that was developed by the Japanese at the beginning of the 20th century. As noted above, longlines were used in many commercial fisheries targeting demersal species but had not been used to target pelagic species, which are mainly found in the water column rather than along the sea floor (Cobb 1906). The Japanese revolutionized this technology by using floats to suspend the longlines at a particular depth, rather than anchors that would keep the lines on the sea floor. This innovation allowed them to use longlines from three to ten miles in length to catch yellowtail, tuna, and other species that were not yet economically valuable in the West (Kalland 1995; Suisankyoku 1915).

Commercialization of Japanese fisheries started prior to their opening to the West in the middle of the 19th century, but Japanese vessels and fishing technologies remained relatively small in scale until the beginning of the 20th century, when the Japanese government started to encourage the adoption of "European methods" (Howell 1995; Suisankyoku 1915; Okoshi 1884). This included purchase and then domestic production of the larger Western-style vessels, mainly using sails and steam. From 1902 to 1912, the Japanese industry added 124 steamers and 669 smacks to an extant fleet of more than 400,000 Japanese-style "junks" (Suisankyoku 1915, 10, 12–13). During the same period, the average size of vessels in the traditional Japanese fleet also increased, as the number of small boats (<18 feet) declined by 23% while the number of medium-size vessels (18–30 feet) increased by 56% and the number of large vessels (>30 feet) increased by 11% (5). The Japanese government also educated fishers to use new technologies and developed shipyards where larger "European-style" vessels could be built. Much as in other historically dominant fishing countries, these innovations completely altered the fishery, increasing the potential impact on fish stocks while reducing the marginal or per unit costs of production substantially. This shifted the equilibrium level of production out and widened the profit disconnect.

Industry Stratification Fleets from many other countries adopted steam power at the turn of the 20th century (J. D. Campbell 1884; Yen 1910; Hurd and Castle 1913; J. T. Jenkins 1920). This further contributed to the stratification of the fishing industry and the widening of the profit disconnect. Large steam-driven fishing vessels were expensive and required far more capital than wooden smacks or schooners (J. T. Jenkins 1920; Teuteberg 2009). Data on fishing capital during this time are scarce, but information from the Scottish trawl fishery is indicative of the larger trend. While the number of fishing vessels decreased from the late 1800s to the early 1900s, the total tonnage of fishing capacity increased considerably. Furthermore, the value of the Scottish fleet rose from £1,712,349 in 1887 (more than US$140 billion in 2010 dollars) to £6,035,952 in 1913 (more than US$440 billion in 2010 dollars), even though the number of vessels declined by more than a third (J. T. Jenkins 1920, 13).[2] Because operating a steam-powered vessel required substantial capital, many individual owners were replaced by corporations or limited partnerships during this period.

Sails and individual ownership remained common only in small-scale, near-shore fisheries that did not compete directly with steam power. Thus, the stratification of the fishing industry that began with capital-intensive whaling, cod, and herring operations increased in the era of the steam engine. Much as their ancestors had done, coastal fishers utilized small vessels, called "cutters," that were well adapted to local conditions. A cutter cost between £160–210 in the United Kingdom in 1914 (US$16,600–21,800 in 2010 dollars). Although this was a large sum at the time, it is small in comparison with the cost of building a steam trawler, which averaged around £6,000–7,000 before World War I (US$624,000–728,000 in 2010 dollars; J. T. Jenkins 1920, 20). The gear used by coastal fleets was relatively inexpensive as well, and often was produced locally or by the fishers themselves. In contrast, the cost of rigging a ship with a full set of otter trawls was about £600 (US$62,400 in 2010 dollars), three times the cost of a cutter (22).

This trend in capitalization continued after World War I. By 1920, a single deep-water steam trawler could cost about £9,800 (US$420,000 in 2010 dollars). Fully outfitting it with gear, provisions, and coal, ice, or salt for fish storage could cost as much as £9,000 per year (US$386,000 in 2010 dollars), much of which would have to be raised in advance of a long trip. The return on this investment was much larger catches—between 45 and 54 tonnes per vessel, and in one case exactly £10,704 (US$459,000 in 2010 dollars) worth of fish (J.T. Jenkins 1920, 21-22). These numbers are for British vessels, but the pattern was similar in other countries (Teuteberg 2009). Faced with heavy competition and declining profits, small-scale fishers lobbied for government protections during this period. They were unsuccessful politically (see chapter 6), but also took economic action, innovating to remain competitive against industrial fleets. By the early 1900s, many of the old sail-powered vessels were transitioned to internal-combustion engines in order to compete with their steam-powered rivals (Gibbs 1922; Sverrisson 2002). Small-scale fishers also adopted mechanical winches, much like those used by the big trawlers, to haul in nets or lines (J. T. Jenkins 1920; Suisankyoku 1915).

These technological improvements deepened the core problems of overexploitation and overcapitalization in many regions and added new layers to the stratification of fisheries around the world. In developed countries, those with ample capital could invest in distant-water fleets, while those with less could modernize their coastal operations. Elsewhere, fisheries remained at subsistence levels, with very low capital requirements. Many of these fisheries were sustainably managed for centuries through local institutions. However, this state of the world changed in the early 1900s, when capital and fishing technologies began to flow from historically dominant fishing countries into developing countries. The profit disconnect that started in the age of steam widened considerably. Costs declined for small, medium, and large scale fishers and the fishing industry spread throughout the world as petroleum power catalyzed further expansion of fishing effort.

3.1.3 Petroleum Power

World Wars I and II, as well as the Great Depression, dampened growth and innovation in fishing effort in the 1920s-1940s, but these effects were temporary. By 1948, global landings were slightly higher than they had been in 1938 and much higher than they had been before World War I (see figure 3.5). Furthermore, the world's fishing capacity continued to increase well beyond pre-World War I levels as fishers increased both the size and the efficiency of their fleets (Holt 1978). Petroleum was pivotal in this period for three reasons. First, petroleum-fueled engines were smaller, cheaper, lighter, more powerful, and more convenient, allowing for reduced costs of production in fisheries and many other industries. Second, petroleum was used to make nylon nets and lines that were lighter, stronger, and more durable than cotton or hemp. Third, the boom in petroleum power fueled global trade and economic growth, allowing fishing companies to take advantage of terms of trade by moving operations to countries with the lowest costs of production. Combined, these innovations catalyzed the creation of modern factory-fishing vessels, which in turn allowed for the third wave of geographic expansion described in chapter 2 and substantially increased the profit disconnect in many fisheries globally.

The shift from steam to petroleum power began toward the end of World War I. Fishers primarily appropriated technologies that were already in use in other sectors, but some endogenous advances in gear technologies were also necessary. Large-scale (> 500 grt) oil-powered shipping vessels were in use at the turn of the century for non-fishing purposes and were common

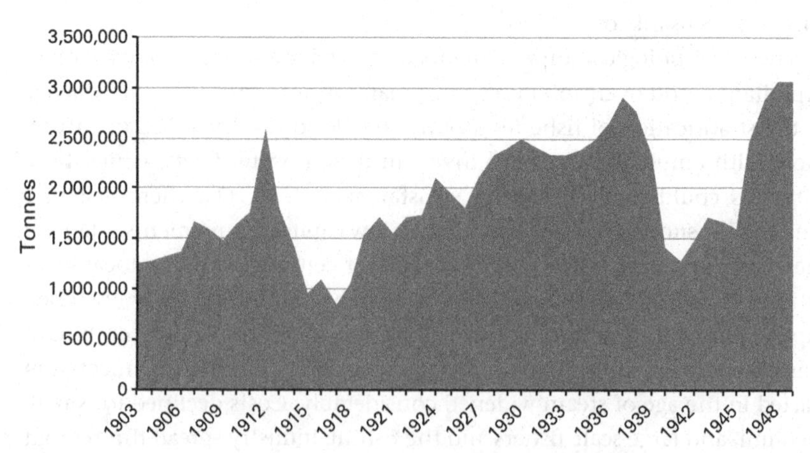

Figure 3.5
Total landings by European fleets, 1903–1949. Source: ICES 2012.

by 1925 (Juda 1996, 57). Japanese fishers started experimenting with diesel-powered fishing vessels in 1927, but the technology did not catch on until the late 1940s (Sverrisson 2002). A few fishers from other countries also experimented with oil-burning steam and diesel engines before World War II, but oil-powered motors were mainly used as auxiliary engines on sail-powered vessels or as primary engines on small vessels (<100 grt) until the postwar period. At that point, high prices and volatility of supply in the coal markets drove fishers with access to large amounts of capital to replace their old coal-fired steamers with either oil-fired steam engines or diesel engines. With government assistance, less wealthy fishers invested to replace their converted sailing vessels with gas-powered motor boats (Robinson 1996, 210–211; Royce 1987, 31; Chapman 1966, 11).

Coastal and Offshore Fleets Evidence for the rise of petroleum power and the decline of steam can be found in figure 3.6, which shows the definite reduction in steam capacity for selected countries from 1930 to 1960. For these nine major fishing countries, the number of steam vessels reported fell from over 3,500 in 1930 to about 1,000 in 1955. The 1960 total is incomplete because data were not available for France, but the downward trend is clear for all countries and it is reasonable to believe that these trends were the same in other industrialized fishing countries (B. A. Parkes 1966, 29).[3] Two developing countries, Morocco and South Africa, reported steam-powered vessels in 1960. This reflects a broader shift in fishing effort from historically dominant to developing fishing countries that started in the interwar period, largely as a result of increasing exploration by fishers from historically dominant fleets and escalation in the globalization of trade. China became the third-largest producer of fish in this period, and fleets also began to industrialize in India, Persia, Burma, and Korea, some with secondhand steamers and others with petroleum-powered vessels (US Congress 1947a).

Available data suggest that two major trends in investment expanded production in coastal areas after 1950. First, like capital-poor fishers from developed countries in the first half of the century, capital-poor fishers from developing countries augmented their sail-powered or oar-powered vessels with outboard or inboard motors to reduce their travel time and increased the distance they could go to find fish. For instance, in 1960, 97% of the powered vessels reported by Ghana were canoes modified with inboard or outboard motors (FAO 1962, table G-1). The People's Republic of China also encouraged its fishers to switch from sail to petroleum power. From 1956 to 1963, the number of motorized vessels increased from 56 to

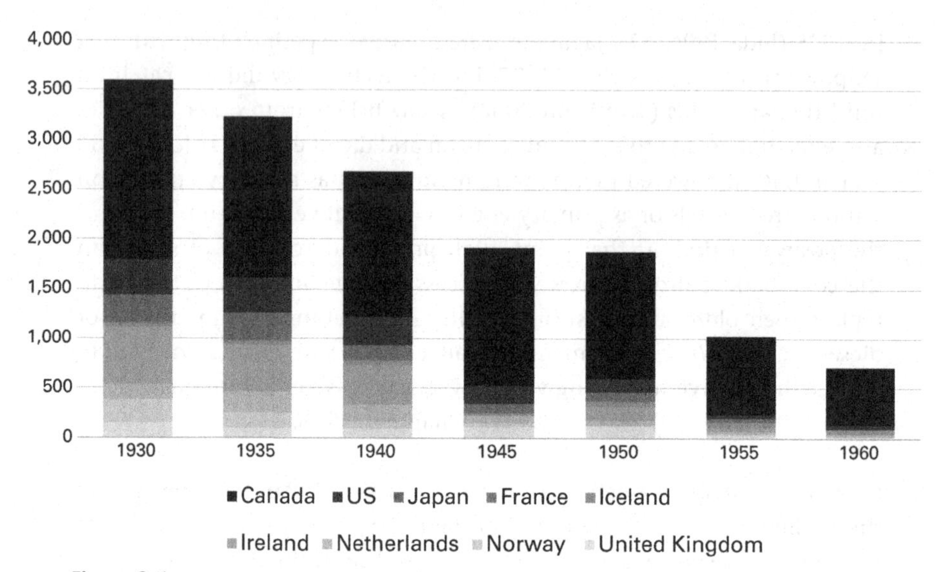

Figure 3.6

Count of steam-powered vessels reported by selected countries in five-year increments from 1930 to 1960. Compiled from FAO Fisheries Statistics Yearbooks (FAO 1947, 1952, 1954, 1960a, 1962). No data were reported for France in 1945 or in 1960.

1,200 (Muscolino 2009, 182). Unfortunately, exact numbers are not available for most other countries, but data on average tonnage (where available) and notes in various reports suggest that many countries harbored growing fleets of powered canoes, sailboats, and other small craft in the early 1960s (FAO 1962, Appendix G).

At the same time, a few developing countries reported additional fleets of larger motorized vessels on par with the fleets of developed countries. It is difficult to say where the capital for such vessels originated, but it is most likely tied to exploration by fishers from historically dominant fleets. In some countries, like Ghana, expansion took the form of foreign direct investment. Seeking new fishing grounds, Japanese fishing companies started building up a fleet of motorized pole-and-line vessels in Tema, Ghana, in the late 1950s (Miyake, Miyabe, and Nakano 2004); they may have provided for the addition of motors to local vessels previously used for subsistence fishing as well. Japan also established fleets in its colonies (Korea, Taiwan, and Manchuria) before World War II and may have contributed to reconstruction of those countries' fleets in the postwar period (Suisankyoku 1915; Muscolino 2009). US fishing companies also invested in fleets overseas, including in Macau, the Philippines, and other territories captured during the war (Finley 2011). Capital flowed from colonial powers to new fleets in developing countries like Morocco and South Africa, which reported rapid increases in steam- and petroleum-powered vessels after World War II. Again, the need to expand away from overfished fishing grounds also led to the propagation of new, more efficient fishing technologies around the world.

Globalization further increased investments in innovation through economic development programs and through the expansion of international financial institutions. For instance, in South America several governments subsidized the development of domestic fleets in the 1950s and the 1960s. Their primary goal was to capture the benefits from fisheries resources in their newly claimed sovereign territory, usually the area within 200 miles of their coastlines (see chapters 5 and 6). Peru opened up the international fishery for Peruvian anchoveta, one of the largest and most valuable stocks in the world (Holt 1978). By 1960, the Peruvian fleet reportedly included more than 700 commercial fishing vessels, of which 268 exceeded 50 feet in length (over 100 tons gross) and 10 additional vessels longer than 100 feet (over 300 tons gross, on average). Brazil, Chile, Mexico, and Venezuela also pursued development through commercial fishing in this period (FAO 1962, table G-3). In addition, developed countries subsidized the rebuilding of their fleets after World War II, contributing to the expansion of existing size classes and to the building of new vessels capable of fishing in far-distant waters.

Distant-Water Fleets As coastal fishers expanded in numbers and effectiveness through motorization and mechanization, distant-water fishers searched for ways to increase the amount of fish they could catch and preserve on long voyages. Before World War II, Japanese fishers started using a variation on the mother-ship configuration established by European schooners hundreds of years earlier. In the new Japanese system, however, the mother ship was a floating cannery that would process and store fish provided to it by fishing vessels that traveled with it. This system allowed the Japanese to fish for salmon off Alaska before the war and then to expand their fleets throughout the oceans once restrictions were lifted in the postwar period (Finley 2011). In fact, the 22 steamers that Japan retained in 1960 were probably cannery mother ships, given their average gross tonnage (4,744 tons (gross); FAO 1962, table G-4).

The United States also utilized the mother-ship approach, starting with a 390-foot converted steamer that was repurposed by the Alaska Southern Packing Company in 1939. This steamer was almost four times the size of a typical high-seas fishing boat (100–150 feet long at this time; see figure 3.7). The floating cannery project was successful, and with the high demand for fish protein after the war, the US government backed the reconfiguration of several World War I era naval vessels by private companies, including the 423-foot *SS Mormacrey*, which was transformed into the *Pacific Explorer* in 1946. This reconfigured vessel fished for salmon in the North Pacific and for tuna in the tropical Pacific, specifically targeting areas that had been utilized by the Japanese before the war. With its attendant purse seine fleet, it could catch and carry 3,450 tonnes of fish in a single trip. Two other vessels with smaller carrying capacity (907 tonnes) were also fitted out at the same time (Finley 2011, 46–47).

Another innovative approach to reducing costs of production in distant-water fisheries was to build fishing vessels that could both catch and process huge amounts of fish. This method was developed in Britain in the 1950s to expand trade in frozen fish, which was an increasingly popular alternative to the canned product produced by Japanese and US mother-ship fleets. Christian Salvensen & Co. Ltd. commissioned and then launched the first factory trawler in 1954. It was a modern version of the older factory whalers and cod busses used by the Basques, Norwegians, and Dutch. Named the

Figure 3.7
A US floating cannery with freezer storage, circa early 1900s. Source: Cobb 1921.

Fairtry, the first factory trawler was a 245-foot-long ship that displaced 2,800 tons of water.[4] It could catch and process 27 tonnes of cod in a day and could hold up to 544 tonnes. Onboard processing included cleaning, filleting, and freezing the catch, creating a longer-lasting product compared to the existing method of preservation, which involved removing the heads of the fish and packing the bodies in ice. The *Fairtry* could also produce and store up to 45 tonnes of cod liver oil and 90 tonnes of fish meal. Fishing in the North Atlantic, this vessel was so successful that her owner soon commissioned two more factory ships, with quite a few adjustments to improve performance (Robinson 1996, 216). Before long, other countries ordered similar ships to plow through the cod, herring, and other valuable stocks in both the Atlantic and the Pacific (Gulland 1974). By the early 1960s, fishing vessels of 500–1,000 tons (gross) were common throughout Europe and vessels over 1,000 tons (gross) could be found in many countries (FAO 1962, table G-5).

Producing the *Fairtry* and other factory trawlers required significant amounts of capital and multiple technological innovations in the setup of the trawling gear. First, developing ways to haul in gear while moving was important for trawlers and required innovation that would allow deployments of nets from the rear of the vessel (Robinson 1996). Second, to fill the holds of these larger vessels quickly, fishers also had to engineer bigger gear. Though some design changes were necessary, like the double cod end for factory trawlers, the shift to monofilament nets and lines was also very important. This started with the development of nylon and other plastics for various military and civilian uses during World War II. Fishers appropriated these technologies, investing in ropes, lines, and nets made out of artificial fibers because they found that gear and rigging made from these materials was stronger and more durable than those made from cotton, hemp, or silk (Robinson 1996). Furthermore, monofilament nets and lines were more effective because they were harder for the fish to see and avoid. This was particularly important in gillnet and driftnet fisheries, but also benefited purse seine fleets. Thus, even small-scale fishers benefited from the introduction of plastics to the industry (Royce 1987).

With the use of monofilament technology and the related modifications to vessel size, stability, and machinery, the maximum size and efficiency of other types of net-based and line-based gear increased substantially. For instance, in the 1870s longlines could be as much as eight miles long, with hundreds of hooks (Holdsworth 1874, 137). Almost a century later, with the use of monofilament technology, they could be more than 60 miles long, with 3,000–4,000 hooks (Ward and Hindmarsh 2007, 502). The impact on trawling was just as impressive. In the 1870s, when sail was still the primary method and gear was mostly hemp or cotton, the maximum size of a trawl haul was about one tonne (Holdsworth 1874, 80). A steamer of the early 1900s using an otter trawl net could catch about four times as much as a sailing smack (Robinson 1996, 112). The *Fairtry* could catch about 27 tonnes a day in the 1950s, which made it about eight times as effective as a steamer and 30 times as effective as a smack (216). Today's double trawlers, which have nets on each side of the vessel, can catch as much as 90 tonnes per day when fish are abundant (Royce 1987, 31).

Access to Capital Of course, factory vessels with appropriate gear technologies are more expensive than their predecessors and so are only available to those with considerable access to capital. For instance, it cost US$5 million to retrofit the *Pacific Explorer* in 1946 (Finley 2011, 63). That would be more than US$43.6 million in 2010 dollars (Williamson 2014). Less than ten years later, construction of the *Fairtry* cost £1,000,000—more than US$31.1 million in 2010 dollars (Warner 1997; Officer 2013). Similar vessels commissioned by the Soviet Union would have cost about US$3.2 million each to produce in the United States at the time—a total of US$21.6 million in 2010 dollars (CIA 1959; Williamson 2014). In free market economies, such vessels could be purchased only by large companies with substantial capital. In most cases, governments subsidized the purchase of these megafleets and at times also provided assistance to cover operating costs. In centrally planned economies, governments purchased large numbers of these vessels, crowding out private investment (Armstrong 2009; see chapter 6).

Expansion of factory fleets was also facilitated by secondary economic actors. Fish processors, desiring a steady flow of product, often chose to purchase their own vessels rather than rely on independent fishers (Hamilton et al. 2011). Some processors also provided loans to fishers to allow them to buy their own vessels and gear, or to cover operating costs. Many examples of such behavior occurred throughout history, and merchants or wholesalers, including the Hanseatic League and the Dutch Herring Councils, frequently used the same practice to guarantee access to supplies of fish. Other examples were observed in the 18th and 19th century in China, Japan, and India, though technologies supported by fish brokers and merchants at the time were not mechanized (Muscolino 2009; Kalland 1990; Day 1873). In the 20th century, many large US fleets, including the tuna and sardine fleets of California and salmon fleets of Alaska, were purchased with capital raised by canneries (Cobb 1906; US Congress 1947b). Some fishing corporations, like Christian Salvensen & Co., expanded their operations from fishing to processing, but generally canneries and merchants accumulated capital that they would then use to purchase their own vessels or to provide loans to fishers.

Although capital requirements limit entry into large-scale fisheries, over time resale markets provide wider access to not-quite-new technology. The vessel prices noted above are at initial sale. When new vessels are built, old vessels are often sold rather than scrapped. The lower prices for used fishing vessels enable fishers who lack access to sufficient capital to purchase the latest equipment to purchase more up-to-date vessels and gear. This is particularly important in developing countries, where capital tends to be limited. For example, in 1969, the *Fairtry* was transferred from the original owner to the K M Corporation, and the vessel was reflagged in Panama as the *Joy 2*. It was not decommissioned until 1985 (Aberdeen City Council 2013). The resale price is not recorded, but current resale prices for similar vessels range from US$7.5 million (for a vessel built in 1984) to US$16 million (for a vessel built in 1994), both much lower than the inflation-adjusted price of the *Fairtry* (Atlantic Shipping 2013). However, prices in resale markets vary considerably in time and space, depending on the condition of the targeted fisheries. Any new discovery that leads to higher profits in a fishery also generates increased demand and higher prices for new and used vessels.

All in all, the process of investment for innovation in vessel and gear technologies sped up the action cycle for fisheries around the world, widened the profit disconnect in most areas, and exacerbated the problems of overexploitation, overcapitalization, and ecosystem disruption. Because of high capital requirements, new technologies also led to the stratification of the fishing industry, with fishers in different niches using different types of technology to reduce costs of production in spite of declining stocks and intensifying competition.

3.2 Fish-Finding Devices

In addition to developing faster vessels, mechanized gear, and factory fishing operations, fishers also increased their efficiency by developing better methods for finding fish. This increases the profit disconnect and facilitates expansion of fisheries in two ways. First, better fish-finding technologies make it easier to find fish in spite of declining stock sizes, thereby dampening the cost signal. Second, these same technologies allow fishers to expand the depth and breadth of their searches, helping them to find and exploit new fisheries. In fact, most distant-water fisheries would not be feasible without the advanced fish-finding technologies described next. The history of technological advances in fish-finding devices is somewhat independent of innovation in vessel and gear combinations, so it would be difficult to present an integrated historiography. That said, the general pattern of escalating innovation through cycles of growth does hold for this type of technology. This section covers patterns of innovation in fish-finding technologies with discussion of implications for costs of production, capital requirements, and stratification of the fishing industry.

One of the most important traditional ways to find fish that are dispersed through large areas, or that cannot be seen from the surface, is to use information about their habits and habitats to predict their locations. From the earliest days, fishers recognized the need to record and analyze such data. Seasonal movements, temperature ranges, and food preferences all can help fishers to find the best locations for harvesting a specific species at a given time. Such information was coded into the traditional ecological knowledge that shaped many ancient fisheries (Shackeroff, Campbell, and Crowder 2011; Cash et al. 2003; Lauer and Aswani 2010; Sethi et al. 2011; Berkes and Folke 2002; Berkes 1999). As fishers became wealthier, many gained literacy and began using charts and logbooks to keep track of patterns in ocean conditions and fish abundance for specific areas and species. Beginning in the 1800s, marine science added significantly to the store of knowledge about the relationships between fishes and their habitats. Though governments and academic institutions fund many marine research programs, fishers themselves often invest in science that will help them to find new fishing grounds or to follow fish when they are hard to find.

Correlations between visible surface disturbances and the less obvious underwater presence of fish have also been extensively used to find fish or other species that congregate near the surface. For example, whales, dolphins, and some species of fish are highly visible because of their breaching behavior. Whalers looked for this and the telltale spouts of their prey for many centuries (J. T. Jenkins 1921). When close to the surface, schools of fish disturb the pattern of waves, often creating a "flat spot" or a "boiling spot," depending on the level of activity underwater. Boiling spots usually occur when a school is being attacked by predators, which drive it to the surface, sometimes forcing their prey to leap out of the water. Seabirds provide an even more obvious indication that fish are present. They will often converge on a school in large, visible, and raucous groups. Fishers have followed conglomerations of seabirds to rich fishing grounds for much of history. Other associations are more mysterious. For instance, in the eastern Pacific dolphins associate with schools of yellowfin tuna. The dolphins are not feeding on the tuna, or vice versa, but the two species commonly travel together, and thus fishers can find tuna more easily because they can spot the dolphins (Felando and Medina 2011).

Using lures is another traditional method of finding fish. Bait, the most common lure, is mostly used in line fisheries. As noted above, baitboats chum the water, attracting fish to their boats. Longlines of various types use baited hooks. Most bait is "trash fish"—fish that is low in price or can't be sold. However, fishers are careful to choose bait known to be preferred by the species they are targeting. Fishers also sometimes use lights to attract nocturnally active species, such as squid and mackerel. Torches were used originally, but eventually were replaced by powerful electric bulbs and glow in the dark sticks (Suisankyoku 1915; Okoshi 1884; FAO 2012d). Association with floating objects is also important in some fisheries. Although the reason for this association is still uncertain, artisanal fishers noticed the relationship and began building artificial floating objects as early as the 17th century. In the 1990s, tuna fishers in the eastern Pacific began using manmade fish-aggregating devices (FADs) on an industrial scale because of political controversy over the killing of dolphins in the fishery. The method was extremely successful both in increasing catch and in reducing the costs of production. It is now used in other regions as well (Bromhead et al. 2003).

Fishers who target species that are easily spotted at the surface (surface fisheries) have found many ways to increase their search range. The spyglass of old gave way to high-powered, deck-mounted binoculars that were originally developed for military use. Many vessels now use spotter planes and helicopters to improve visual location (pioneered by Harold Medina in the 1950s), and use satellite and radio technologies for geo-positioning so that a skipper can quickly get to a school of fish once it has been found (Felando and Medina 2011). Some commercial seiners that travel too far out to sea to rely on land-based helicopter services now carry a helicopter on board. Even though the capital costs are high, a helicopter increases efficiency so much that smaller seiners either pool their resources to purchase a shared helicopter or rent helicopter time. At a lower price point, GPS is commonly used in commercial fleets, and skippers use both radio- and cell phone-based communication to keep in touch with each other. Encryption is now important to prevent valuable information about the location of a school from being appropriated by a rival.

Other technologies are used by fishers targeting species that are not as easily spotted at the surface (groundfish and deep-water fisheries). Sounding, an old method for determining the depth and other characteristics of the bottom, was used by fishers targeting groundfish (bottom dwellers;

Gibbs 1922). When fish like cod were very abundant, fishers could also detect schools by sensing vibrations caused by the fish hitting the weights on their longlines (Huxley 1884). Using technologies developed during World War I, British fishers began experimenting with echolocation systems in the 1930s. Twenty-five years later, after considerable refinement of the technology by the US military during World War II, both sonar and radar were in widespread use and had become essential to fishers around the world (Finley 2011). Early sonar systems required a lot of skill to interpret, but today computer microprocessors are able to analyze the information from echo devices quickly and precisely.

Fishing vessels of all size classes now have radar and sonar capabilities that differentiate the sizes of individual fish within the search radius. These "fish finders" are widely available around the world at a variety of prices. Cheaper versions can be purchased for a few hundred US dollars and provide information for depths of 500–1,000 feet. For higher prices (US$3,000–$18,000), commercial fishers can purchase systems that will map the size and location of fish at depths of more than 9,000 feet (sonar) or identify schools and large agglomerations more than 100 miles from the ship (radar; see, e.g., Furuno 2013; Hondex 2013; NFUSO 2013; Simrad 2013). Until recently, there was a substantial tradeoff between range and resolution in sonar, but new "chirp" technology allows for high resolution even at great depth and distance (see, e.g., Raymarine 2013).

Some fishers also utilize GPS devices to track their gear, and many use satellite imagery to locate likely fishing spots. For instance, free-floating gear like FADs were first tracked via radio signals, which could be monitored by everyone, but now FADs are located using encoded GPS signaling systems. Fishers also attach sonar arrays to FADs so they can assess the size and composition of associated schools from a distance. Satellite imagery of cloud cover was first provided to fishers by governments to improve safety by increasing available information on weather conditions. Now, satellite imagery of important environmental conditions such as sea surface temperature and chlorophyll concentration are freely available through several national or joint private-national institutions, including NASA and JPL in the US, JFIC in Japan, and CSIRO in Australia (NASA 2013a; NASA 2013b; JFISC 1999; CSIRO 2013). Although some fishers are very good at analyzing this information themselves, there are also private services that supply fishers with analytical software that compiles and interprets satellite imagery for specific fisheries (see, e.g., SeaView Fishing 2013; DigitalGlobe 2013; Ocean Imaging 2013; SeaStar 2013; SpaceFish 2013; Catsat 2013).

In combination with the vessel and gear innovations described in section 3.1, these new fish-finding technologies catalyzed an enormous increase in the effectiveness of fishing effort, widening the profit disconnect substantially. Measures of the exact effects on costs are not available, but Royce (1987, 33) estimated that the biggest and best fishing fleets of the 1980s were 1,000 times more effective than subsistence fishers using manually operated boats and gear and 100 times as effective as modern small-scale coastal fishers. Even the less advanced versions of distant-water fishing trawlers, seines, and longlines were 100–300 times as effective as subsistence fishing methods. Given that many new devices have been introduced since the 1980s, it is safe to say that fishing efficiency and effectiveness continue to improve even though marine capture production stagnated in the 1990s. I will return to this issue after a brief discussion of aquaculture as a potential innovation in fish production.

3.3 Aquaculture

Aquaculture is one of the most important sources of increased fish production in recent years. Harvests in capture fisheries leveled off in the 1990s, and the growth in fish production since that time is largely due to aquaculture. Most aquaculture operations are based in freshwater systems (60%) rather than in marine systems (40%), and growth is higher in freshwater production as well (FAO 2012d). Furthermore, fishing and aquaculture require two very different skill sets, and aquaculture techniques evolved separately from fishing technologies (see chapter 6). Therefore, it is difficult to consider aquaculture as an endogenous response in the fisheries action cycle. Instead, I treat it as an exogenous factor that can have substantial if unpredictable impacts on the price signal. This section briefly covers the development of aquaculture and related implications for capture fisheries.

Freshwater aquaculture dates back at least as far as 5000 BCE (Yen 1910, 370). There are also prehistoric examples of lagoon-based aquaculture systems in which community management distributed endowments and entitlements similar to those adopted for capture fisheries (Huxley 1884, 7). In the 19th century, entrepreneurs and scientists began working on hatchery and seeding programs, often with government support (see section 6.2). Seeding involves raising fry in a lab setting and then introducing them into the wild environment. This has been done successfully with white sea bass off the California coast and with salmon in various parts of the Pacific, but is not economically feasible for large stocks of marine species (Hervas et al. 2010). Mariculture has been much more successful than seeding alone.

Definitions for mariculture are diverse, but it generally includes any form of human intervention designed to increase the productivity of a marine population. Two common forms of mariculture are salmon farming and bluefin tuna ranching. Farming involves raising fish from hatcheries; ranching entails the fattening of wild-caught fish. Both ranching and farming can increase the biomass harvested, but ranching may reduce overall fecundity, exacerbating stock decline (Volpe 2005).

These mariculture techniques are controversial for several other reasons. First, economically viable species tend to be large, carnivorous fish that require substantial protein input, usually from capture of smaller marine species that would provide more human nutritional value if consumed directly. Second, concerns exist about concentrations of effluent waste around mariculture operations, which result in depletion of oxygen and general habitat degradation. Third, because of the close proximity of fish in pens, antibiotics are commonly used to reduce disease mortality. This creates several problems, including introduction of antibiotics into the marine environment and the proliferation of antibiotic-resistant diseases in wild stocks. Fourth, parasites also breed much more successfully in pen environments and can easily transfer to wild stocks. This is particularly problematic for fry and small fish, which can easily succumb to just one or two parasites. Fifth, where saltwater aquaculture is carried out on land, as with shrimp farming, considerable destruction of coastal ecosystems like mangrove forests can result, which, in turn, leads to degradation of coral reefs and related marine ecosystems. Finally, recent attempts to introduce transgenic fish into mariculture operations raise the fear that these genetically modified fish could escape and completely alter wild populations through competition or interbreeding. To cope with these problems, entrepreneurs started developing an approach called integrated multi-trophic aquaculture (IMTA). There is also increasing interest in cultivation of tilapia and other vegetarian species to reduce pressures on wild populations (see Naylor and Burke 2005 and Troell et al. 2003 for a good overview of these issues).

In terms of the AC/SC cycle, both mariculture and aquaculture increase the global supply of fish products, but impacts on price signals are not always predictable. In some cases, growth in aquaculture production increases the supply of substitutes and thereby drives down prices for targeted species. An example would be the decline in the demand for swordfish in the United States in the 1990s, which was at least partly caused by the greater availability of cheap farmed salmon (Webster 2009). Interestingly, salmon culture did not drive down prices for wild-caught salmon. Instead, separate markets developed for farmed and wild-caught fish, with a significant price premium for the latter (Schlag and Ystgaard 2013). Similarly, ranched bluefin tuna does not draw the same high prices as wild bluefin, even though all bluefin in the market are wild-caught (Webster 2009). These effects depend heavily on consumer perceptions of product quality as well as attitudes related to social status and environmental conservation. Indeed, as the next chapter shows, the shaping of these attributes through marketing is a critical determinant of demand for both farmed and wild-caught fish. In any case, as currently practiced, aquaculture magnifies core problems in some fisheries and also creates environmental issues of its own, so it is not yet a viable solution in the fisheries action cycle.

Looking toward the future, it is difficult to predict what new innovations may occur in global fisheries. It is possible that gradual technological improvements will continue, with prolonged impacts as new technologies percolate through the many strata of the global fishing fleet. Transformative innovations are more difficult to anticipate. It appears that economies of scale are exhausted, particularly given the state of global stocks (World Bank and FAO 2009). Fish finders are already highly sophisticated, but searching for fish still involves substantial uncertainty, particularly in open-ocean fisheries. Ironically, the next great innovation in fishing technology will probably be driven by larger environmental concerns, such as climate change and the depletion of fossil fuels. Much as petroleum-powered engines replaced steam engines when the cost of coal increased, renewable energy sources may replace fossil fuels as prices rise and people express growing concern about the effects of greenhouse gases.

Fuel price increases are the most likely driver of the development of substitute propulsion technologies. Unlike the temporary price hikes associated with the OPEC oil cartel in the 1970s, and in spite of recent steep declines, real price increases that started in the early 2000s can be expected to continue for the long term (EIA 2013). In addition, carbon taxes and other regulations designed to reduce greenhouse gas emissions have increased the cost of fuel in some countries and may spread further as the world seeks to cope with climate change (EIA 2012). High fuel costs reduce the profit disconnect temporarily, but fishers are already investing in improving the fuel efficiency of their operations (World Bank and FAO 2009; M. E. Riddle 2003). Innovations include better fish-finding technology to reduce search and travel time, more efficient engines, and changes in vessel operation to increase fuel economy. I have also spoken with several fisher-entrepreneurs who are looking for ways to harness renewable energy to replace their diesel-powered engines. If this occurs, it would be yet another transformative innovation in global fisheries, and could lead to even greater increases in fishing capacity, fish catches, and, ultimately, the overexploitation of fisheries resources. Indeed, given that fuel accounts for about 50% of the operating costs for commercial vessels and that new vessels are counted as assets rather than liabilities, the effect of this innovation on the action cycle and the profit disconnect could be quite profound.

Aside from renewable energy, not many other transformative innovations can be expected to occur in fisheries in the near future. Indeed, there have only been a few major changes in fishing technologies since the 1970s, and those were in fish-finding devices rather than in vessel-gear combinations. Nevertheless, even gradual innovation can dampen the costs of production, widening the profit disconnect and worsening the core problems of the action cycle. Exploration is approaching global limits. Innovation is leveling off. The next chapter tackles market expansion.

4 Opening New Markets

The third type of economic response predicted by the AC/SC framework is the opening of new markets to dampen price signals. This response has short-run and long-run aspects. In the short run, price signals can limit effort by driving some fishers and secondary economic actors out of business. As explained in chapter 1, for any market that is in equilibrium, a surge in supply generated by exploitation of new fishing grounds or innovations in fishing technology will result in a decline in price as long as demand remains constant. This price signal is amplified by the stepwise nature of growth in supply, because production often increases for multiple products within a fairly short amount of time. The short-run effects of sudden price drops associated with rapid expansion of supply vary depending on the state of the economic system. In most cases, a steep decline in price causes profits to fall as well. When this occurs, some fishers and other industry actors are forced to exit, which reduces supply to some extent. Those who remain in the fishery work to increase demand by opening new markets. This response facilitates expanded production and mitigates the price signal in the long run. Alternatively, when supply increases as a result of a substantial reduction in costs of production, as happened in the Industrial Revolution, the effects of lower prices on profits may be countered by an increase in quantity demanded, as many more consumers are willing and able to buy fish products at the new, lower price. This trend does not prevent industry response to the price signal; fishers, processors, and marketers will still seek ways to increase demand and thereby increase the price at any given level of production.

In the longer term, we can expect to see cycles of market expansion that loosely parallel the cycles of effort expansion described in previous chapters. This process is also driven by exogenous changes in demand associated with population growth and economic growth. However, it is important to note that the fishing industry would not be able to take full advantage of these exogenous drivers without concomitant changes in processing, shipping, and marketing. Furthermore, processors, marketers, and other secondary actors are much more active participants in this portion of the action cycle than in cost-driven responses such as exploration and innovation in fishing technologies. They have more expertise in product development and marketing, and they often have considerable access to capital for investment in new facilities or advertising campaigns. Large processors and marketers are also insulated from the costs of overexploitation because they procure primary product from many different fisheries. That is, increasing costs of production due to declining biomass in any one fishery can be avoided by switching to other fisheries, so large producers and marketers are less sensitive to the cost signals described in chapter 1. Similarly, these processors are able to control ex-vessel prices, because they can buy from different fishers, but they are sensitive to the prices that consumers are willing to pay for their products. This gives some secondary actors both the incentives and the resources to respond to the price signals associated with stepwise increases in supply by opening new markets. Even so, advances are still limited by existing technologies, so this response is circumscribed by exogenous catalysts and endogenous inventions that alter the structural context.

This chapter covers three overlapping histories related to the opening of new markets. While fishers are still important in the analysis, most of the focus is on secondary economic actors. Section 4.1 describes how processing evolved from simple preservation using artisanal methods to a global industry that is constantly creating new products to appeal to the increasing appetites of consumers. In particular, processors can increase demand generally by developing more palatable and less perishable products and can increase quantity demanded by reducing costs of production through efficiency improvements. Section 4.2 shows how shipping innovations complemented the process of opening new markets by allowing faster transport of fish products and by incorporating refrigeration technologies that facilitated trade in less processed products that generally bring in higher prices. Finally, section 4.3 reviews the development of marketing practices used to convince consumers to buy ever-increasing amounts of "new and improved" fish products. Given past trends in exogenous drivers of demand and the endogenous drive to mitigate price signals, it is likely that opening new markets will continue to mask the relationship between core problems and fishing effort well into the future.

4.1 Processing

Fish are highly perishable. This fact limits the geographic scope and scale of markets for fresh fish products. In early fisheries, processing was primarily a method of preserving large catches that could not be consumed immediately. With the commercialization of fishing, processing was also used to preserve catches during transport to local population centers. Eventually, long-term preservation techniques were developed that allowed for widespread trade in fish products. Modern preservation technologies are well developed, and processors tend to focus on factors such as flavor, convenience, and novelty to appeal to increasing numbers of consumers. Processors also have found many ways to utilize "trash fish" that are not in high demand by consumers, diversifying the number of different species that fishers are able to profitably exploit. This helps to explain the increased diversity of fish catches and the growth in the volume of fish production described in chapter 1. Processors also innovate to find ways to reduce their own costs of production, including efficiency improvements and elimination of waste through production of non-food products from offal. Although lower costs do not always benefit fishers, they do amplify the profit disconnect for processors, who are already insulated from cost signals. Subsection 4.1.1 covers the evolution of the processing industry for food products, subsection 4.1.2 describes advances in non-food products, and subsection 4.2.3 documents how stratification shaped the growth of the modern processing industry.

4.1.1 Food

Processing fish for human consumption is an ancient practice. Curing is one of the oldest methods of fish preservation, and was common in Egypt, Sumer, China, and many other early societies (Adams 1884a; Kalland 1990). Artisanal fishers and artisan producers still make and sell cured fish products today. However, all methods of curing are labor-intensive and there are only a few cured products that have a long shelf life and are palatable to consumers. Thus, large-scale production and trade in food-fish products was limited until the invention of canning and freezing in the 1800s. These technologies allowed for production of better-tasting products that lasted longer and were cheaper than either fresh or cured fish. All of them helped to expand markets for fish products, accommodating increasing supply but also widening the profit disconnect.

Curing There are many types of curing, including drying, salting, pickling, and smoking, but two specific cured products were most important in the rise of large-scale European fisheries: salt cod and pickled herring. Salt cod was favored by many consumers because it tasted better than most other salted fish products and it also lasted for months. The processes used for salting cod are generally the same in most areas and time periods, with some local variations. The colonial New England cod fishery is typical. Cod was either "dry cured" or "green cured." For a dry cure, fish were usually cleaned, split, salted, and then stored aboard ship after each haul so they would not deteriorate at sea.[1] Once the ship returned to shore, the fish would be repacked, resalted, and eventually sun dried into tough, durable slabs that could be stacked and stored like wood. Dry curing was used when fishers traveled long distances to fishing grounds. In contrast, green curing, in which the fish were cleaned, salted, and dried on land rather than at sea, was used on shorter trips to nearby fishing grounds. Usually these fish were smaller as well (Gibbs 1922; Innis 1940). Because it was a faster process, green curing did not wring as much water out of the fish as dry curing. Green-cured fish therefore were considered to be of lower quality and did not last as long. Colonists would keep some of the best dry-cured fish and send the rest to Spain, where prices were highest. Lower-quality green-cured fish usually went to the West Indies or elsewhere in North America (Pearson 1972).

Herring was the other widely traded fish product from the Middle Ages through the Industrial Revolution. Fish caught mainly in the North Atlantic were pickled or preserved in brine for shipping throughout Europe. Pickling predates the maritime rise of the Hanseatic League in the 11th century, but the Hansa consolidated and ultimately monopolized production for centuries. Moving to control an already profitable industry, the League set up large facilities in a few key ports to process the herring harvested for them by local fishers. Scania, in Denmark, was the most famous of these ports. In these towns, the League built wharf-side neighborhoods where people lived and worked to process the herring, performing every necessary activity other than fishing, from cleaning the fish to building the storage barrels. The League also used its military power and economic influence in other trades to ensure that the fishers and curers in these towns had access to all necessary materials, including wood for vessels and barrels, and salt for preserving the fish. Both wood and salt were scarce and valuable at the time. Furthermore, the Hansa supplied inspectors to ensure a uniform, high-quality product, and other officials to enforce the League's rules and settle disputes (Zimmern 1889).

As described in chapter 3, the Dutch developed improved methods to capture herring, increasing the efficiency of their fleets over the Hansa-sponsored Danes. However, the Dutch would not have been able to break the Hanseatic monopoly on herring without the development of a new pickling technique that resulted in a better-tasting and longer-lasting product. According to most sources, a man named William Beukelsz (or

Beukelius, or Belkinson depending on the source) developed just such a technique in the middle of the 14th century. The Beukelsz method required gutting the fish as soon as they were caught, packing them with salt into tightly closed barrels, and dowsing them with fresh sea water or "pickle" every two weeks until they were sold and shipped to consumers (Beaujon 1884, 11).[2] Knowledge of the Beukelsz technique was not the only advantage possessed by the Dutch. To use the process, they also had to invest in large, decked vessels, where fishers could clean and salt the fish as they were caught. Many, including the Dutch themselves, credit their rise as a great fishing power to the Beukelsz method; however, there were also exogenous forces at play, including the general role of the Dutch as a major military and trading power during the period (Beaujon 1884; J. T. Jenkins 1920; Adams 1884a).

British fishers began to dominate the international market for pickled herring in the 1700s. They appropriated the Beukelsz method from the Dutch in the 1600s, but for more than 100 years they could not break the Dutch monopoly. This is because the Dutch used branding strategies and negotiated market restrictions in other countries to monopolize the international herring market. Even with substantial government support, British herring did not start to eclipse the Dutch product until the Industrial Revolution, when pickled herring became a cheap staple rather than an expensive luxury. With substantially reduced costs of production at all stages from capture to distribution, British processors supplied the new middle classes of Europe with a cheaper product of only slightly lower quality than Dutch brands (see section 4.3). Quantity demanded increased rapidly, soaking up the expanding production of British fishers who were innovating to reduce costs and increase catches. Rather than adapt to changing circumstances, the Dutch herring boards—groups of wealthy fishers and processors who were the *de facto* regulators of the fishery—simply continued their existing practices, both in fishing and in processing (Beaujon 1884; J. T. Jenkins 1920).

Of course, cod and herring were not the only species of fish that were sold and cured; they were merely the most valuable species and the largest fisheries of the time.[3] In Europe and in North America, halibut, plaice, mackerel, menhaden, and salmon were also important species in markets for cured fish. In Asia, other marine organisms, including bonito, cuttlefish, abalone, and sea cucumber, were cured in several countries. Most of these products were destined for domestic consumption, usually in large cities or towns some distance from fishing ports. However, small quantities of the most valuable products were traded internationally. For instance, records indicate that during the Edo period (1600–1868 CE) Japanese merchants sold roasted sea cucumber, dried shark fins, and dried abalone to customers in China (Kalland 1995, 167). Because curing by drying or salting remains one of the cheapest and easiest ways to preserve fish, it is still used by artisanal fishers in many developing countries (FAO 2012d). Valuable trade in specialty cured products such as dried shark fin, pickled herring, and smoked salmon continues today.

Although the limiting effects of lower prices were fairly small in the early 1800s, continued expansion in the supply of fish products due to exploration and innovation periodically narrowed the profit disconnect toward the end of the 19th century. Trade in fish products was still limited by the shelf life of cured products, and demand was restricted by their limited palatability. Therefore, countering the price signal required the development of new processing methods. In response, two processes were appropriated for this purpose: canning and freezing. Both methods generally produce a more palatable product than curing, and properly canned fish can last for years with no special handling. In conjunction with the expansion of refrigeration technology, demand for frozen fish products increased substantially in many parts of the world. Indeed, without these innovations, demand for fish products would be much lower and the scope of the fishing industry as a whole much smaller than it is today.

Canning Canning technologies were first invented to preserve high-priced meat products but were eventually modified for fish. Essentially, canning requires that the fish be cleaned, sliced or minced, packed into cans (sometimes with oil), and sterilized through heating to high temperatures. Cans are also sealed to prevent recontamination with the micro-organisms that generate decay. Interestingly, knowledge of canning processes predated scientific theories that correctly attributed decay or "putrification" to cellular processes rather than to "vital humors" in the 1830s. Numerous scientists advanced cellular theory but discoveries by Louis Pasteur and John Tyndall were crucial for the canning industry. They showed that heat could be used to sterilize organic matter and that removing oxygen could prevent the regrowth of bacteria (Kilbourn 1884, 219–220).

Europeans were early pioneers in the canned seafood industry, but the majority of mechanical innovations in canning occurred in the United States. Experiments in canning seafood started around 1820. Canned sardines originated in France in 1824 during a boom in harvests and quickly gained popularity throughout Europe and the Americas (Fichou 2004). Oysters, lobster, and salmon were also canned, though this industry did not take off until the 1840s. Most of these early canneries focused on producing tinned seafood but would also preserve fruits and vegetables when catches were low. Like curing, the process of canning was highly labor intensive at this time. Everything was done by hand, including making the tins, processing the fish, and sealing and labeling the final product (Cobb 1919).

By the 1880s, canning was important in most of the major fishing countries, though Britain was well behind the United States, France, and other European countries, and Japan was just beginning to experiment with canning fish for export (Okoshi 1884; Walpole 1884b). The United States was the largest manufacturer of canned seafood, producing over 31 million cans of salmon, 4 million cans of lobsters, and hundreds of thousands of cans of other fish products for domestic consumption and international export. This reflects the size of the US fishing industry, the power of the US economy in the period, and the scope of demand from growing US populations. However, canned goods still accounted for only a small portion of total US fish production, particularly by value (Goode 1884). By this time, canneries were operating on both the East and West Coasts of the US and were just starting production in Alaska (McFarland 1911; Cobb 1906, 1919). In the early 1900s, China and other countries began producing canned fish products, and by the 1930s there were canneries throughout Southeast Asia (Yen 1910; Barnett 1943).

Mechanization of the canning process substantially increased the efficiency of production. This shifted the supply curve to the right, reducing the prices of canned products but increasing the quantity that could be sold. It also increased cannery demand for raw materials, which kept ex vessel prices for fish high, reinforced or expanded existing profit disconnects in fisheries, and facilitated fishery-level exploration and efficiency improvements. Mechanization began in the late 1800s. Processors first invested in machines that would automate the process of making cans, then developed machines to process the fish (McFarland 1911). Perhaps the most important machine was the "Iron Chink" (figure 4.1). Invented in Washington State in 1903 for processing salmon, this machine completely cleaned and "slimed" the fish, replacing (mostly Chinese) workers who had previously done the task for fairly low wages (Cobb 1921, 115). Other inventions of the early 1900s included a "cutter" that cut the fish to the size of the cans, a "filler" that filled the cans with fish cut to size, a "double crimper" that attached the lids to the cans without solder, and various types of elevators and hoppers that moved the fish from the docks to the factory floor and then from machine to machine. Larger, more efficient boilers, steamers, and cooking processes were also developed. Much of this innovation occurred in the Alaskan salmon canneries, where labor was more expensive and the amount of fish pack was greater than in the lower 48 states. Together these innovations enabled canneries to produce 15–20 times more in the 1920s than they had in the late 1800s (113).

The efficiency of canneries continued to improve in the 20th century, and the industry continued to expand both in scope and scale. By 1938, 7% of global landings of fish were processed in canneries. After a slight decline during World War II, canned production increased rapidly. The amount of landings used in canneries worldwide quadrupled from 1948 to 1976, and overall production of canned seafood in major fishing countries more than doubled from 1955 to 1976.[4] During this period, production of canned lobster and salmon declined, but production of canned tuna, which became widely popular in the United States after World War II, grew tremendously. Production of canned anchovies increased as well, particularly in the 1960s, after Peru started exploiting the huge stock of anchoveta off its coasts. By 1976, production of canned tuna, sardines, anchovies, and other pelagic species was more than twice the amount of all other types of fish (FAO 2012a). Interestingly, as canned production increased, production of cured fish also increased slightly, reflecting the symbiotic relationship between

Figure 4.1
A bank of "Iron Chinks." Source: Cobb 1919.

expansion in the supply of fish and growth of demand for processed fish products globally (see figure 4.2).

Freezing Canning of seafood products was a solid economic success, but as shown in figure 4.2, freezing rapidly eclipsed all other methods of preservation in the second half of the 20th century. Initially, frozen fish, particularly herring, was used as bait in other fisheries, rather than as food. However, there were several roadblocks to the extension of freezing to fish destined for human consumption. First, until the invention of refrigerated train cars in the 1870s it was difficult to transport frozen fish to inland markets. Second, few retailers had facilities to display frozen foods, and few consumers had means to store them (Frohman et al. 1966). Third, public perception was that freezing reduced the quality of the fish—which may have been true at the time, because flash freezing had not yet been invented and consumers did not know how to cook frozen fish (Kilbourn 1884, 222).

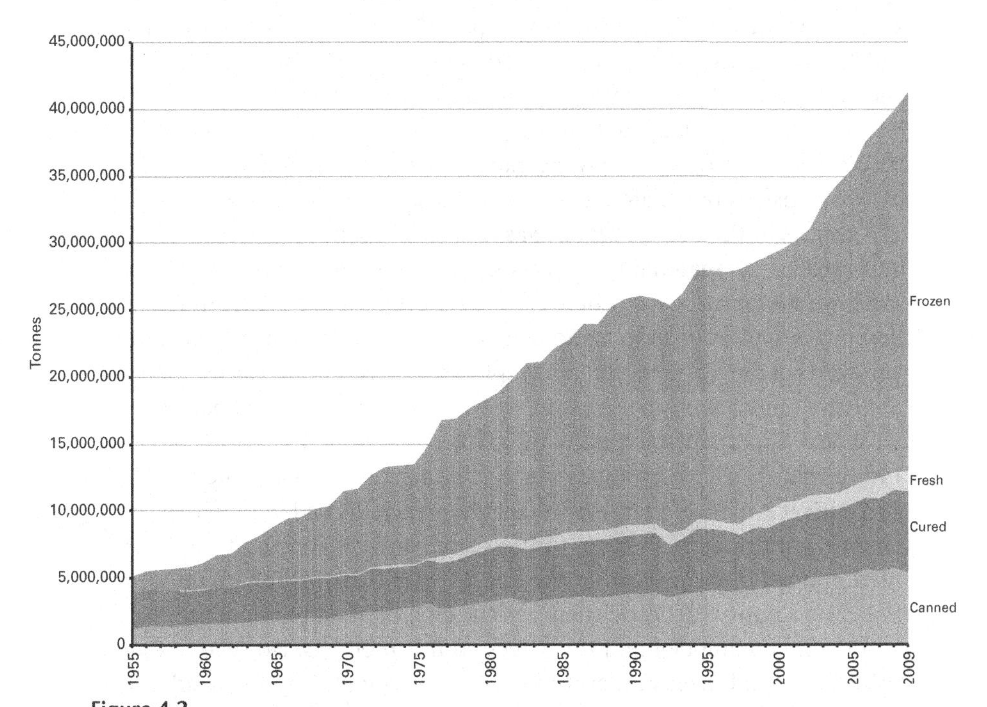

Figure 4.2
Production of processed fish products for use as food, 1955–2009. Sources of data: FAO 1960b for 1950–1960; FAO yearbooks from the respective year for 1961–1975; FAO 2012a for 1976–2009.

In spite of these early barriers, the food-processing industry pursued new methods of refrigeration to expand markets temporally and geographically. The first machines for freezing food were developed in 1877. Fish processors started freezing salmon and other important food fish in the 1880s. They specifically adopted freezer technologies to store fish in periods of glut, when there was not enough time to can or cure all of the catch before decomposition set in. Once the glut had passed, the fish would be unfrozen and processed as usual, smoothing out natural periods of surplus and shortage and stabilizing prices for processed products (Cobb 1921; Teuteberg 2009). Smelt and other small fish could be frozen naturally in the winter, but many processors invested in cold-storage plants, so that they could freeze fish year round and also could freeze larger fish more quickly, improving palatability of the final product. In 1913, a Danish fish importer developed flash-freezing technologies that reduced freezing times and also improved the quality of frozen fish products (Teuteberg 2009, 205). Later, superchilling was developed to freeze fish on board fishing vessels. In processing facilities, flash freezing with liquid nitrogen was used early on, but quicker and cheaper freezing methods were developed in the late 1960s, including use of blast cooling in tunnel freezers. All of these methods are still used today (Berk 2009; B. A. Parkes 1966).

With the development of refrigeration technology, production and trade in frozen fish as raw material for the processing industry skyrocketed. Frozen fish from the United States was in high demand in Europe, where it was used as raw material by processors who could not get sufficient product from European fleets (Cobb 1919, 1921). This helped smooth out geographic variations in supply and demand that were created in large part by the cycles of overexploitation and exploration described in earlier chapters. Similarly, changes in fishing and freezing technologies on board vessels were also important for expanding markets. By the turn of the 20th century demand for fish products was increasing substantially due to exogenous drivers. To meet increasing demand, processors had to purchase larger amounts of fish. New factory fishing vessels and other distant-water fleets used onboard freezing technologies to expand their fishing grounds, and there was a positive feedback between the growth in supply from these fleets and the increasing demand for processed fish products. However, existing fisheries for cod, haddock, and other popular whitefish were already overexploited in this period and could not fill this new freezer and processing capacity, so fishers and processors turned to new species, including pollock, turbot, and albacore tuna (FAO 2012a).

Direct consumption of frozen fish products by consumers did not begin until after World War II, when frozen foods in general became more common. Here again, exogenous innovations catalyzed a major change in the resources available in the structural context for fisheries. There is no record of retail sales of frozen fish before the war, but by 1948 about 5% of global landings were used to produce frozen fish products that were directly purchased by consumers (FAO 1960b). Growth in "prepared" frozen fish products picked up in the 1950s and the 1960s, thanks in part to Clarence Birdseye's invention of methods for freezing fish fillets rather than whole fish. Other drivers of the shift included changes in consumer beliefs about the palatability of frozen food products generally and the expansion of cold storage in supermarkets and homes. These technological changes were part of a general trend, particularly in developed countries, in which supermarkets quickly replaced old-fashioned general stores and greater affluence allowed middle-class consumers to purchase new appliances like refrigerators (Teuteberg 2009). To stay competitive in this changing marketplace, many fish processors began developing and marketing new products (see Section 4.3).

As a result of these exogenous and endogenous changes, production of prepared frozen fish products increased from 1.2 million tonnes in 1955 to 28.3 million tonnes in 2009. Products like frozen fish sticks were popularized earlier; consumption of these products in the US jumped from 3,400 tonnes in 1953 to 36,000 tonnes in 1963 (Frohman et al. 1966). By 1970, frozen products accounted for 30% of all fish products sold directly to consumers. By 1980, frozen fish was at 40% and by 2003 it was at 50% (FAO 2012a; estimates for years before 1976 were compiled from FAO 1960b, 1962, and 1977). Like canned products, frozen products are usually at the lower end of the price spectrum, with some variation. Frozen fish products are consumed mainly in developed countries, but consumption has also been increasing in developing countries in recent years (FAO 2012d). Freezing is by far the most prolific means of preservation for human consumption, and many different types of marine organisms are now sold as frozen seafood products. Indeed, without these technologies, it is likely that demand for fish products would be much lower than it is today, simply because the palatability and the variety of products available would be less enticing to consumers.

Flavor and Appearance Despite the many advances in seafood preservation since the 1960s, the principles of preservation have not changed much. Heat is still used to sterilize products and cold is still used to slow the growth of micro-organisms, though processors are always developing or copying more efficient methods (Garthwaite 1997; Horner 1997). Chemical additives are now used liberally for many purposes at several stages of the production processes. For instance, sulfides are applied to shrimp and other crustaceans at sea to prevent brown spots that occur with oxidation. Potassium sorbate and sorbic acid are also used as preservatives in highly processed seafoods like surimi (Taylor and Nordlee 1993). Several types of modified atmospheric packaging are also common now, including vacuum packing (in which all air is removed) and other processes in which air is replaced by a combination of gases such as carbon dioxide, nitrogen, and oxygen. Plastic is commonly used as packaging in this type of processing because it is both flexible and non-permeable (A. R. Davies 1997, 201).

These new methods of preservation stimulated the creation of new markets for several types of processed fish products. First, note the growth of "fresh" processed fish products shown above in figure 4.2. These have been filleted, cleaned, or otherwise processed but not frozen, cooked, canned, or cured. Vacuum packing and refrigeration are common methods of preservation for these types of products.[5] Better preservatives and chilling facilities also catalyzed growth in the surimi industry. Based on a centuries-old Japanese product called kamaboko, surimi is a generic product produced from many different species of whitefish—usually Alaskan pollock and other lower-priced species.[6] To make surimi, the whitefish is minced, washed, and then boiled or steamed. The resultant mass naturally gels and can be mixed with flavoring, colorant, and other ingredients to make fish cakes, imitation crab meat, and similar foods. These products are commonly produced and consumed throughout Asia, and some are familiar in other parts of the world where growing populations and economic development drive increasing demand for these products (Kalland 1995; Hall and Ahmad 1997; Pietrowski et al. 2011).

Several important similarities and differences are evident between processing and fishing. On one hand, technological innovation in food processing was both exogenously driven and endogenously modified. Processors, like fishers, engineered for efficiency to compete with one another and with other food producers. They also took advantage of the many innovations generated during the Industrial Revolution. On the other hand, there is substantial potential for economies of scale in processing, much more than in fishing, because processors do not face the same logistical problems and physical limitations that fishers deal with on the oceans. Furthermore, processors can purchase fish from many fishers, often from different countries, thereby reducing the inherent risks of volatility in capture fisheries production. Before examining the capital implication of these comparisons, it is necessary to take a look at the other side of processing, production of non-food fish products.

4.1.2 Non-food Uses

Many people utilized fish products for non-food purposes at different times in history. Before the advent of industrial processing, fish would be used for fertilizer when it could not be consumed directly or preserved in a palatable form. In the 1800s, some fishers targeted herring, menhaden, and other species to produce oils for industrial applications, including the manufacture of paints and similar products (McFarland 1911). Fertilizers, fish meals, and fish oils could also be produced as by-products of food fisheries. Where rendering plants were not readily accessible, offal often was discarded at sea or dumped on land (Cobb 1921; Catarci 2004). However, certain by-products, like cod liver oil, were quite valuable. As the name suggests, processors would remove the livers from the rest of the offal and process them separately to obtain the oil, which was first used in the manufacture of paints and leather products in the Middle Ages and which later became a dietary supplement (Ackman 2003).

Whales were among the first living marine resources to be commercially targeted for non-food purposes. They were captured specifically for their oil as early as the 11th and 12th centuries in Europe. Whale oil was originally used as a fuel in lamps. Over time, Europeans began using whale oil to lubricate machinery and in the manufacture of candles, soaps, paints, and varnishes. "Whalebone"—the bristles that baleen whales use to filter their food—was also highly valued as the raw material for corset stays, the hoops in hoop skirts, and other items that required rigid flexibility. In the mid-1700s, Dutch exports of whalebone to England were valued at £500 per ton.[7] The commercial value of whalebone declined in the early 1800s, fluctuating between £60 and £300 per ton (approximately US$4,000–25,000 in 2010 dollars). Although whales were not targeted for it specifically, ambergris was an important by-product of the fishery, because it was used in the production of expensive perfumes (J. T. Jenkins 1921, 41).

Growing demand and increasing prices for non-food products had a significant impact on the development of the whaling industry. As explained in chapter 3, there were two types of whaling: shore based and sea based. The shore-based whalers would tow their whales to shore, where they could be processed for both food and non-food uses. In Europe during the Middle Ages, there was not much demand for whale meat, but there was considerable demand for whale oil, so whales were targeted mainly for non-food uses. Eventually, the Basques and their imitators developed methods for processing whales at sea, collecting the blubber, and reducing it to oil aboard their ships. Whalebone and ambergris were collected, but the rest of the carcass was discarded because it was not valuable enough to warrant the large amount of time and space required to transport it back to land for sale. Nevertheless, in the Americas and in other areas where whales could still be harvested from shore, whale meat was used as food for humans and cattle, and the carcass was processed into fertilizer and bone meal, which was also an important agricultural input (J. T. Jenkins 1921). Although prices started to decline in the 1700s, the whaling industry remained important until the mid-1900s, when the proliferation of synthetic substitutes led to a sharp downturn in demand for whale-based products and growing concern about declining whale populations generated political actions to curb the fishery (see chapters 5 and 8).

Other types of marine living resources were also targeted for non-food uses. For instance, the Japanese started using dried sardine meal as a fertilizer more than 1,500 years ago (Kalland 1995, 1). Since they did not have large sea-worthy vessels, fishers or entire villages would wait for the seasonal appearance of schools of sardines near shore. Once the schools appeared, everyone in the village would race to set a shore-based purse seine net to catch as many sardines as possible before the school dissipated. Some of the sardines were consumed fresh, but the catches were so large and the quality of preserved fish so low that most would be dried and sold at low prices as fertilizer. Another species, the sand lance, was also utilized in the same manner in Japanese villages. In the 17th century, with the development of better boats and nets, Japanese fishers started targeting herring for the fertilizer market. Because Japanese entrepreneurs consolidated labor and capital to create fish fertilizer manufacturing centers, some believe that this was an important step toward the industrialization of Japanese fisheries (Howell 1995). It certainly generated a positive feedback between growth in supply and demand, much as in the symbiotic the relationship between freezing technologies and distant-water fleets described above.

As fisheries industrialized, the volume of offal increased substantially, as did the number of fish species utilized for non-food products. Species targeted directly for meal and oil included herring, sardines, mackerel, and anchovies (clupeids and scombroids) as well as blue whiting, sprat, hake, and pout (gadoids). Again, there was not much demand for these species as food fish, but they could be captured in high volumes at low cost, so fishers looked for non-food markets for their catch. In addition, as gear became larger and less selective, by-catch from food fisheries was also processed for non-food uses (FAO 1986, sec. 2.2). Much of the now highly valuable stock of Atlantic bluefin tuna was depleted in the 1960s, when it was not in demand as an expensive food fish and instead was used primarily in the production of animal feed (Webster 2009, 175). Sharks and rays, as well as any other "trash fish" that has a low economic value, may also be used to manufacture fish meals and oils (FAO 1986, sec. 2.2). These products are still used as additives in fertilizer and animal feed but are no longer common in industrial applications.

Another use for high-quantity, low-value fish is for bait in other fisheries. In fact, "natural" freezing methods were first developed in conjunction with the marketing of herring as baitfish in the mid-1800s. By this time, huge quantities of herring were harvested off the coast of the United States as well as in Europe and numerous substitutes for pickled herring were available, so supply was high relative to demand, ex-vessel prices were quite low, and the profit disconnect was narrow. Looking for a new market for his catch, a US captain named H. O. Smith experimented with natural freezing in the winter of 1854–55. He froze a catch of 80,000 herring by exposing them to the natural temperatures of the season, and then loaded them into an insulated hold on his ship for transport to Gloucester, Massachusetts. In port, he successfully sold his frozen catch as bait to schooners targeting cod and other groundfish. Given Smith's success, the practice was quickly emulated by other herring fishers and a new market was established (McFarland 1911). These innovations ensured that increasing demand for high-priced fishes spurred a growing demand for low-priced fishes as bait, increasing the profit disconnect at both ends of the price spectrum.

Similarly, population growth, economic development, globalization, and industrialization drove indirect demand for fish products for use in agriculture (including aquaculture). To tap these markets, processors developed industrial methods for production of fish-based protein additives for animal feeds in the 1960s. Processors also increased production of fish meal and fish oil for use in the growing pet-food industry (Naylor and Burke 2005; Skewgar et al. 2005; FAO 2012d). New non-food uses for marine living resources continue to emerge, including multiple applications in the cosmetics and pharmaceutical industries, as additives in non-fish food-processing industries, and in gelatin, leather, and handicrafts (FAO 2012d). In FAO datasets, all of these uses fall under the category "not elsewhere included" (nei), which is a small but important part of overall production because it allows the industry to utilize raw materials that would otherwise either go to waste (offal and discards) or be left in the ocean (low-priced fish like herring). These actions widen the profit disconnect for certain segments of the industry, allowing production to increase well beyond the levels that would be predicted in a basic common pool resource scenario.

Although production increased for all product types from 1955 to 2009, the proportion of meals, oils, and nei production declined from about 36% in 1955 to about 20% in 2009 (FAO 1960b, 2012a). There are many reasons for this change, but one major factor was an increase in demand for fish as food due to global population growth and economic expansion. This trend, combined with marketing to introduce consumers to new products, allowed processers to earn high profits by canning, curing, or freezing sardines, anchovies, and other fish for human consumption, rather than turning them into fertilizer or fish meal (Garcia and De Leiva Moreno 2001). Recently, however, with declines in fish stocks resulting in increasing competition over raw materials, processors are again working to reduce waste and to find ways to utilize all parts of the fish. In 2010, 36% of world fish meal production was obtained from offal. This was a substantial increase over the 2000 level (4%), but much of this production is still derived from species targeted specifically for non-food uses (FAO 2012a,d). Data are not available on offal versus targeted production of fish meal prior to 2000, but, given that economic conditions were similar to modern times, it is likely that the proportion of offal-based meals was much higher in the late 1800s and the early 1900s.

4.1.3 Stratification and Growth

The fish-processing sector is similar to other industries in many ways, but there are a few key differences. As the discussion of offal suggests, processors go through cycles of growth and recession, as per Schumpeter ([1946] 1976). Growth is driven by increasing returns to scale and the potential for innovation in highly concentrated industries that can leverage substantial capital for investment. To some extent, this scenario parallels the development of fishing fleets. Larger processing operations, like larger fishing vessels and fleets, required greater amounts of capital. However, unlike fisheries, there were few economic or environmental niches for small processors to exploit, and so the stratification observed in capture fisheries is inverted in the processing industry. Certainly, subsistence and artisanal fishers still process their own catch, but in commercial fisheries only a few artisan curers and canners remain. The industry is completely dominated by large multinational companies that have access to considerable capital and can control large shares of the marketplace. In fact, the processing industry is highly oligopolistic, and the lack of competition can undermine efficiency, resulting in considerable waste. Nevertheless, processors do adapt over time, and high prices for raw materials or low demand in the market can easily spark spates of innovation and efficiency improvements (Clark and Munro 1980; Mansfield 2003; McDonald 2000; Gopinath, Pick, and Li 2003).

The concentration of capital has several additional costs and benefits. It gives processors monopsony power in the purchase of fish from fishers. Thus, ex-vessel prices for fish that are targeted for processing are set internationally, and fishers are price takers in the industry (Clark and Munro 1980). On the other hand, the natural volatility of capture fisheries production is problematic for processors because they need stable flows of resources to maximize their efficiency. Freezer facilities can smooth out the feast-and-famine cycles that are common in fish production, but the scale of modern processing facilities is so large and so highly specialized that other methods are necessary. Diversification of sources is the primary approach used by processors. By purchasing fish from many different parts of the world, they can still obtain raw material even during regional shortages (see, e.g., Asche et al. 2012; Burke and Phyne 2008). However, this very flexibility insulates modern processors from cost signals related to overexploitation and over-capitalization in specific fisheries. As catch from one fishery declines, processors can switch to suppliers in areas where the costs of production are lower rather than pay higher prices for raw materials.

In fact, industrial processors, like commercial fishers, expand the profit disconnect by establishing facilities in different countries to take advantage of lower operating costs. As figure 4.3 shows, production of processed fish products in developing countries started increasing in the 1980s and continued to increase, while production in developed countries peaked in the late 1980s and then went into a gradual decline. By 2009, developing countries produced 63.5% of all processed fish products, three times the percentage in the 1980s (FAO 2012a). This change in production is not equally distributed among all developing countries but rather is concentrated in a few, mainly in Asia and the Americas (FAO 2012d). China is by far the largest producer, with 25.8% of all processed fisheries production in 2009 (FAO 2012a). Among developing countries, Thailand is currently the hub of several major processing sectors, including canned tuna and frozen shrimp (Miyake et al. 2010; Goss, Burch, and Rickson 2000). Thailand and Indonesia produce 4.7% and 4.2% of all processed fish products, respectively. This is less than Japan and the Russian Federation but more than that of the United States. Peru, Chile, India, Vietnam, Myanmar, and Taiwan also account for 2–4% of production, on par with several European countries. Most other developing countries produce less than 1% of processed fish

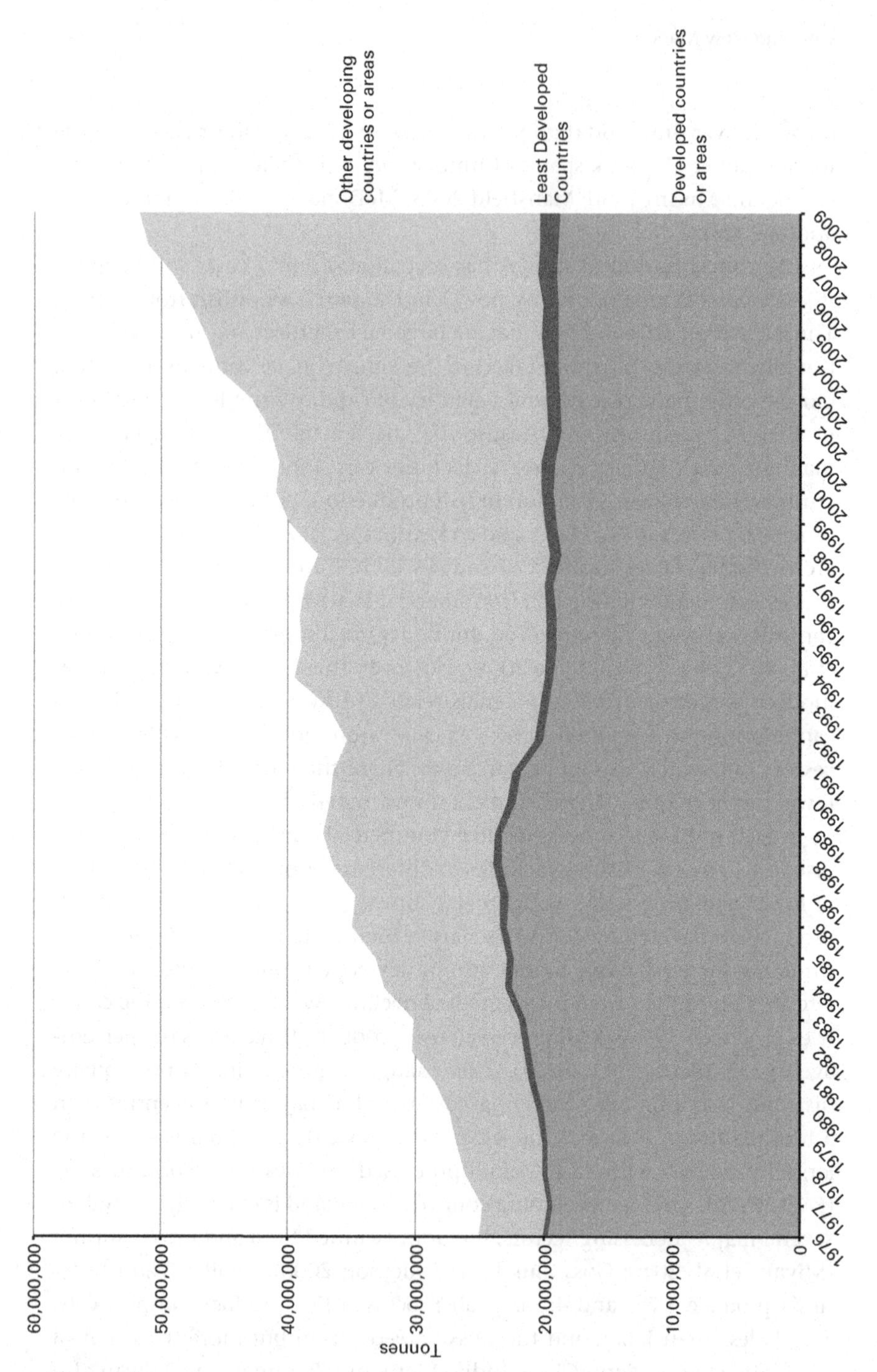

Figure 4.3

Production of processed fish products, 1976–2009. Source: FAO 2012a.

products, and the least-developed countries have very few processing facilities (FAO 2012a).[8] Much as with mechanization, globalization of the processing industry keeps costs of production—and prices—low, so that higher volumes of products can be sold. This helps soak up the increasing capture production and aquaculture production described in chapter 1.

This internationalization of the processing industry—in the production of both raw material and finished goods—could not have occurred without many concomitant changes in fish production, shipping, and marketing. Without these economic and technological changes, processors would not have been able to expand to their current size, because scarcity would have forced ex-vessel prices to rise, thereby reducing quantity demanded in what is primarily an industry focused on production of large quantities of lower-priced products. The opposite is also true, insofar as fishers would not have been able to open up new markets for their increasing landings of lower-priced fish without the rise of industrial processing facilities. For both processed and non-processed fish products, improved shipping technologies and marketing were also necessary to the expansion of the industry.

4.2 Shipping

Improved methods of transportation were important catalysts for the expansion of the world's fisheries. First, as fishing voyages increased in length, fishers needed to ensure that their catch would not deteriorate before they returned to land. Processing at sea was developed for this purpose, but fishers also invested in new transportation technologies so they could get fresh fish to markets faster and in better condition. In general, people are willing to pay more for higher-quality fish, so faster transport methods increased both the value of the catch and the range of markets where fresh fish could be sold. Second, improved preservation during shipping was necessary so that the industry could bring both fresh and frozen products to inland and international markets. This effectively increased the demand for these products regionally and globally and therefore mitigated the price signal associated with the growth of supply. Lastly, better shipping technologies further reduced costs of production, even for processed products like canned goods that need little preservation. This cost reduction helped to prolong the economic life of global fisheries by further increasing the profit disconnect.

As with fishing and processing technologies, this historiography reveals both endogenous innovations developed by industry entrepreneurs and exogenous changes that fishers, processors, and marketers appropriated to improve transportation of their products. But there is also a distinction here between changes that were financed by the fishing industry, which therefore increased capital costs, and public investment in transportation infrastructure, such as railroads, airports, and dock facilities, which spread the burden of cost over larger populations. Most of the policy implications of investment in infrastructure will be covered in Part II. These include government subsidization of fisheries through infrastructure investment and industry influence on governments due to their private investment in domestic infrastructure. This section describes the technological evolution of shipping methods and the related impact on the growth of the industry.

Many of the historical advances in marine transportation were covered in chapter 3, but a few other important changes that improved transport of fish to markets or processing facilities merit attention. Fishers often appropriated and then modified designs for bigger, faster ships developed for navies or the merchant marine, increasing their own speed and reducing their travel time. Factory vessels enabled fishers to process fish on board so the catch would be preserved on long trips. Yet prices for fresh fish were often higher than those for processed fish, so fishers supplying markets for fresh fish developed methods to keep fish alive during transport from the fishing grounds to port (Drouard 2009). From early times, fishers from many countries fitted their undecked vessels with pens that could be filled with seawater and would keep most species alive for a few days (Suisankyoku 1915). Such modifications are still used by artisanal fishers. Similar technologies are also used to store bait fish in modern commercial and recreational vessels (FAO 2012d).

At an industrial scale, fishers initiated the use of welled boats, which had pierced hulls to allow seawater to circulate in certain holds, keeping the fish alive until they reached shore and could be sold fresh. It is difficult to discern the origin of this technology. Holdsworth (1874) claims that the first welled vessels were developed in Harwich, England, in 1712. He provides specific information on the number of vessels constructed by various companies and the impacts of the vessels on the fisheries of England. However, Beaujon (1884) speculates that the Dutch must have used welled boats much earlier due to the distance between their fishing grounds and home ports. He also cites a law passed in 1777 that prohibited sale or transfer of welled vessels to foreigners. Although their origin is in doubt, it is clear that welled smacks were important for the expansion of North American and European fleets that supplied fresh markets into the 19th century, as they are mentioned in many different texts. In fact, this technology catalyzed the commercialization of the lobster fisheries of North America by facilitating the transport of live lobster to major urban centers (Acheson 1997). Like other technological advances, welled vessels cost more than their "dry" counterparts, but the difference was not as great as that between sail and steam vessels. For instance, in the late 1800s a welled trawler cost about 25% more than a conventional trawler (Holdsworth 1874, 141).

In the 19th century, steam power improved transportation from port cities on the coasts to inland regions and fostered the creation of new markets overseas. Few of these innovations were developed specifically for fisheries. For instance, the steam power that helped fishers increase their harvests also increased the speed at which fish could be transported to market. Rather than sail to the nearest port and then transport the fish over land, as was most efficient with sailing vessels, steamers could travel to the port of sale by sea, often in the same amount of time or faster (Blake 1884). In fact, in Britain steam-powered carriers were first employed to transport fish from the fleets at sea to port locations in 1868. That was eight years before steam was used to power fishing vessels. Furthermore, many of the earliest steam-powered fishing boats were converted tugboats or carrier vessels that were still used to haul fish when not engaged in fishing (10, 26).

Starting in the 1800s, railroads provided a faster and cheaper alternative for the distribution of fresh and preserved fish products on land. Before the railroads, fresh fish was not available to most consumers living more than a few miles from the coast (Gibbs 1922; Kalland 1995). Processed fish could be delivered using pack animals, but this method was both slow and expensive, so preserved fish—like meat in general—was a luxury good in inland markets (Beaujon 1884). Many authors note the impact of railroads. According to Day (1884, 9), in India transport of dried and salted fish more than doubled in only 10 years (1872–1881) because of the Grand Trunk Peninsular Railroad. Smitt (1884) relates how Swedish fisheries grew as a result of improved access to markets via railroads. Cobb (1921) describes how the transcontinental railway facilitated the growth of fisheries along the Pacific coast of North America, particularly for cured salmon and (later) canned salmon, which could be shipped via rail to major population centers on the east coasts of the US and Canada and around the Great Lakes. Rail transport was also important in the transformation of European markets for herring and other cured fish, which were historically produced in low quantities and sold for high prices but became low-priced, high-quantity during the Industrial Revolution.

The speed of railways opened up inland markets, greatly increasing demand for canned and cured fish, but the addition of refrigeration opened up markets for fresh and frozen fish all over the world (Beaujon 1884). Ice was the earliest method used to refrigerate fresh fish. In 1854, the British pioneered the use of ice to preserve fish for the rail journey from the major port of Yarmouth to the Billingsgate fish market in London, a distance of about 100 miles (J. T. Jenkins 1920, 145). They also commissioned the first refrigerated rail car in 1879. This was an exogenous innovation, as the goal was to transport meat rather than fish, but the technology was quickly appropriated for fish and many other perishable goods (Blake 1884, 38). The Chinese used natural ice to preserve fish as early as 1840, if not before. Rather than keep the ice on board the fishing vessel, ice boats would follow fishing fleets, taking in the catch and then ferrying it back to land, much as the Dutch tender ships had in previous eras (Muscolino 2009, 31–32). In Europe, the Germans were probably the first to use ice boxes aboard fishing vessels, starting in 1868 (Teuteberg 2009, 197). British vessels began using ice on board in the 1870s, and the practice spread to India and the Americas by the turn of the century (Holdsworth 1874; Day 1884; Cobb 1906). By the 1920s, most large fishing companies had installed ice factories in their port facilities (Gibbs 1922). Refrigerated holds and cold-storage lockers were also common on larger-scale commercial fishing vessels by this time.

Combined, the technologies of steam, rail, and refrigeration facilitated a substantial increase in the production of live, fresh, or chilled fish, which is now about 40.5% of total fisheries production (FAO 2012d). Even more than cured fish products, fresh fish was a luxury good outside of fishing ports prior to the mid-1800s. However, with new methods of transportation, fresh fish became an affordable source of protein for the growing middle class. Holdsworth (1874) describes a substantive transformation of markets for fish products in Britain and related effects on fishing practices. Demand for fresh fish increased so much that fishers began to sell "offal fish" such as haddock in the fresh markets instead of either discarding it at sea or selling it at low prices to curers (89). On the other hand, prices for "prime fish," such as live plaice brought in by welled boats, declined as the supply of chilled plaice increased. In this case, as the middle classes began consuming more and more plaice, the upper classes no longer viewed it as a status symbol, and prices for live plaice declined even as prices for chilled plaice rose (91). This trend continued, and in spite of some price declines the fishery for plaice was one of the largest and most valuable in Britain in the early 20th century (J. T. Jenkins 1920).

Trade in fish products, whether fresh, frozen, canned, or salted, also increased due to the proliferation of standardized shipping containers in the middle of the 20th century. US entrepreneur Malcom McLean introduced the concept in 1954, when he started the Sea-Land Corporation.

McLean's innovation was that, instead of moving goods from ships to trucks piecemeal, his company would use standardized boxcars that could be filled with goods and transferred from trucks to boats and vice versa. Although this approach had considerable potential to reduce the costs of transportation, particularly by minimizing labor costs at port, McLean's business struggled and ultimately went bankrupt. However, the idea took hold, and today most shipping is now based on standardized containers that can be interchanged between trucks, trains, and boats (Roland 2007; Levinson 2006). This both reduces the potential for spoilage during transfer of products and reduces the costs of production for fishers, processors, and marketers, widening the profit disconnect for much of the fishing industry.

In the early 1970s, marketers began transporting Atlantic bluefin tuna via airplanes. This transformed the bluefin fishery by providing Atlantic fishers with access to the sushi and sashimi markets of Japan, where prices for high-quality air-freighted bluefin were 20 times the price of the low-quality bluefin that was canned or used in animal feed. Because of this change, some recreational fishers began targeting giant bluefin commercially, and a number of commercial fishers entered the fishery (E. H. Buck 1995, 6). Today, air freight is commonly used to transport several high-priced species, including bluefin and bigeye tuna, salmon, crab, and lobster. Many air-freight companies have specialized branches to handle live, chilled, and frozen seafood, and lobsters are often shipped overnight via UPS, DHL, and similar companies.

These new and improved air, land, and sea shipping methods spurred a tremendous increase in the global fisheries trade in the second half of the 20th century. Figure 4.4 illustrates the growth from 1957 to 2009. Although early data (1957–1975) are somewhat circumspect because the FAO did not yet have a single established method of data collection, the overall trend represented in the data is still valid.[9] Trade in fisheries products expanded almost sevenfold over the period. By 2010, 10% of agricultural trade and 1% of total global trade was in fish and fishery products (FAO 2012d). The value of trade increased as well, from about US$1.2 billion in 1960 to US$15.5 billion in 1980 to US$109 billion in 2010. In real terms, the value of trade in fisheries products increased twelvefold from 1960 to 2010 (FAO 1960a, 2012a,d; World Bank 2012). That the value of trade grew even more quickly than the magnitude of trade reflects the quality improvements through new processing techniques described in section 4.1, as well as exogenous growth in demand for fish products internationally. As described in the next section, diversification of consumer preferences was also a necessary

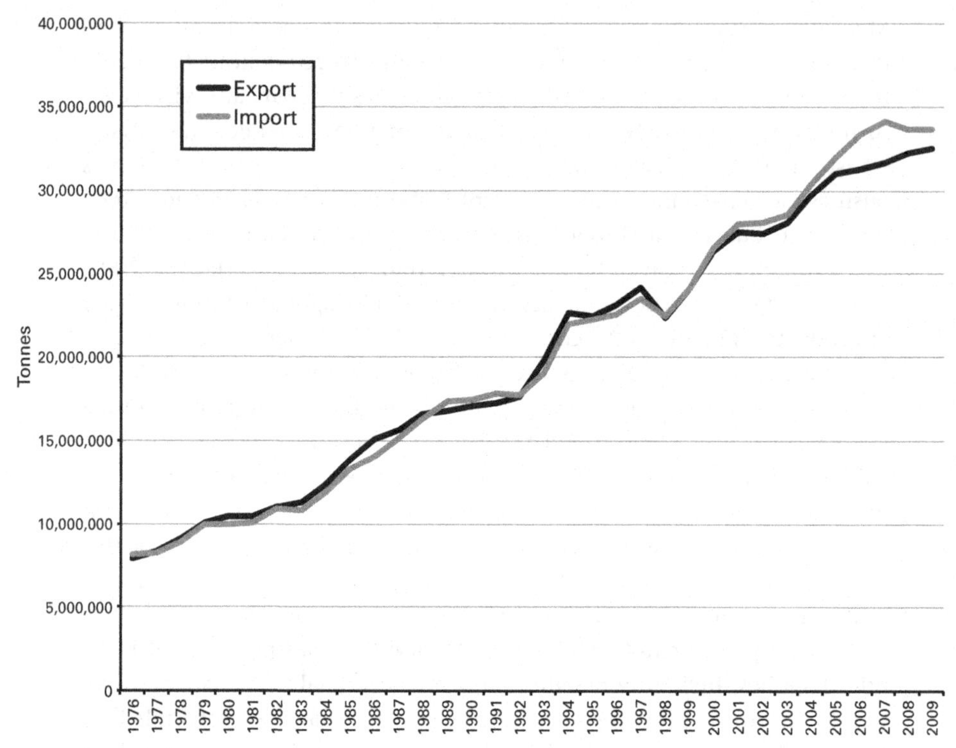

Figure 4.4
Volume of trade in fishery products. Sources of data: FAO 1960a for 1955–1961; FAO yearbook for respective year for 1962–1975; FAO 2012a for 1976–2009.

component in the growth of both domestic and international markets for fish products.

4.3 Marketing

Even with better processing and transportation, the fishing industry could not expand markets without also convincing consumers to buy new goods. Many factors affect consumers' willingness and ability to pay for fish products. As noted several times, population growth and economic growth can both increase demand for fish products, all else equal. Cultural factors can also increase or reduce consumption of seafood products. Indeed, much of the demand for pickled herring and salted cod in Europe during the Middle Ages was driven by the Vatican's injunctions to avoid eating meat other than fish on Fridays and holy days. Similarly, a number of fish products like shark-fin soup and bluefin tuna, are viewed as status symbols in some cultures, which substantially increases consumer willingness to pay as long as access to these goods is limited to social elites (note the transformation of plaice from luxury good to staple product described above). The availability of complements and substitutes may also affect demand, as evident in the decline of the whaling industry as consumers switched from whale oil to petroleum products for lighting and lubricants.

In addition to these exogenous factors, fishers and processors try to manipulate demand in several ways. First and foremost, they work to generate products that appeal to consumer tastes. This drove much of the innovation in shipping and processing described above. Other attempts to manipulate demand can be divided into three categories: quality control, quantity control, and diversification. As described in subsection 4.3.1, when fish production and trade was relatively low, processors established municipal and national brands to justify high prices by maintaining exceptionally high-quality products. Modern brands are still used for product differentiation and related price premiums. With the introduction of canning and freezing, quality control was also necessary to prevent episodes of food poisoning, which can depress demand for certain fish products. Subsection 4.3.2 documents how the rise of capitalism and related industrialization shifted competition to quantity rather than quality. During this period, temporary local gluts in established markets became common. Producers responded by advertising their products in new markets to stimulate demand. Lastly, subsection 4.3.3 shows how, in the 20th century, marketing campaigns were used to further expand demand by convincing consumers to buy new products (e.g., canned tuna and frozen fish sticks), diversifying production and consumption because known species were already overexploited and the fishing industry needed to switch to new species.

4.3.1 Quality Control

Reputation building and networking are important in fisheries marketing, particularly because fish and fish products are so perishable. When markets were all relatively small in scale and most sales were local, fishers (or fishwives in some cultures) would cultivate networks of clients by forging personal relationships with them. Reputation was important then as now, but easily monitored because fish could be inspected for freshness by buyers and prices were difficult to conceal in tight-knit communities. This dynamic still exists in small isolated fisheries, but with globalization, few fishing communities remain insulated from the outside world. As trade in fish and fish products expanded, merchants started acting as middlemen, buying fish from the fishers and selling it to the public. Merchants would therefore cultivate networks of fishers as well as networks of consumers and would have to manage their reputations on both ends (Muscolino 2009). Personal connections are still important in many locations, including the wholesale fish market in Tsukiji, Japan—the largest market for fish products in the world (Bestor 2004). However, many other approaches to marketing evolved with the stratification of the fishing industry. These approaches were important in the rise of large-scale commercial fishers in Europe and in the expansion of markets associated with the Industrial Revolution.

High Price, Low Quantity In Europe, commercial processors of pickled herring began branding their products in the 1400s. Branding was more important for this product than others of the time because the herring were packed into barrels so a buyer could not fully inspect the purchase at the point of sale. Less scrupulous vendors sold barrels packed primarily with low-quality fish but topped with a few layers of high-quality fish to fool buyers. Furthermore, because the trade in pickled herring extended throughout Europe, sellers moved around and cheated buyers were often unable to prosecute them. Distance also made it difficult to monitor the reputations of sellers or to obtain accurate information about prices. Brands were therefore used to establish a widespread reputation for quality and honesty; they were linked to city of origin rather than any specific company. As mentioned previously, under the Hanseatic League the Danish town of Scania earned a reputation for producing the highest-quality pickled herring in Europe. Their brand was strictly controlled by the Hansa, who provided inspectors to monitor the processing of the fish and assess its quality before the brand was applied (J. T. Jenkins 1920).

The Dutch followed similar practices during their ascendency in this market. Although some national-level legislation was enacted to ensure the quality of barrels as early as 1519, town brands were first used to indicate the quality of the fish (Beaujon 1884, 14–15). Beginning in the 1500s, Dutch herring producers also invented new brands to market their products to specific regions or consumers. For instance, barrels packed for exportation to France were usually labeled "Rouen brand" or "Cologne brand." Herring caught during different seasons or on particularly auspicious days were labeled to appeal to consumers—for example, there were brands for St. James Day, Lady Day, and Elevation Day (50). These brands were used in addition to town and national brands, providing for product differentiation within the larger context of quality control. Competing towns were known to tamper with the brands of their rivals in order to either gain from a higher-value symbol or to undermine the perceived quality of their competitors' products (44).

As noted previously, the Dutch were monopolists like the Hansa, and so protected their access to European markets with military force and by negotiated treaties as well as through branding. As part of their attempt to break the monopoly power of the Dutch, the British began branding—and subsidizing—production of Scottish pickled herring for export to Europe in 1718. These early attempts failed, and several iterations of "investment societies" for pickled herring went bankrupt during the 18th century (J. T. Jenkins1920). However, the brand itself survived as an indicator of quality, and British herring eventually won out over the Dutch product because their prices were lower and the quality was still acceptable to consumers. Broader changes were also important in this period, including the shift from mercantilist to capitalist ideology throughout Europe. A general influx of relatively cheap goods into Europe and the rise of a middle class that could purchase them also helped to break the Dutch monopolies. Finally, Dutch regulations designed to maintain the quality of the brand kept costs and prices high, making competition difficult in open markets (Beaujon 1884).

Low Price, High Quantity Eventually, capitalism replaced mercantilism, and municipal brands were superseded, for the most part, by corporate brands. This left quality control in the hands of corporations and retailers. At the beginning of the Industrial Revolution, rapid population growth, migration, and urbanization undermined producer accountability. Adulteration to reduce the costs of food production was common throughout Europe and the Americas. Rotten food was sold to the desperately poor and food poisoning was common. Canned meat products, including fish and lobster, caused major food poisoning events into the early 1900s (Collins 1993). Many outbreaks occurred in North America and Europe, all of which reduced consumer trust in canned fish products.

Most of the earliest forms of food poisoning that were associated with fisheries products stemmed from either histamine or botulism, which result from bacterial activity in fresh or preserved products, respectively (Collins 1993; Cobb 1919; Catarci 2004). Consumption of raw fish products, such as sushi and uncooked shellfish, can expose consumers to pathogens like *Salmonella, E. coli,* and *Staphylococcus aureus*. Outbreaks of illness due to these diseases and to less-well-known pathogens, such as *Vibrio parahemolyticus,* occurred in Japan from the 1950s to the 1970s, and were also common in other developed countries. A number of major outbreaks during this period reignited government and consumer concern regarding the safety of seafood products (D. P. Sen 2005; Su and Liu 2007). Around the same time, scientists discovered that fish can be tainted with heavy metals and persistent organic pollutants. These chemicals can have long-term impacts on neurological development (Moon and Choi 2009; "Dangers in Tinned Tuna" 1970; Catarci 2004). Certain algal blooms can also produce neurotoxins, which may be stored by shellfish and can cause illness if consumed by humans (FAO 2012d).

All of these adverse effects on human health can have negative impacts on demand for fish products. In fact, the deaths of even a few individuals can drastically reduce demand for a product, regardless of the source. For instance, in 1892 the British Admiralty banned tinned herring because of the death of a single officer from ptomaine poisoning. That well-publicized event also dampened other markets for canned products in the period (Anonymous 1892, 981). Sixty years later, two women died as a result of eating canned tuna tainted with botulism in the United States. In response, US officials inspected canned tuna all over the country and found tainted cans in Ohio, Michigan, and Georgia (The Tuna Scare 1963, 101). That incident, which was linked to only one particular label from one company, resulted in a 35% drop in demand for canned tuna throughout the United States. The cost to the industry was heavy, as several plants were closed and workers were laid off. In response, representatives of the tuna industry invested US$10 million (about US$56 million in 2010 dollars) in a marketing campaign to convince reluctant consumers that canned tuna was a safe and healthy product (Tuna Back in Favor 1963, 98; Williamson 2014).

These incidents are just a few of many that galvanized seafood producers to invest in both quality control and general marketing campaigns (Taylor 1986, 94–99; Wong et al. 2005, 563–571; Brown et al. 2001, 105–116; Weber et al. 1993, 451–454; Burns and Williams 1975, 1–6). In Europe and North America, fishing industry organizations began working with governments to ensure a minimal level of quality control to prevent food-borne illnesses much as they had in the 1800s. Many in the industry realized that public response to an outbreak was not limited to a specific brand or even a particular product. As the industry spread out into other parts of the world, the development of quality control mechanisms followed similar patterns (Horowitz, Pilcher, and Watts 2004; Cobb 1921; Mansfield 2003). Today the World Health Organization and the UN Food and Agriculture Organization work with private NGOs like the International Organization for Standardization to ensure safe handling and processing of seafood, particularly in developing countries (FAO 2012d). Outbreaks of food poisoning related to fish products still occur, but their effects on demand appear to be localized and temporary, thanks largely to quality control within the industry and related marketing campaigns to maintain consumer confidence. As described below, industry actors also use marketing to increase demand for known products, like cod and haddock, so that they could increase supply without dampening prices.

4.3.2 Quantity Control

Under the monopoly control exercised by the Hansa and the Dutch, quantities of fish sold remained relatively low in Europe. The transition to capitalism and the Industrial Revolution triggered a shift to high-volume production. This increased the importance of advertising as a means to convince consumers to buy more and different fish products. In the Middle Ages, European shop owners sometimes hired town criers to advertise for them and sometimes employed "touters" to stand outside their shops to entice buyers to step inside (Sampson 1875). In the 1500s printers began placing advertisements in books and pamphlets, and in 1628 the first newspaper advertisement appeared in a French publication (62). Into the early 1900s, print advertisements generally provided information about products or wares. Figure 4.5 provides an example of such an advertisement for a company selling many different types of fish products. This ad, published in the program for the 1883 World Fisheries Exhibition in London, is similar to many others of its day.

While increasing demand driven by the growth of the middle class soaked up increasing supply in the early 1800s, in the late 19th and early 20th centuries the advent of steam power, the spread of mechanization, and the opening of new fishing grounds caused stepwise increases in supply that generated extreme volatility in markets for some fish products. High prices for pickled herring, canned salmon, and even fresh fish like shad attracted capital investment, but the resultant increase in supply glutted existing markets, driving prices to extremely low levels. This, in turn, caused exit from the industry, resulting in decreased supply and higher prices that attracted new entrants, thus repeating the cycle. This pattern occurred in Europe and the Americas, and in each case surviving producers worked hard to expand consumption of their products, either overseas or in inland markets (see, e.g., Jenkins 1920; Cobb 1906). With new markets, demand increased and prices returned to relatively high levels. With the concomitant increase in profits, new entrants plunged back into these markets, often repeating the cycle yet again.

The case of salmon canning in Alaska is representative of this dynamic. Following the success of salmon canneries in the continental United States,

Advertisement. 67

JOHN MOIR & SON,

LIMITED,

LONDON, ABERDEEN, SEVILLE, AND WILMINGTON

(DELAWARE, U.S.A.).

Purveyors to H.R.H. the Prince of Wales.

Exhibit from their London, Aberdeen, and American Factories, the undernoted Fish, preserved in glasses and tins—

HERRINGS.—Fresh Herrings, Kippered Herrings, Red Herrings, Devilled Herrings, Herrings in Shrimp Sauce, Tomato Sauce, and Fennel Sauce; Pickled Herrings, Herrings à la Sardines, Digby Chicks, Potomac Pickled Herrings, Fried Herrings, Spiced Herrings.

HADDOCKS IN GLASSES.—Fresh Haddocks, Findon Haddocks, Haddock Roes.

SALMON IN GLASSES AND TINS (Scotch).—Fresh Salmon, Collared Salmon, Salmon Cutlets, Salmon Cutlets in Sauce, Salmon Cutlets in Oil, Pickled Salmon with Vinegar, Salmon Cutlets in Indian Sauce, Kippered Salmon.

COD FISH IN GLASSES.—Fresh Cod, Compressed Cod, Cod Roes, Cod Fish Balls, Smoked Cod Roes, Cod Sounds.

LING FISH IN GLASSES.—Fresh Ling.

TURBOT IN GLASSES AND TINS.—Fresh Turbot, Turbot Cutlets.

POTTED FISH IN GLASSES AND TINS.—Lobster, Salmon, Shrimp Bloater Paste, Anchovy Paste, Potted Turtle Meat, Potted Crab, Potted Findon Haddocks.

MACKEREL.—Fresh Mackerel, Kippered Mackerel, Mackerel in Sauce.

SOLES IN GLASSES.—Soles, Fried Soles, Fillets of Soles.

LOBSTERS IN GLASSES AND TINS, also Curried.

OYSTERS IN GLASSES AND TINS.—Fresh Oysters, Pickled Oysters, Curried Oysters.

ANCHOVIES IN GLASSES.—Gorgona, ditto in Oil.

FISH SOUP IN GLASSES AND TINS.—Fish Soup white, Terrapin Soup, Green Turtle Soup for Invalids, Real Turtle Soup clear and thick, Lobster Soup, Oyster Soup, Mulligatawny Fish Soup, Bouillabaisse, or French Fish Soup.

VARIOUS FISH IN GLASSES.—Green Turtle, Sturgeon, Sturgeon Cutlets, Potomac Herrings in Jelly, Chesapeake Bay Rock Fish, Delaware White Perch, Delaware Shad in Jelly, Florida Red Snapper, New England Brook Fish, Brandy Wine Suckers, Delaware Perch, Susquehanna Yellow Perch, Delaware Rock Fish in Jelly, Eastern Shore Common Terrapin, Chincoteague Sound Diamond Back Terrapin in Jelly, Red Snapper Cutlets, Red Snapper Cutlets in Jelly, Red Snapper Cutlets with Oyster Sauce, Broiled Shad, Dried Sprats, Russian Caviare, Soft Speldings, Whitebait.

SHELLS.—Diamond Back Terrapin.

SAUCES.—Fish Sauce, Essence Anchovies, Essence Shrimps, Essence Lobster. Indian Sauce for Salmon Cutlets.

CURRY POWDER.—Specially prepared for Fish Curries.

SARDINES.—Sardines in Oil, Sardines in Butter, Sardines boneless, Sardines with Tomatoes.

Figure 4.5

Advertisement in program of World Fisheries Exhibition, London, 1883. Source: William Clowes & Sons 1884.

two companies established canneries in southeast Alaska in 1878. Although only one of the companies survived more than a few years, canning was quite successful in Alaska. The industry weathered the recession of 1886 and by 1889 36 new canneries were built in the territory. Production shot up to over 720,000 cases of canned salmon (see figure 4.6). This flooded existing markets, causing prices to plunge. In only four years, the number of canneries was reduced by more than half, although total production declined by about one third. From this low, production again began to increase as canners improved their advertising and found their footing in growing markets throughout the United States and Europe. Prices remained relatively high, though still quite volatile. Production increased steadily to more than 2.2 million cases in 1903, but then declined again. That decline too was due to downward pressure on prices caused by an overshoot of existing demand (Cobb 1906, 23). Production increased again during the food shortages associated with the two world wars, but then declined for a variety of reasons, including reduced stock sizes and catch rates, regulations on the amount of fishing effort, increased competition from canned tuna, and higher demand for fresh and frozen salmon versus canned (Finley 2011; Wick 1946). Cycles similar to those illustrated in figure 4.6 occurred in many fisheries around the world.

Throughout this period of expansion, advertising was important because existing markets could not accommodate the rapid growth in production. Canners pioneered many marketing techniques in a quest to overcome public reluctance to purchase canned goods. Originally, cans were lacquered to prevent rust and other unsightly damage. This started in Europe and North American canners followed suit, since they primarily targeted European markets at the time (Cobb 1921). Labeling started in the 1800s but early labels were small and contained only limited information, such as the name of the company and the type and quality of the product. In 1870, the first full-size paper label was introduced in US canneries, though the labels themselves were mainly ordered from printers in Europe and Asia. This innovation gave canners the opportunity to use bright colors and imagery to generate brand recognition and loyalty. Sardine canners were the first to use cartoons on their labels and many others followed suit (122).

Full-size labels became standard in the early 20th century and some were used in campaigns to create markets for new products rather than extend markets for existing products. "Chicken of the Sea" is arguably the most famous of these brands, at least in the United States. This brand was created by the Van Camp Seafood Company, which started canning tuna in 1914 and began to market it commercially in the US in 1917. At the time, US

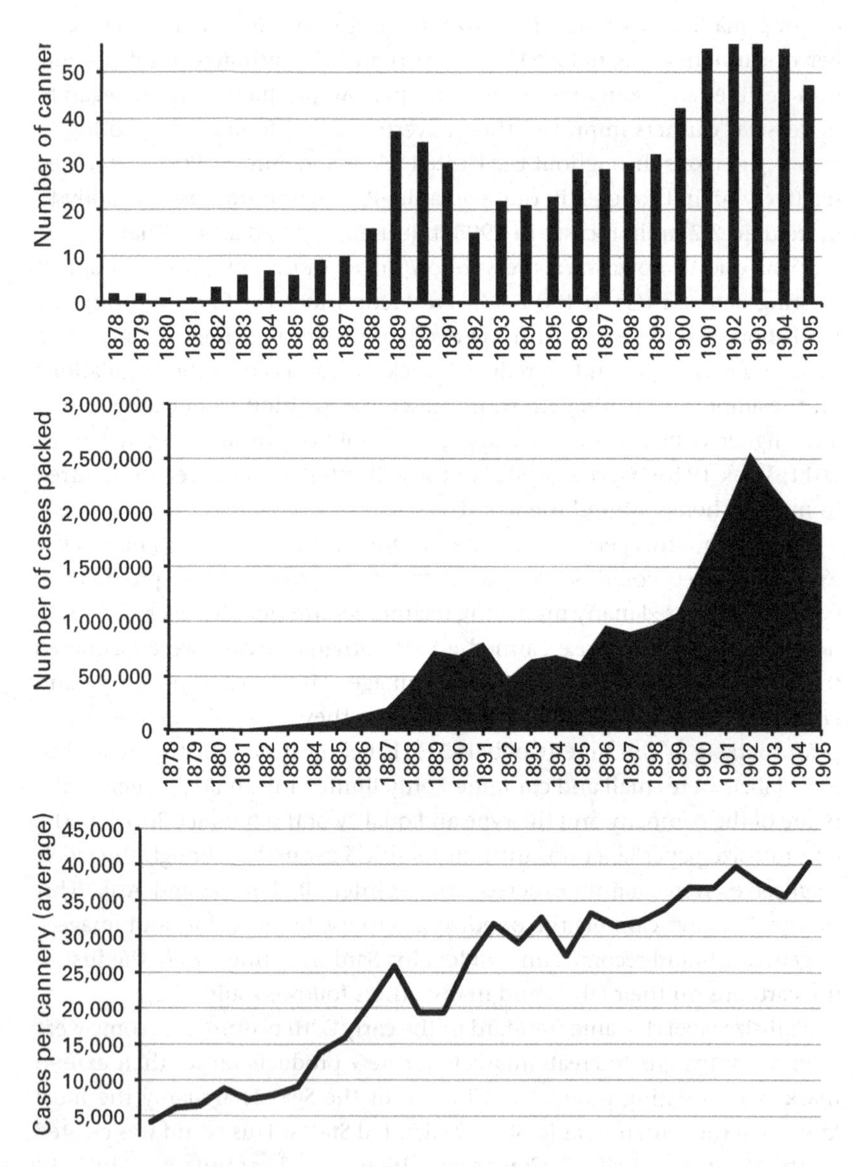

Figure 4.6

Cycles of production in Alaskan salmon canneries, 1878–1905. Source: Cobb 1906.

consumers eschewed canned tuna because it had a brownish color, but Van Camp packed albacore tuna under the "Chicken of the Sea" brand, capitalizing on the fact that albacore is whiter when cooked than other tuna species. This convinced many US consumers to try the new product. In the 1960s, Van Camp developed a new mermaid logo for the Chicken of the Sea brand and pioneered the use of radio and television marketing methods to sell fish products (Chicken of the Sea 2013). Figure 4.7 shows a print advertisement used in a Chicken of the Sea campaign during the 1940s. Note the strong contrast to figure 4.7, which has no imagery or story, but rather focuses on product information.

The introduction of canned tuna to US consumers is just one example of how advertising was used to create markets for new fishery products. After World War II, fish processors made concerted efforts to bring new products

Figure 4.7
A 1946 advertisement for canned tuna. Source: Vintage ad 1946.

to consumers (Fisheries of North America 1966). These included joint marketing campaigns launched by formal industry lobbying organizations such as the National Fisheries Institute, which was formed in 1945 (US Congress 1947a, 44). At the same time, fish processors employed professional public-relations officers to help them develop and market new "prepared foods" that could simply be heated in the stove or microwave. Many of these products were advertised as time savers for busy homemakers. Frozen fish sticks, breaded shrimp, and frozen fillets are just a few examples. As supermarkets replaced the old general stores, packaging and display became more important for marketing fish products. Like many other producers, fish processors started targeting "impulse shoppers" with bright labels and in-store advertising. They also worked with ad agencies to produce television commercials, and they developed complicated, integrated marketing strategies (Frohman et al. 1966). With globalization, this trend in marketing expanded to many other countries, and now canned tuna and many other fish products are advertised to consumers all around the world.

4.3.3 Diversification

Although expansion of markets for traditional fish products was important, advertising was also used to create new markets for fish that previously had been considered by-catch or trash fish. For instance, in the early 1900s, British fishers found that they were catching large amounts of a fish called saithe with their herring harvest. At first all of this by-catch was discarded but through clever advertising, merchants created a demand for the fish in foreign markets. Within a few years, saithe—which we now call pollock—became an intentionally targeted species (J. T. Jenkins 1920). Today, higher-volume substitutes for historically popular fish are not well known because they are marketed to the public as generic processed foods like fish sticks or breaded fillets. For example, the fishery for Alaskan (or walleyed) pollock is now one of the largest in the world but few people have heard of the species. Since the fish has a mild flavor and holds its texture when processed, it was a good replacement for cod and other overfished whitefish. Because consumers are unaware of the species composition of generic fish products, the transition to pollock allowed fishers to reduce costs, increase supply, and widen the profit disconnect without any need for new processing techniques or additional advertising. More broadly, this diversification of species harvested to make similar products masks the core problems of the fisheries action cycle and obscures both cost and price signals from consumers, who often have little information about the species of fish they are eating or the region in which it was caught.

Table 4.1 shows the overall change in the species composition of the top 10 US fisheries in the 1880s and in 1946. Note that menhaden—a small, oily fish used for fertilizer and animal feed—remained number one in volume of production in both periods. However, cod, one of the most important species historically, moved from second to tenth place, because overexploitation of cod led to decreased landings from historically targeted stocks while diversification and exploration increased landings of other species. Pilchard, rosefish, mackerel, and shrimp were not widely known in US or European markets prior to World War I. The same is true of tuna, which, with salmon, was one of the most important canned fish products of the 20th century. Fishers and processors had to convince consumers to purchase these new products in very large quantities in order to keep prices high relative to costs. In fact, the total value of US fisheries production increased tenfold during this period, while the volume of production increased by only about 40% (US Congress 1947a, 11). Again, this was due in part to population growth and economic growth, but it would not have been possible without successful advertising campaigns.

Many of the "higher-end" species now served in restaurants also were once considered "trash fish" but were transformed into valuable products by marketing campaigns. For instance, in the 1970s television's French Chef, Julia Child, introduced the world to a very "ugly but tasty creature called the monkfish." She cited high prices for "our usual fish" as a reason that "our fishery people" had paid more attention to their catch of late and started targeting monkfish as another source of fresh fish. Monkfish

Table 4.1
Top ten species harvested by US fleets, by weight, 1880s vs. 1946.

1880s		1946	
	Pounds		*Pounds*
1. Menhaden	495,000,000	1. Menhaden	900,000,000
2. Cod	294,000,000	2. Pilchard	500,000,000
3. Oysters	153,400,000	3. Salmon	390,000,000
4. Mackerel	131,250,000	4. Sea Herring	235,000,000
5. Alewives	52,000,000	5. Tuna	223,000,000
6. Salmon	45,000,000	6. Rosefish	180,000,000
7. Sea Herring	42,820,000	7. Haddock	156,000,000
8. Haddock	42,800,000	8. Shrimp	150,000,000
9. Lobster	28,880,000	9. Mackerel	106,000,000
10. Mullet	21,600,000	10. Cod	105,000,000

Source: US Congress 1947a, 8.

was cheaper than cod, haddock, sole, or swordfish, and with the help of Julia's recipe it became popular, almost doubling its price in three years (Hanes 1980). Less celebrated chefs popularized fish like skate and orange roughy, marketing them to patrons as the availability of fresh fish fluctuated. In the 1980s, an entrepreneur transformed Patagonian toothfish from an unwanted by-catch into one of the world's most profitable species by renaming it "Chilean sea bass." The subsequent marketing campaign swept the United States and then Europe, and today Patagonian and Antarctic toothfish are so valuable that "fish pirates" harvest them illegally in spite of international attempts to protect dwindling stocks of these long-lived, slow-growing species (Fallon and Kriwoken 2004).

Figure 4.8 shows how the world's biggest fisheries shifted from one species to another from 1950 to 2010. Note that Atlantic cod and Atlantic herring were the dominant stocks in the 1950s. These fish were popular in North America, in Europe, and in the Soviet Union. Landings of each species topped 2 million tonnes in the period, four times greater than any other species. However, as stocks of these well-known species dwindled, fishers moved on to new fisheries, including those for anchoveta (Peruvian anchovy), yellowfin and skipjack tuna, Alaskan pollock, European pilchard (sardines), and chub mackerel. Some of these species were already familiar, but others were not. In either case, the industry had to engage in marketing to ensure that prices would not plummet with the increase in supply. Subsequently, as some of these newly targeted species declined, fishers increased their harvests of other substitutes, such as largehead hairtail. By the time they did so, markets for large-scale commercial fisheries existed over a wide geographic range, including countries in Asia, South America, and Africa. Growth of demand in China, in particular, drove increased landings of largehead hairtail and several other species (Caddy and Garibaldi 2000).

But figure 4.8 does not show the total increase in the number of stocks exploited commercially in the 20th century. Table 4.2 demonstrates that marine capture production expanded as a result of increased catches of known and accepted species but also as a result of greater exploitation of relatively "new" species. Indeed, the number of known species targeted increased almost threefold from 1950 to 2010.[10] Furthermore, the number of stocks exploited commercially (that is, at industrial levels of production) also increased substantially. Landings of all of the top ten stocks listed in figure 4.8 were greater than 1 million tonnes in 2010, and for four of them landings topped 2 million tonnes. Some ecologists refer to this increase in the number of species harvested as "fishing down the food chain" (Pauly et al. 1998). Others show that fishers are actually targeting the highest-priced

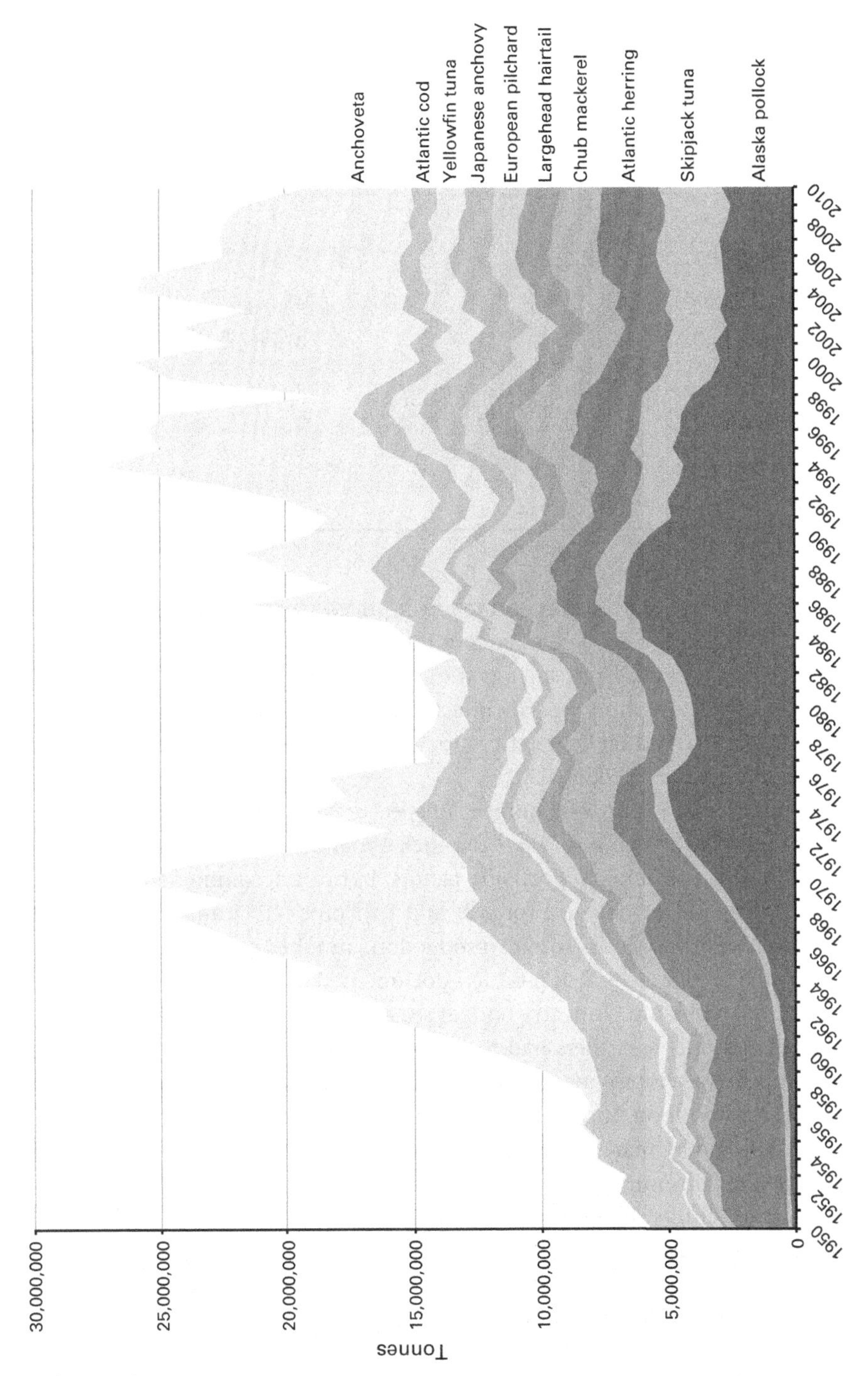

Figure 4.8

Trends in marine capture production for major species, 1950–2010. Source: FAO 2012a.

Table 4.2
Number of known species landed, by level of production

	Number of known species landed	
Tonnes produced	1950	2010
1–10,000	370	979
10,000–50,000	93	181
50,000–100,000	16	45
100,000–1 million	32	87
1–2 million		5
2–3 million	2	3
> 4 million		1
total > 1 tonne	513	1301

Source: FAO 2012a.

species first and then moving on to lower-priced species (Caddy and Garibaldi 2000). These perspectives are not mutually exclusive, partly because of ecological characteristics, such as the fact that larger fish are slower growing and therefore more heavily affected by fishing, and partly because of present-day consumer preferences for larger species, such as salmon and tuna (Stergiou and Tsikliras 2011).

The increase in the total number of species exploited commercially may be driven by the common pool resource dynamic, as well as exploration, innovation, and other supply-side factors, but it is accommodated by relevant increases in demand for fish and fish products around the world. In fact, the changes in overall production, number of stocks or species exploited, and trade in fish products documented in this chapter reflect the growth of the fishing industry both at sea and on land. To sell their increasingly large catches, fishers had to expand the markets for their fish. They did this by developing new methods of processing and preserving fish for food and non-food purposes, by appropriating better transportation technologies, and by marketing their products globally. The result is that more people are consuming more types of fish products in more places around the world. This is shown by the steep growth in total consumption of seafood products depicted in figure 4.9. Because of this increasing demand, ex-vessel prices are higher on average than they would be without this growth in consumption of fish products.

In contrast, figure 4.10 shows that trends in per capita consumption of seafood are more complicated than the absolute number reported above. In

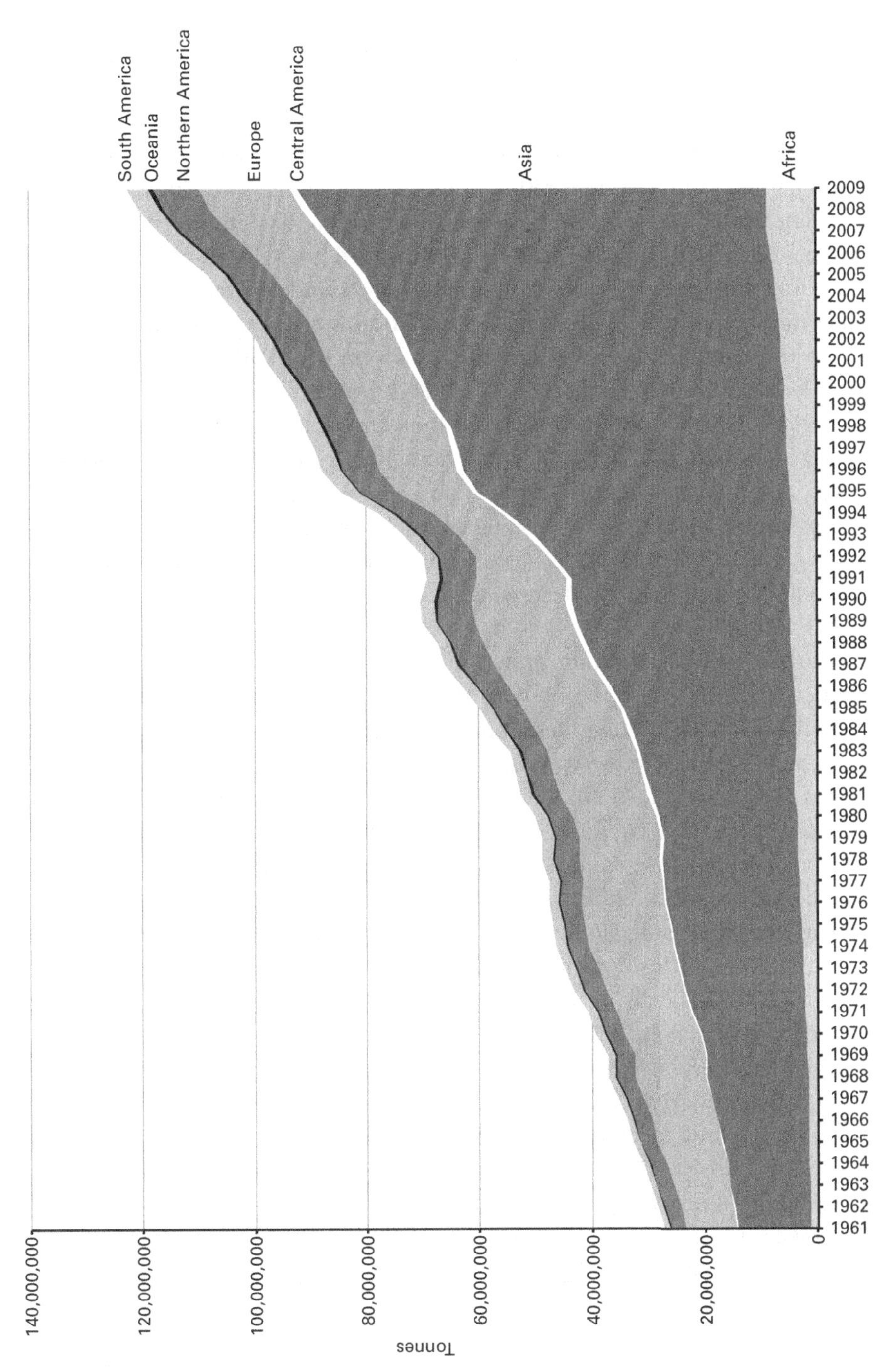

Figure 4.9

Global consumption of seafood by region, 1961–2009. Source: FAO 2012b.

Europe, in Central and South America, and in Africa, per capital consumption of fish products leveled off in the early 1980s. European consumption declined per capita in the early 1990s—probably as a result of closures of important historical fisheries, which led to a decline in supply of popular species—and only recently returned to previous levels. In contrast, North America and Oceania consumed less fish per capita than Europe in the first three decades of the time series, but continued on an upward slope and are now the largest per capita consumers at 71 and 67 grams per person per day, respectively. Per capita consumption increased tremendously in Asia throughout the period, but particularly in the 1990s, and is still on a relatively steep upward trajectory. This is probably because of economic development in China and in other industrializing countries. Indeed, as long as economic growth and population growth continue, fishers, processors, and marketers will continue to advertise fish products to consumers. If exogenous biophysical limits prevent the supply from increasing to meet the growing demand, prices will rise even more quickly than they did in periods when supply was decreasing.

Figures 4.9 and 4.10 demonstrate the difficulty of differentiating between different causes for the observed changes in demand for seafood. The many local-scale changes described in this chapter cannot be detected in the aggregate data represented in these figures, at least in the short run. That said, these changes in processing, shipping, and marketing had lasting long-run effects on the fishing industry. Without such innovations, fishers would face the economic costs of the tragedy of the commons much earlier and more often. Furthermore, given current trends it is likely that exogenous and endogenous drivers will cause demand to increase in the future, and that this dynamic will continue to mitigate the price signal and expand the profit disconnect in spite of the fact that the fishing industry is approaching global limits on fish production. Indeed, without successful governance response, it is possible that fish products will again be a luxury commodity that can only be purchased by wealthy elites. Some of the most overfished species already fall into these categories, but contractions in supply associated with overexploitation combined with exogenous and endogenous increases in demand could transform staple products such as canned tuna or surimi into rare and valuable commodities. Part II describes the co-evolution of fisheries governance and delves into the effects of these economic trends on the future political economy of global fisheries.

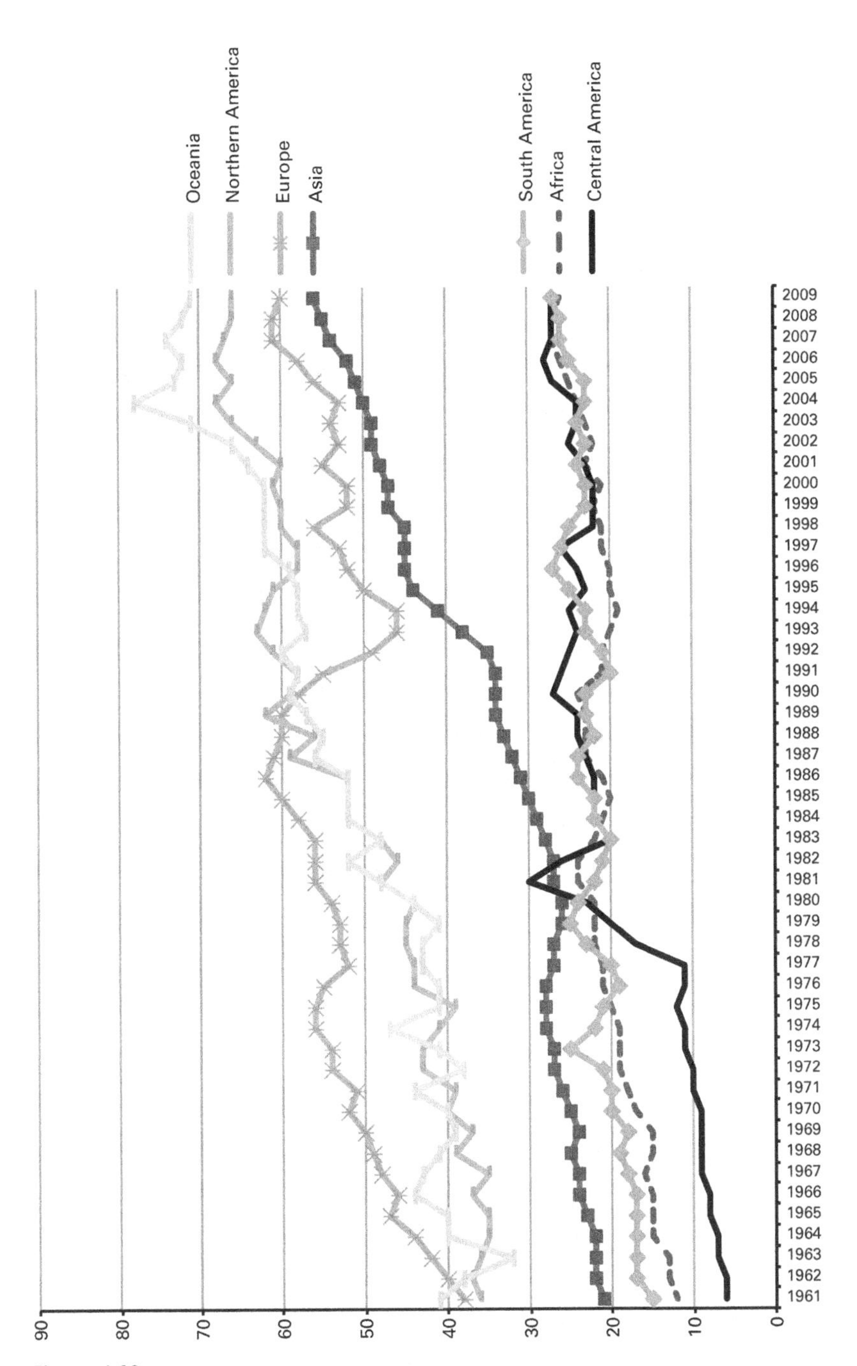

Figure 4.10
Per capita consumption of seafood, 1961–2009. Source: FAO 2012b.

II The Politics of Response

5 Exclusion

Given the expansionary effects of the profit disconnect described in Part I, governance is clearly necessary to control fishing effort and related core problems in the action cycle. However, when governance is responsive rather than proactive, controls on effort are usually delayed until problem signals are quite strong. As explained in chapter 1, small, isolated communities are usually sensitive to problem signals and so engage in early management response, but global expansion of fishing effort and markets for fish products described in Part I profoundly increased the profit disconnect in fisheries all over the world, thereby dampening the economic problem signals associated with catch, costs, and prices globally. Politically, this dynamic is compounded by the power disconnect, which occurs when actors who are sensitive to problem signals are marginalized in the decision making process. On the other hand, adding governance to the AC/SC analysis also introduces new problems, including conflict between fishers and power struggles between decision makers. Political problem signals associated with conflict over resources often trigger response much earlier than economic problem signals. Part II delves into the co-evolution of responsive governance and economic expansion. This chapter covers the politics of direct exclusion and related effects on fisheries governance. Indirect exclusion through application of restrictive management measures will be covered in chapter 7.

Although in theory marine fisheries are open access, in reality the establishment of access rights and exclusion in general, are the most common political responses in the fisheries action cycle (Wilson 1982). Domestically, the decision to exclude is usually a response to the problem of conflict over rights of access. As fishing grounds become more crowded, fishers may work to lay claim to specific stocks or areas. They will respond strenuously when others are seen to impinge on established access rights. The structural context is a key determinant of fisher response and their ability to influence decision makers. In systems where violence is tolerated and where fishers have little access to decision makers, self-enforcement of access rights through threats and violent acts is common. Where sea tenure or other forms of access rights are recognized by governance organizations, fishers can appeal to decision makers to enforce exclusion. In cases where broader governance institutions favor open access, fishers may either establish informal access rights or they may seek to exclude others through the regulatory process. Informal institutions are easiest to maintain when governing bodies lack capacity or incentive to control fishing effort, so changes in the size and capabilities of governments are important exogenous factors. The distribution of political and military influence among different groups of fishers can also strongly influence exclusionary outcomes.

Power is an important determinant of exclusion at the international level as well. Here, there is a distinction between domestic and international politics. As a result of the expansion of offshore and distant-water fleets described in chapter 2, conflict between fishers from different countries increased greatly. While coastal fishers demanded protection from foreign fleets through nationalization of fishing grounds, distant-water fishers wished to protect their access to fisheries near the coasts of other countries. The balance of power between these interest groups—or put another way, the growing power disconnect that accompanied the rise of distant-water fleets—had profound impacts on national claims regarding access rights to marine resources. Negotiations over international access rights were also shaped by the power disconnect between fishing countries or, in the formal terminology of international relations, *fishing states*. Powerful states, most of which were historically dominant fishing countries, favored bilateral settlement of disputes over access rights. Less powerful states, most of which had small or developing fleets, favored a multilateral setting for negotiations so they could gain power through coalition formation. Ultimately, these domestic and international struggles for control over access shaped the International Law of the Sea, which formalizes exclusionary measures in coastal areas and on the high seas.

This chapter reviews the evidence on direct exclusionary responses by industry actors and decision makers at different levels of analysis, using the AC/SC framework. Section 5.1 briefly describes historical approaches to political problem signals associated with the incursion of outsiders and threats to power, including various forms of sea tenure practiced in different parts of the world. Section 5.2 shows how monopolization of hard power by sovereign states from the 17th through the 19th centuries altered the structural context of fisheries, reducing piracy and making fishing safer and less expensive while also laying the groundwork for future regulations. Section 5.3 covers the establishment of the 3-mile territorial sea as a contested political attempt to exclude foreign distant-water fleets from coastal fishing grounds. Section 5.4 covers the establishment of the first international bodies designed to manage shared fish stocks. Section 5.5 explains how continued conflict over control of marine resources led to the bifurcated system of domestic regulation of species within the 200-mile exclusive economic zone (EEZ) and to international management of highly migratory, straddling, and transboundary stocks of fish. Domestic management within the EEZ will be covered in subsequent chapters. The role of exclusion in the creation of international regulatory bodies called regional fisheries management organizations (RFMOs) is covered in sections 5.5 and 5.6.

5.1 Endowments and Entitlements

There is evidence of the local depletion of fisheries in the ancient world, particularly in inland waters and in intertidal zones. However, most documented governance regimes regulated the distribution of benefits from the resource without placing any direct limitation on catches. This observation is supported by research on traditional arrangements with long-standing historical foundations (Ostrom 1990; Bromley 1992). Prevention of conflict among fishers is a common cause for distributional rules and norms, but these institutions were also used to reinforce existing power hierarchies, which in turn tend to be associated with profit and power disconnects in local areas of rapid economic growth. In some instances, norms based in traditional ecological knowledge were utilized to anticipate and therefore benefit from fish migrations or to reduce targeting of immature or spawning fish (Kalland 1990).

As mentioned in chapter 2, one of the earliest known examples of fisheries governance is the guild-based system of ancient Sumer (around 2300 BCE). This was a hybrid sea tenure system in which temples shared governance responsibilities with collectively organized fishing guilds. Government at the time was largely theocratic, and the temples authorized guilds to fish in certain areas with specific types of gear, provided them with boats and other equipment, settled disputes between fishers or with buyers, and punished any non-guild-members caught poaching fish. In return, guilds were responsible for providing the temples with the appropriate quantity and quality of fish they demanded. Any fisher who was unable to pay his quota would accrue a debt, which could be paid in fish, in barley (akin to currency at the time), or through military service and menial labor. A fisher and his entire family could be sold into slavery for non-payment of these debts. Distribution was the primary purpose of this system, which provided fishers with endowments (the right to appropriate resources) and temples with entitlements (the right to some benefit from resource appropriation). In spite of the limits on entry and additional costs placed on fishers, which should have minimized the tragedy of the commons, population growth and economic prosperity eventually drove fishing effort to unsustainable levels. Indeed, even as fish stocks declined, demands from temples increased, causing many fishers to go into debt (Kreuzer 1978b). This was an important power disconnect, as the temples controlled the size of the tribute from fishers but did not pay the costs of overexploitation and could even profit when indebted fishers, their goods, and families were sold to pay off debts.

Evidence of fisheries institutions used by other ancient civilizations is scattered but supports predicted goals of resource appropriation and conflict minimization. Egyptian rulers claimed some entitlements, including ownership of all living resources found in inland waters (Adams 1884a). The only written Roman record found regarding fisheries is associated with their law for wild animals and plants of all kinds, on land or in the water. These were proclaimed *res nullius* (property of no one) until they were appropriated through harvest or capture (Radcliffe 1921). Recent interpretations suggest that this concept was quite different in Roman times than in modern times and that it was used primarily to justify the appropriation of lands and other resources that were not "owned"—though they might be used by people from other nations—rather than to establish common property resources as such (Fitzmaurice 2014, 52–53). With increased population and development, crowding also increased in the European seas, causing growing concern regarding conflict between fishers. In the early common era, Emperor Tiberius (14–37 CE) consolidated disparate maritime rules into the Rhodian Laws, which were also confirmed by subsequent emperors through the reign of Septimus Severus (193–211 CE). These laws provided basic rules to ensure maritime safety, including a ban on fishing with torches to attract nocturnal species like squid. A subset of rules governing conflicts between fishers prohibited violent altercations and set up a graduated structure of punishment for rule breakers (Adams 1884a).

Documentation is sparse, but from what is known, rulers around the world increasingly regulated fisheries in order to appropriate a share of the benefits from the resource and to keep the peace among fishers. In feudal Europe, the aristocracy could demand tribute from fishers as well as from peasants, claiming dominion over water as much as land. Inland waters were restricted with licenses or given over to sport fishing for the privileged, driving commercial fishermen further out to sea (Adams 1884a; Teuteberg 2009). The claims of a monarch could also extend to coastal fisheries. For instance, according to the French Rôles d'Oléron, written in the 11th century at the behest of Eleanor of Aquitaine, marine fisheries were open access (article XXXIV) except for certain "great fishes" and "oil fishes" (possibly whales), to which the king might have some rights, depending on the custom of the country (articles XXXVII–XLII). The Rôles d'Oléron also laid out rules for resolving conflicts aboard a single ship and when ships worked together to harvest large schooling fishes like mackerel or herring. These formal laws did not prevent fishing communities from establishing informal endowments and entitlements (Adams 1884a; Admiralty and Maritime Law Guide 2002).

Formal fisheries laws in Asia also fluctuated over time, moving through periods of closed and open access. In China, fishing communities tended to be isolated and generally governed themselves (Yen 1910, 370). During the Tang Dynasty (618–907 CE), fisheries were formally designated as commons resources, but fishers were still largely left to their own devices (Makino 2011, 22). Little is known about early fishing communities, but it is likely that they established their own management regimes. Clear evidence exists of the local use of *de facto* fishing rights in some parts of China's coastal areas during the Qing Dynasty (1644–1912 CE), including assigned fishing grounds for specific groups, rules for governing general use of fisheries resources, and dispute settlement procedures (Muscolino 2009). Later in the period, increasing population growth and industrialization forced Qing rulers to work with local communities to reduce conflict between growing numbers of fishers over dwindling resources (5). By the 1800s, there were many well-established fishing guilds in China with customary rules of self-governance (J. D. Campbell 1884).

Early Japanese fisheries regulations were based on Chinese models from the Tang Dynasty (618–907 CE). Established under the Taiho Code of 701 CE, these laws designated marine resources as open access. However, for many centuries rulers of Japan largely ignored fisheries, leaving fishing communities to govern themselves through informal local distribution of access rights. This system continued until the Edo period (1600–1868), when the Tokugawa dynasty established nationwide rules of sea tenure, largely to help feed and employ growing populations but also to appropriate benefits from the export of fish products to China (Makino 2011, 22). In this period, fishing rights were established with the same status as property rights on land. Primary ownership of the sea and its resources belonged to feudal lords, who granted endowments to fishers in return for a portion of the catch or other entitlements. Access was governed in many ways, including limits on what type of fishing could be undertaken by which fishers and at which times. In addition, the system established conflict resolution mechanisms, including a hierarchy of arbiters, and set taxes and fees that should be paid to various participants throughout the hierarchy. In some cases these fees could be onerous, and, like anything that increases the costs of production, probably had some limiting effect on fishing effort (Kalland 1995).

Documented instances of fisheries management in India date back to the reign of Chandragupta Maurya (322–398 BCE). His kingdom covered most of the subcontinent, and the law gave the king ownership over fisheries and established various taxes, fees, and tolls to ensure that the monarch received his due from fishers for use of his property. Fishers were required to provide dried fish for sustenance of the king's armies when they traveled and fresh fish if they camped near fishing grounds (Agarwal 2007, 38–41). Around 246 BCE, a king named Ashoka (also known as Priyadarsi raja) took the throne. A convert to Buddhism, Ashoka issued edicts establishing closed days to fishing for important species during the spawning season (48–54). This is a rare example of a measure designed to conserve resources rather than distribute benefits or reduce conflict. Otherwise, it appears that fisheries management in India followed patters similar to other parts of the world. Research from the late 1800s showed that fishing communities on the subcontinent were known to stoutly defend access to fishing grounds over which they claimed ownership. A hierarchy of chiefs would settle disputes and would levy fines or fees on fishers both for infringement of norms and as compensation for their work. Some of these positions were hereditary while others were elective or determined informally, usually on the basis of wealth and social status, and thereby reinforced existing hierarchies (Day 1884, 17–18, 20).

What is known about early fisheries management in Africa suggests that ownership could be communal or assigned to individuals, but in either case rights of access were well defined through formal and informal institutions. For example, kinship ties are documented as important determinants of access rights in Anlo society in Ghana from the 15th century on. The Anlo, well known for their fighting prowess, protected their fishing territory from outsiders and bought or conquered fishing grounds held by neighboring groups (Akyeampong 2001). In the Kazembe Kingdom in what is now Zambia, which was probably founded in the 19th century, ownership of fishing rights in specific lakes and lagoons was well defined through political and spiritual institutions. Rulers controlled access to their fishing areas and demanded remuneration for access from fishers. In turn, these rulers engaged in rituals thought to bring bounty to the fisheries, and also gained political power by enticing migrants to their areas, thereby increasing the size of the population under their control (Gordon 2006, 52–53).

Little is known about pre-colonial fisheries governance in the Americas. There is evidence of large, well-organized fisheries during the Inca period in Peru (Coker 1910). By the time of colonization, most coastal Native American peoples had established rules regarding fishing territories, techniques, and distribution of catches. For example, Newell (1997) describes a wide range of fisheries management institutions in pre-colonial British Columbia. Although indigenous peoples in the area did not conceive of fish or fishing grounds as property per se, groups did establish formal use rights that could be held by an individual, a family, a village, or a clan. These rights were transferable through inheritance and could be loaned or leased out. In several instances, chiefs or other important members of a group distributed use rights as a form of patronage, claiming the first harvest from any fishing ground as tribute for the favor.

As this brief survey shows, historical fisheries management tended to focus on distribution of rights and benefits as well as conflict resolution, rather than some concept such as sustainability. In most cases, fishing was costly and labor intensive, so there was no profit or power disconnect except in a few cases associated with large population centers. That said, it is also true that resource users compiled substantial information on the biology and the ecology of the fisheries they exploited, and that this traditional ecological knowledge (TEK) also was a part of the management process. Berkes (1999) and others show that TEK can be deeply ingrained in traditional management systems and may be a crucial factor in the sustainability of long-standing collective action regimes (Shackeroff, Campbell, and Crowder 2011; Cash et al. 2003; Lauer and Aswani 2010; Sethi et al. 2011; Berkes and Folke 2002). As fishers spread out to new fisheries and expanded their harvests through industrialization in the 19th century, governments also sought to extend their reach, and many traditional commons approaches to fisheries management were marginalized both economically and politically. In their stead, governments created formal management systems that required domestic and international recognition of the power of sovereign states (Finley 2011). Although this step was necessary because the scope of fishing operations expanded beyond a single community, it also increased the power disconnect because in most cases a few large corporations had greater political access, alignment, and leverage than many small fishing communities.

5.2 Sovereign States

When applying the AC/SC cycle it is important to look for exogenous factors that alter the structural context as well as endogenous factors associated with the action cycle. The rise of sovereign states had a tremendous impact on fisheries governance. The extension of state power at once undermined community management practices and facilitated incursions by large fishing corporations and other outside actors that were well insulated by the profit disconnect and had little interest in the sustainability of local resources. At the same time, international conflict over access to rich fisheries resources was one of many factors that drove sovereigns to consolidate their control over the use of force and establish rule of law. This generated a potent feedback effect that allowed large-scale commercial fishing corporations and similar actors to increase their political power and influence along with their wealth, widening both the profit and the power disconnects. However, as people recognized the legitimacy of the state's monopoly over the use of force, they also accepted the state as the protector of shared resources such as fish stocks, which ultimately altered the state's role from appropriator to manager. This section covers feedbacks between the expansion of fisheries exploitation, responses focused on the exclusion of outsiders, and the rise of the modern international system based on sovereign states. Implications for domestic governance are covered in subsequent chapters.

Violent conflict over fisheries resources was common before the modern era but was most destructive after the proliferation of cannons, guns, and related firearms in the 17th century. The industrialization of violence started in Europe, where governments were relatively weak and individuals who had access to capital armed themselves, either for protection or to forcibly appropriate resources from others. The oceans and remote coastal areas were particularly lawless, and fishers could be both victims and perpetrators of violent acts. Piracy was rampant and while fishers frequently engaged in piracy as an additional source of income, they were also frequent victims of independent pirates or of the government-sanctioned pirates known as privateers. During wars or times of international tension, instances of tit-for-tat violence between fishers from different groups or nationalities occurred as well. This heightened violence increased costs of production as fishers were forced to invest in arms, and, for most, the benefits of preying on others were counterbalanced by the increased risk of falling prey themselves. As such, violence served as a limit on increasing fishing effort but eventually led to the strengthening of formal governance institutions.

The distinction between public and private military power was blurry in the Middle Ages. Here again it is useful to start by examining the operations of the Hanseatic League, which dominated production of pickled herring and trade more broadly in Europe in the 14th and 15th centuries. As a proto-state, the League was created as a pact of merchants seeking to defend themselves against pirates. By pooling capital and other resources, merchants belong to the League were able to purchase expensive armaments and to use sanctions and other economic measures to punish any group that attacked League members. As their military and economic power increased, the Hansa themselves turned to piracy for a time, but gradually gave it up for respectability and comfort (Gosse 2007). Even so, the League continued to use violence and negotiation to maintain their monopoly on pickled herring, which, at the height of their power, was worth more than the manufactures of France or England (Adams 1884a; Johnston 1965). They even went to war against the king of Denmark in the early 13th century when he charged high fees for access to the herring fisheries and the port facilities of Scania (Zimmern 1889). In an ironic twist, the Hansa hired a group of pirates to help combat the Danish forces. Once the Danes were defeated, the League tried to rein in these same privateers and was only successful at very high cost (126–127). Eventually, the Hansa lost their monopoly over trade from the Baltic to the British and Dutch, and their power dwindled considerably (Palais 1959, 856).

During their ascendency as the primary producers of pickled herring, the Dutch followed similar heavy-handed practices, though an interesting division of labor emerged between the government and the fishing industry, which was represented in these matters by a body known as the Grand Fisheries Council. In the late 14th century, the dukes of Burgundy, who then ruled the Netherlands, started charging Dutch herring fishers for protection. This special tax, called "last money," was paid first for protection from pirates and then for protection from foreign navies as the Dutch went to war with Spain, Britain, France, and Sweden at various times in their history (Beaujon 1884, 14, 16). Indeed, all of these governments allowed their own privateers to harass Dutch herring fishers, and Dutch privateers—including some fishers—returned the favor. This type of tit-for-tat behavior continued well into the 17th century and was costly for all involved. At times, even with the last money the Dutch government could not afford to send enough naval ships with the fleet, so the Grand Council spent additional funds to hire their own armed escort (18–19). Indeed, the herring industry subsidized many of their government's wars, investing funds in navy ships and armaments that might otherwise have been spent on fishing vessels and gear.

The herring grounds of the northeast Atlantic were not the only fisheries plagued by pirates during the golden age of piracy (ca. 1650–1730; Mabee 2009). The Grand Banks off Newfoundland were a prime target. Pirates would attack the fishing fleets there to restock their own supplies of food and drink, including fish, which would sometimes be sold for profit as well. They would also replenish their ranks with fishers, who, being able seamen, were either enticed or coerced into joining them. Although Mediterranean fisheries were not as large scale at the time, fishers and all others setting forth on that sea were subject to capture and enslavement by pirates of the Barbary Coast. Indeed, while sacking of Spanish and Portuguese vessels laden with treasure from the New World was a common undertaking for all pirates operating near Europe during this time, the pirates of North Africa made their fortunes by catching and selling Europeans and others as slaves. The capture and trade in slaves was also a prominent feature of pirate activities in Asia, Africa, and the Americas. Smaller in scale, fishing vessels were not targeted for plunder, but fishers could be captured to be sold into slavery (Gosse 2007).

Throughout Europe, piracy was illegal and yet tolerated in the 17th and 18th centuries because of the broader structural context. In particular, the combination of weak states and mercantilist philosophy generated the perfect conditions for the large-scale piracy described above. In contrast to free trade liberalism, mercantilism dictates that sovereigns should do everything in their power to maintain a positive balance of trade (more imports than exports) and thereby enrich the kingdom. The norms of the times allowed the use of both public violence (warfare) and private violence (piracy and privateering) to appropriate goods through colonial conquest and direct battle with rivals. Sovereigns also sought to control trade by attacking foreign ships, including fishing vessels, and confiscating all goods on board. Navies were relatively weak at this time, so sovereigns would implicitly or explicitly employ pirates or privateers to supplement their forces. Usually this was done under a "no plunder, no pay" policy, so the direct expense of hiring these privateers was low (Gosse 2007, 116). However, since they were subject to the predation of foreign and domestic pirates and privateers, merchants and fishers often had to pay for their own protection, either by arming their vessels or by paying for an armed escort, like the Dutch fishers, so the indirect costs of the privateer system were high (Mabee 2009).

According to Thomson (1994, 19), piracy declined in Europe in the 18th century as sovereigns sought to consolidate and legitimize the power of the state. Monopolization of the use of force at sea was an important part of this transformation. Rather than pay directly for their own protection, merchants agreed to be taxed to pay for stronger, more effective royal navies. In return, sovereigns committed to protect their merchant fleets at sea. Of course, the Dutch had already tried this several hundred years earlier, with only temporary success, but there were several key structural differences in the 1700s. Thomson views this transition through an international relations lens and so treats sovereigns as the primary drivers of change. However, at least where piracy was concerned, changes in domestic perceptions of the practice and social costs of piracy were also important. Though many local government officials in England, France, and the Netherlands had once had profitable ties to pirate gangs or had even gone to sea as pirates themselves, the benefits of both legal and illegal commerce dwindled with general escalation in what could be called piratical effort. That is, just as in fisheries, individuals entered piracy because they saw it as an opportunity to profit, but with unlimited entry those profits were driven down to near zero. At the same time, many people were forced to shoulder the indirect costs of piracy, either by paying for protection themselves or through economic and personal losses, so political pressure to quell the practice increased, as did demand for protection from sovereign governments (Gosse 2007).

Establishing sovereignty over the seas around Europe did not end piracy. Pirates and privateers were active all over the world by the 16th century. When the structural context was favorable pirates developed "combines" that were like small navies, both in terms of firepower and in terms of organization and discipline (Gosse 2007). Here again, concentration of power in the hands of pirate bands was attributable to increased trade and the industrialization of violence, but was also a direct result of mercantilist European policies. For instance, in the mid-1600s privateers associated with the Dutch East India Company formed a combine with Chinese pirates who had taken over the island of Formosa (now known as Taiwan). Their goal was to plunder mainland China, which was wealthy in highly valued goods but refused to trade with the West at the time (Andrade 2004). French and English colonial powers also employed privateers to break Spanish or Portuguese dominance in India and the Americas. Because the Caribbean was an excellent place to waylay Spanish vessels transporting silver and gold from the New World to the Old World, many combines were formed in the region. When European sovereigns cracked down on piracy in European waters, they displaced many European pirates, swelling the ranks of preexisting pirate groups in other parts of the world.

Combating piracy and establishing sovereignty was even more difficult outside of Europe, because in many cases colonization destroyed existing governance institutions. However, the growth of piracy in colonial areas placed an increasingly heavy burden on trade, just as it had near Europe, so imperialist powers eventually went to war with pirates to protect trade routes and establish public monopolies on the use of violence. Establishment of the rule of law at sea was dependent on the rule of law on land, so increased naval power only succeeded when complemented by conquest of coastal areas. Indeed, ending piracy was a common excuse for the invasion and colonization of many parts of the world. France invaded Algeria in 1820 and went on to conquer much of North Africa, in part to lessen piracy in the Mediterranean (Gosse 2007, 69). Britain and the United States sent considerable military power to Africa, India, the Middle East, and the Caribbean to deal with pirates and to extend their imperialist influence. The Chinese government worked to reduce piracy in the seas around China, particularly to put down the pirates of Formosa and their Dutch counterparts, reinforcing their hold on power in the region (Andrade 2004).

By conquering pirates on land as well as at sea, sovereign states established relative peace on the high seas by the middle of the 19th century, just when fishing effort started to expand substantially in both scope and scale. Military effort was not the only vector of victory over piracy; the introduction of steam power increased capital costs for pirates as well as for fishers, creating barriers to entry. Nevertheless, the establishment of standing navies and of coastal policing organizations such as the US Coast Guard (founded in 1789) extended the rule of sovereign states to many coastal and marine areas (US Coast Guard 2013). These government entities protected fishers as fleets expanded in the 19th and 20th centuries, reducing risk and improving the business climate. They also provided the institutional foundation for the management of marine fisheries that followed in the wake of fisheries overexploitation. Finally, the system of sovereign states that emerged with the monopolization of violence by governments shaped negotiations over the law of the sea and the management of international fisheries.

5.3 Territorial Seas

Privateers and pirates operated in a time when access to resources was highly contested among governments and exclusion of foreign fleets was politically popular. For much of history, shared fishing grounds were a common cause for legal and military disputes between countries. These conflicts were fueled by pressure from fishers, who clamored for the protection of near-shore fisheries from foreign incursions or, conversely, lobbied their governments to ensure continued access to fisheries near foreign shores. Related claims of sovereignty over coastal areas of between 60 to 100 miles from shore were common in Europe during the Middle Ages. Disputes often broke out between local rulers when claims overlapped. Some of these disputes were settled peacefully, but others resulted in war (Fulton 1911). Similar claims and conflicts occurred in most other parts of the world. Generally, "ownership" of marine resources was established through force rather than law, though at times laws were enacted to legitimize territory gained through military might. This section describes early attempts to establish sovereign control of fisheries resources. Each attempt to exclude altered the structural context for specific fisheries and the cumulative effect of these alterations shaped the evolution of broader exclusionary institutions through international law.

Modern legal discourse on sovereignty over the sea and its resources is usually traced back to Grotius' 1609 work *Mare Liberum*, which advocated open access to the seas as part of a larger debate between the Dutch and the English about access to herring stocks off Scotland, to stocks of whales near Spitzbergen, Norway, and to the East India trade (Fulton 1911). As noted previously, precedent for this approach goes back at least to Roman times, when fish as well as all other wild animals were considered *res nullius*, or property of no one—at least until they were caught (Radcliffe 1921). However, for the Romans *res nullius* provided legal rational for the appropriation of "unowned"—though not necessarily uninhabited or unused—terrestrial and maritime resources. This modern interpretation of Roman law differs substantially from that of Medieval legal scholars like Grotius, who used *res nullius* as a justification to maintain open access to commons resources (Fitzmaurice 2014). In his 1635 response to *Mare Liberum*, the Englishman John Seldon suggested that states should favor closed coastal fishing zones. He argued that a nation that did not extend its jurisdiction would be giving up resources to foreign fishers and, ultimately, enriching foreign governments. Seldon's ideas supported the stance of King James I of England, who had commissioned his work and who encouraged considerable privateering against Dutch fishers who fished near British coasts (Beaujon 1884; J. T. Jenkins 1920). It is important to note that Seldon's treatise anticipated the tragedy of the commons argument put forth by Hardin (1968), but at the time there was no evidence that the shared resources in question were overexploited. James I was responding to the political problem signal of conflict over access rights and also was concerned about threats to his political power in the region. The Dutch favored the opposite policy, but for similar reasons.

There were several practical effects of this theoretical debate. In the same year that Grotius published *Mare Liberum*, James I proclaimed that no foreign vessels would be permitted to fish in English waters without a license from his government. Assertions were made that the purpose of this measure was to conserve coastal resources, but it was fairly clear that James' true purpose was to exclude the Dutch fleet from the herring grounds off Scotland and to secure more of the benefits from the fishery for domestic interests. James was unsuccessful in that the Dutch and many others refused to comply with his terms and Britain lacked capacity to enforce the edict (Juda 1996). The issue recurred as the two kingdoms struggled for control over the seas throughout the 17th century (Johnston 1965). Exogenous factors in the structural context played an important role. During the first half of the 17th century, when the Netherlands was in jeopardy from Spain, the Dutch placated the English but once the Spanish threat receded, the Dutch were more willing to fight for access to important fisheries and trade routes. The resulting Anglo-Dutch Wars lasted from 1652 to 1684. England eventually prevailed, devastating the Dutch fleet and taking *de facto* control over trade routes and fishing grounds in the North Sea (Fulton 1911).

One reason the British won out over the Dutch was that they took advantage of changes in the military capabilities of ships and weaponry. Many other governments and private navies did the same, and by the end of the 17th century the need for protection from pirates and foreign navies had led to general acceptance of the "cannon shot rule," which gave states dominion over the oceans for the distance they could defend with shore-based cannons. Although no true wars were fought over fisheries specifically after the Anglo-Dutch conflict, access rights continued to be contested by states. By the 18th century, Britain and France were the key players, competing for access to the cod stocks of the Grand Banks. Each country sought to maximize its fishing fleet's production of dried cod, which Catholic Spain would pay dearly for in gold bullion. Concessions and rights of access to these fisheries figured into the Treaty of Utrecht (which ended the War of Spanish Succession in 1713), the Treaty of Paris (which ended the Seven Years' War in 1763), and the Treaty of Versailles (which ended European hostilities associated with the American Revolutionary War in 1783; Fulton 1911).

With independence, the United States also weighed in on international fisheries law. Access rights to the cod fishery were a significant sticking point in negotiations for the Treaty of Paris that ended the American

Revolutionary War in 1783 (Fulton 1911). In 1793, the government of the United States was the first to introduce the 3-mile territorial sea. Previously, the cannon shot rule had been included in many treaties, but without a specific measure of distance (573). The US claim was not consistent, as the government often claimed greater areas, but the idea was taken up by France and Britain. Ignoring the increasing number of states with fleets fishing in the North Atlantic, those countries continued to focus on bilateral negotiation of international fisheries regulations. By 1839 they had worked out a convention that fully defined the 3-mile territorial sea and established a commission to develop rules to prevent disputes between the old types of longlines and the new sail trawlers, which were quickly becoming dominant. Regulations were approved by the two governments, but they were poorly enforced and they failed to change the practices of the vessels of other nations (Juda 1996; J. T. Jenkins 1920).

As congestion and competition increased in the North Atlantic, legal scholars and natural scientists began to call for the extension of territorial waters and the regulation of fisheries on the high seas. A second commission, set up in 1882, established the first international vessel registry and flag system and set rules for appropriate fishing behavior on the high seas (formally defined as any area outside of the territorial sea). Norway and Sweden took part in the negotiations but refused to ratify the convention because it set the territorial seas at three miles when they had established their own coastal zones at four miles. Though claims of territorial seas of three miles or more were now a *de facto* aspect of international law, lack of participation and enforcement continued to undermine attempts to exclude foreign fishers from these areas. The Dutch, whose fleets had fished their coastal waters for centuries, were becoming increasingly concerned about diminishing domestic catches caused by high levels of production by foreign fleets near their shores, but they had little power to alter the behavior of distant-water fleets (Juda 1996).

The international power disconnect described above was paralleled by domestic power disconnects in states with large distant-water fleets. Tensions between coastal and distant-water fishing fleets were felt in all the major fishing nations of Europe, as well as in the United States, Canada, Russia, and Japan. Around the turn of the 20th century, most fishing was still taking place only a few hundred miles from the coast. Distant-water fleets still stayed close to land, even if not their native shores. On one hand, coastal stocks of fish were being depleted, often with the help of foreign fishers with high profit disconnects. On the other hand, distant-water fishers needed access to stocks close to foreign shores in order to maintain economies of scale. Local fishers were much more susceptible to profit signals at the time. Their power varied from country to country. In the United States powerful salmon-canning interests balanced distant-water tuna fleets, but in Europe coastal fishers were much less powerful than distant-water fleets, even though they were more sensitive to the costs of overexploitation (see chapter 7). For representatives of distant-water fishers, who had considerable influence in all major fishing states by the early 1900s, demands for protection by coastal fishers were problematic. Closure of domestic nearshore fisheries to foreign fleets brought the danger of tit-for-tat closures of much larger tracts of valuable coastal resources near foreign shores. Since the coastal regions around major fishing states were already heavily overfished, this would be a significant problem for distant-water fleets (Juda 1996). Nevertheless, international attempts to control access to coastal resources continued and expanded in the 20th century, as governments worked to balance the competing claims of domestic fishing interests.

5.4 Voluntary Abstention

Although they were not enshrined in international law until late in the 20th century, several international bodies now known as regional fisheries management organizations (RFMOs) were created to manage shared fisheries in the early 1900s. Exclusion of foreign fleets from coastal fisheries was clearly the major driving concern in the negotiations of these first RFMOs. That said, the evidence suggests that states did not reach agreement on cooperative management until economic problem signals were strong as well. In a few cases, economics favored agreement prior to heavy overexploitation, usually when demand was relatively elastic and prices dropped substantially with rapid increases in production. In other cases, severe overexploitation and overcapitalization caused economic hardship that drove the states whose fishers were most adversely affected to make concessions. In line with the responsive governance foundations of the AC/SC framework, members of the first RFMOs exhibited responsive rather than proactive behavior. It is also interesting to note that all three of the RFMOs established before World War II had limited membership (only two to four states) and were generally considered successful in both reducing fishing effort and rebuilding depleted stocks. As Young (1999a) noted, there is not necessarily any causal relationship between these two factors, but the overlap between the organizations is important because it reflects key diplomatic resources in the structural context, notably shared strategies and governance institutions. The United States and Canada were members of all three of these RFMOs, indeed were the only members of two, and so had a major role in shaping these early regimes.

5.4.1 The Fur Seal Commission

Established in 1911, the North Pacific Fur Seal Commission (NPFSC) was the first formal regional fisheries management organization. Seals are not fish, but because they are marine mammals this was still considered an important fishery of the period. Prior to this time, many formal and informal sharing arrangements for international stocks of marine species were numerous but none established an actual regulatory body or formalized the ongoing negotiation of management measures. The NPFSC consisted of representatives from four countries (Canada, the United States, Russia, and Japan) that collectively managed the fishery for fur seals living on and around the Probilof Islands in the northeast Pacific. Most of the larger islands were US territories. Russia controlled some of the smaller islands. Disputes in the area began in the mid-1880s, when Canada (represented as a colony by Britain) began harvesting seals at sea. US sealers, who harvested the animals on land, saw this "pelagic sealing" as a great trespass on their rights of access. The transboundary nature of this problem gave Canada a slight edge in negotiations but maintaining good relations with the US was important to the Canadian government, so, in response to US protests (a secondary political problem signal), these two states temporarily solved their differences in a bilateral agreement. However, the treaty had broken down by the turn of the century due to an increasing profit disconnect and related demands to increase sealing by fishers from all four countries (Young 2010, 152).

Japanese sealers began harvesting fur seals around the Probilofs in the early 1900s. Demand for the seal furs was high at the time, and related high prices generated considerable expansion of effort by sealers from all four countries. Indeed, the profit disconnect widened even further in this period. As would be expected, higher hunting mortality resulted in smaller stock sizes and higher marginal costs of production. This cost signal alarmed US and Russian fishers, who complained to managers in their respective countries, and both states began trying to reduce land-based harvests of seals. Their efforts were undermined by the sea-based fisheries run by Japanese and Canadian sealers who had lower costs of production and had multiple sources of revenue from other fisheries. This profit disconnect dampened their sensitivity to problem signals. International conflict escalated with the decline in stocks, but no lasting agreement was reached until 1911, when the population of fur seals in the area had declined from 4.7 million individuals to less than 130,000 (Juda 1996, 34). By this time, cost signals were high and price signals were no longer negated by increasing demand, which increased US willingness to pay for an international agreement. In exchange for their compliance with the ban on pelagic sealing, the United States agreed to provide Canada and Japan with yearly side payments of

US$200,000 plus 15% of the annual harvest of seal skins (Young 2010, 153). Thus, economic power was as important to the final agreement as economic incentives because if the US had not had the resources to make these side payments, they would not have been able achieve the ban on pelagic sealing or the concomitant establishment of the NPFSC. Indeed, without such an alignment between profits and power, switching from ineffective to effective management would have been delayed much longer in this case.

Although interrupted by war and then made obsolete by exogenous forces, the NPFSC is usually seen as a highly successful RFMO. Its first iteration lasted until Japan withdrew in 1940. During this 29 year period, the fur seal population rebounded to about 2.2 million individuals, a significant improvement on the overexploited state of stocks in 1911 (Juda 1996, 35). After World War II, the problem of pelagic sealing was raised again, and the Convention was reinstated from 1957 to 1984. This second incarnation was also relatively successful; it managed seal stocks at a level that was close to the economic maximum (Peterson 1995, 302). In spite of this high level of effectiveness, the Fur Seal Commission disappeared as a result of political and economic shifts in the structural context. Politically, the establishment of their 200-mile exclusive economic zones reduced US and Russian incentives to remain parties to the agreement, since the fishery was now within their EEZs and they could legally exclude Japan and Canada. Changes in economic demand for fur also caused prices for seal skins to plummet. This reduced the profit disconnect so much that the fur industry declined substantially in all regions. Without pressure from their domestic industries, Canada and Japan did not object to the enclosure of the seal fishery within US and Russian EEZs. Therefore, the signatories quietly allowed the Fur Seal Agreement to lapse when it came up for renewal in 1984 (Young 2010, 162–163).

5.4.2 Management of Pacific Halibut and Pacific Salmon

Two other regional fisheries management organizations were created in the 1920s and 1930s. Both governed important transboundary fisheries in the northern Pacific, and both involved only the United States and Canada. The International Fisheries Commission (IFC) was created in 1923 and renamed the International Pacific Halibut Commission (IPHC) in 1953. This organization grew out of pre–World War II differences between the US and Canada over two major issues. First, US fishers wanted to be able to land their fish in Canadian ports and ship them overland to the United States without paying import tariffs to Canada. Canadian fishers wanted the same exemption in US ports because rail was a much more efficient method of transport than ships at the time. Second, the halibut producers from both countries wanted a closure of the fishery during winter months in order to contract supply and thereby boost prices. By the mid-1910s, halibut production had soared due to technological innovations that had reduced the costs of production and allowed fishers to move offshore to areas where fish were more plentiful. Since no new markets were readily available, fishers saw a closure of the fishery for low-quality spawning fish in the winter months as an expedient political response to this strong price signal (Thompson and Freeman 1930).

Continuing conflict over port access and trade issues prevented agreement on the halibut issue for many years. Finally, in 1922, a new treaty was proposed. It would establish the IFC and give it powers to close the winter fishery as desired by the fishing industry. The agreement was finalized in 1923 and entered into force in 1924 (Thompson and Freeman 1930, 57). A prominent US fisheries biologist, William F. Thompson, was hired to lead the scientific effort of the new International Fish Commission (Smith 1994, 203). Like the Fur Seal Commission, the IFC is generally considered to have been successful from a conservation point of view. In reality, however, it went through several cycles in which management was gradually tightened until stocks were rebuilt, then gradually loosened until a decline was observed, then tightened again. Although the IFC started out as a bilateral agreement between neighboring states, it eventually was used as a rationale to exclude fishers from countries that were not "historical" participants in the fishery during the 1930s (see below). Once the exclusive economic zones were created, such entry was no longer an issue. The halibut commission is still considered a great success, but, as the case of Pacific salmon suggests, things might have turned out differently if the timing had been slightly different.

Negotiations over Pacific salmon started in 1908, well before talks regarding halibut, but all agreements failed until the creation of the International Pacific Salmon Fisheries Commission (IPSFC) in 1937. The nature of the conflict over salmon was inherently different from the halibut problem. Whereas the halibut fishery was relatively homogeneous, with fishers from both countries using similar gear, fishing on the same grounds, and landing in the same ports, salmon fishers were deeply divided both by nationality and by the location of their operations. Salmon are anadromous fish that migrate between the open ocean and freshwater streams. The most important salmon fishery of the early 20th century was based in the Fraser River watershed in British Columbia and in adjacent near-shore waters. Working just outside the 3-mile territorial sea, US fishers could harvest large amounts of salmon as they returned to their native headwaters to spawn, substantially reducing the number of fish available to upstream fishers in Canada. Fishing pressure from all sides led to the rapid overexploitation of salmon stocks, starting with the fish that were highest-priced and easiest to capture. As economically valuable stocks were depleted, lower-priced and less accessible stocks were heavily targeted. By the time the IPSFC was created, many stocks were overexploited and also were under threat from habitat loss or obstructions generated by development along the Fraser River (Johnston 1965).

Although economic problem signals drove much of the political concern for this fishery, exclusion was also a primary driver of the final agreement on Pacific salmon. US and Canadian fishers both feared that Japanese fleets—which overtook the US fleets as the largest fish producers in the world in 1935—would start exploiting Fraser River salmon stocks on the high seas. The presence of Japanese vessels targeting crab in Bristol Bay, Alaska, in June 1937 further inflamed these fears, particularly because representatives of the US fishing industry publicly claimed that the Japanese were also catching salmon in the bay (Finley 2011). One month after this "salmon crisis," the US Congress took action to ratify the fifth incarnation of the agreement that created the IPSFC. This ended a seven-year delay in the implementation of the agreement, which was signed by both parties and ratified by Canada in 1930 (Johnston 1965). This may seem to be a surprising response, since management measures prescribed both by the IFC and the new IPSFC applied to US and Canadian fishers but not to the non-member Japanese. However, the existence of management measures on Alaskan salmon was an important part of the US argument for keeping Japanese fleets out of Bristol Bay, so formalizing management institutions for the Fraser River was a necessary step (Scheiber 1989). The United States used similar arguments to convince the Japanese to refrain from entering the halibut fisheries managed by the IFC as well (US Congress 1932).

A few years after the creation of the IPSFC, Japan was at war with the United States and other Allied countries. After the war, Japanese fleets were limited to a small area known as the MacArthur Zone but were gradually allowed to move further and further away from Japan. By 1952, US and Canadian fishers were again worried about Japanese fleets fishing in "their" waters. Indeed, US fishers had lobbied heavily for the Truman Proclamation of 1945, showing concern about a resurgence of Japanese fleets even before the end of the war. The industry was considerably larger and more powerful in the post-war period, and waged a successful public relations campaign to spur US and Canadian efforts to exclude the Japanese. At the same time, the US faced a dilemma because of its desire to protect the interests of its own distant-water fleets. In 1951, following the advice of State Department employee Wilbert McCleod Chapman, the US advanced what later came to be known as the voluntary abstention principle, which can be paraphrased as follows: If a stock of fish is harvested at or above maximum sustainable yield and is the subject both of extensive scientific research and management for the conservation of the stock, then fleets from countries who have little or no historical interest in the stock should voluntarily abstain from entering the fishery, even if fish can be captured on the high seas (Scheiber 1989, Anonymous 1955).

Keeping in mind that Japan was occupied by the Allied Forces from the end of the war in 1945 until the Treaty of San Diego entered into force in April 1952, it is not surprising that the Japanese accepted the voluntary abstention principle when they signed the agreement for the International North Pacific Fisheries Commission the very next month (Herrington and Chapman 1954). However, as Scheiber (1989) points out, although the Japanese claimed that they agreed to the abstention principle only because of the duress placed on them by North American interests, including threatened boycotts of Japanese goods, they also benefited from the principle. Given their dependence on distant-water fisheries, maintaining access to relatively plentiful stocks off the coasts of developing countries was much more important to the Japanese than exploiting the overharvested stocks of salmon, halibut, and other fish near North America. Later, when disputes over access arose in other RFMOs, Japan, the US, Canada, and other "historical fishing states" based their claims for rights of access to distant-water fisheries on the precedent set by the voluntary abstention principle and its application in the North Pacific.

5.5 Exclusive Economic Zones

Both coastal and distant-water fleets had considerable political power around the turn of the 20th century and decision makers in many dominant fishing states wanted to please both groups at the same time. Thus, they continued to haggle over the size of their territorial seas while also seeking ways to ensure access to distant-water fishing grounds. In doing so, states introduced several new concepts, the most important of which was the exclusive economic zone. Exogenous shifts also had an effect in this period, particularly the rise of food security and economic development as major national goals in many countries around the world. Indeed, after World War II, many more states engaged in negotiation over access rights to global fisheries. This increased the negotiating power of developing countries and coastal states with small-scale fleets, but generally widened the power disconnect at the domestic level by increasing the influence of large commercial interests.

5.5.1 The Truman Proclamation

Between the two world wars, the first attempt to codify an international law for the oceans was undertaken by the newly created League of Nations. At the Hague Codification Conference in 1930, representatives of 36 states met to negotiate the width of the territorial sea. Although a majority (twenty states) favored the 3-mile limit, twelve favored a 6-mile boundary, and once again the four Scandinavian states stuck to their 4-mile limits (Schrijver 1997, 204). There was also talk of adapting an "adjacent zone," next to the territorial sea, in which the coastal state would have well-defined rights to regulate certain issues, including fisheries. Three of the largest maritime powers of the time—and arguably those with the largest distant-water fleets, the United States, the United Kingdom, and Japan—staunchly opposed the institution of any such measures. Instead, they preferred that any disputes between fishing states should be solved bilaterally on a case-by-case basis. Portugal, Iceland, and other states with small coastal fleets strongly opposed this view and favored adjacent zones. As states with limited international influence, these small fishing countries had greater leverage in multilateral settings than in bilateral settings. Given this power balance, no consensus could be reached at that time and war soon ended the negotiations (Juda 1996).

Even before the end of World War II, it was clear that support for extended territorial seas was much stronger than it had been in the interwar period. Exploration and scientific evidence created new incentives for enclosure of larger coastal areas, including the discovery of oil and mineral deposits in the seabed and the development of coastal fisheries by many newly independent states. In addition, issues of food security that had emerged earlier in the century were acute during the war, and the perceived success of economic programs like the New Deal in the United States generated considerable demand for government-sponsored economic development in many countries around the world. As will be described in greater detail in chapter 6, these programs drove rapid fisheries expansion in the postwar period. Indeed, fisheries were seen as a source of both food security and economic development, so government desires to protect their fleets at home and abroad were greatly strengthened by these exogenous trends.

In 1945, a few months after the Japanese surrendered, the United States attempted to exclude foreign fleets from large coastal regions without challenging existing international law, particularly *mare liberum*. This occurred in response to domestic pressures from the salmon industry and other large-scale coastal fisheries organizations that feared the return of Japanese distant-water fleets. Without questioning the 3-mile territorial sea, the 1945 Truman Proclamation on Fisheries stated that the United States intended to establish "conservation zones" to protect coastal areas. If only US fishers were involved, the United States would act unilaterally; if fishers from other nations had historic interest in a fishery, then conservation plans would be negotiated with their flag states. Domestically, US policy makers were trying to balance demands for protection from incursion by foreign salmon fishers in Alaska with the needs of US distant-water fleets (Hollick 1981). It was clear that such an approach would not help either coastal states whose waters had been historically fished by many nations or developing states that had not yet established their fishing interests on a large scale (Juda 1996).

The unilateral nature of the Truman Proclamation set a precedent internationally, and a process of claim and counter claim began that ultimately culminated in the concept of the 200-mile exclusive economic zone. At first the claims of different countries varied widely. In 1952, Iceland unilaterally moved the boundary of its territorial waters from three to four miles from shore, and then extended it again to a 12-mile limit in 1958 (Johnston 1965, 184–185). This was the beginning of the "Cod Wars" between Britain and Iceland. In this and other "fish wars" of the 20th century, a state that had unilaterally extended its jurisdiction would systematically confiscate vessels and imprison the crew of foreign fleets within that area. Usually, the "war" would be between two specific states and "peace" would be negotiated bilaterally, as happened when the United Kingdom recognized Iceland's 12-mile limit in exchange for a 10-year adjustment period in which the UK's fishing effort in the zone could be gradually reduced (189). Here again, context is important, particularly the postwar international norm against violent conflict between states or their representatives (i.e., privateers), which resulted in relatively peaceful resolution of conflicts over marine resources.

Around the same time, Latin American countries also started taking action to protect "their" coastal stocks from exploitation by fleets from Europe and the United States. They also expressed security concerns about Cold War hostilities, particularly the outbreak of war in Korea. At first, these countries took a piecemeal approach. Starting in 1950, Mexico, Honduras, El Salvador, Nicaragua, Ecuador, and Brazil all laid claim to fisheries resources well beyond the 3-mile territorial sea. Most of the claims extended to 200 miles from the coast or beyond, and several states claimed rights to the entire continental shelf. Mexico enforced its claims by seizing foreign fishing vessels within its territorial waters in 1950, and other Latin American states followed suit in subsequent years. This angered US and European fishers, who appealed to their national governments for protection (Hollick 1981). As other coastal states began enforcing their claims using similar tactics, the US and other historically dominant fishing states initiated a barrage of bilateral and multilateral interactions with them through both official and unofficial channels. Feeling the pressure from historical fishing states, Chile, Ecuador, and Peru formed a coalition in support of the 200-mile zone in 1952 (85). Here again, coalitions were used to balance the power gap between historically dominant and coastal developing states.

5.5.2 UNCLOS III

With the proliferation of disputes between coastal fishing states and distant-water fishing states, it was clear that multilateral negotiations were necessary. In 1958, the United Nations held its first Conference on the Law of the Sea (UNCLOS I). Various developing countries in Latin America, in Asia, and in Africa joined the Scandinavian states at this time, expressing their preference for a larger territorial sea. While there was consensus that coastal states should have the right to regulate fisheries in their territorial seas, there was no agreement on the limits of that jurisdiction. Another UN Conference on the Law of the Sea (UNCLOS II) was held in 1960, but it also failed to set any specific international standards. A US-Canadian proposal was presented that established a 6-mile territorial sea with an additional 6-mile fisheries zone, but if failed by a single vote (Schrijver 1997, 204).

Many changes occurred during the 13 years that separated the second and third rounds of UNCLOS negotiations. Fishing capacity increased greatly, especially for distant-water fleets, which spread to every navigable part of the oceans as described in Part I. On the other hand, the number of coastal states also increased with the final collapse of colonial empires. New countries swelled the ranks of the coalitions that favored expansion of the territorial sea and institution of wide exclusive economic zones in which coastal states would have the right to regulate living resources. Furthermore, by the early 1970s coastal states had already unilaterally claimed rights to 4.5 million square nautical miles of ocean, an area three times larger than had been nationalized in 1958. Several of those states backed up their claims by confiscating vessels or otherwise interrupting fishing activities of foreign states within their exclusive areas (Juda 1996, 192). Indeed, according to Peterson (1995, 250), by 1977 the 200-mile EEZ was a *de facto* international norm and fishers and fishing states were forced to adjust accordingly. As access to stocks in distant coastal areas declined, claims for equal protection by domestic coastal fishers began to sound more convincing in states with distant-water fleets. Because enclosure was already widespread, there was no longer a need to worry about setting an exclusionary precedent. In addition, developed countries were eager to legitimize rights of access to oil and mineral wealth located on the continental shelf and to secure airspace and sea routes from arbitrary injunction by coastal states. By 1973, all parties agreed that a new round of UNCLOS negotiations was necessary to deal with these and other altered circumstances.

The UNCLOS III negotiations took nine years, engaging representatives from 144 states or fishing entities, and encompassed not only fisheries jurisdiction but also rights to minerals from the seabed, pollution controls, delineation of national airspace, and access to important trade routes. The convention that was finalized in 1982 gave states the right to claim sovereignty over territorial seas up to 12 miles from their coasts and gave them authority to manage natural resources in an exclusive economic zone that would extend no more than 200 miles from their shores. Rights were not conferred without responsibilities and coastal states were enjoined to protect and preserve the marine environment (Juda 1996). Moreover, coastal states were given a legal duty to ensure that the fisheries resources within their EEZs are exploited optimally. "Optimal management" was not well defined in the treaty, but one clause required states to set a total allowable catch (TAC) for each fishery in order to prevent overexploitation. In addition, if the coastal state's own fishing capacity was insufficient to take the full amount of the total allowable catch, the state was directed to make some of the harvest available to foreign fishers (230).

As described in greater detail in chapter 7, the United States pushed strongly for acceptance of maximum sustainable yield (MSY)—defined as the maximum amount that can be harvested annually without reducing future harvests—as the dominant standard for an optimal setting of TACs. This was part of a coordinated effort by the US State Department to both reduce fishing by foreign fleets in US waters and to continue expansion of US distant-water fleets. This was essentially a continuation of the voluntary abstention strategy described above. By the time of the UNCLOS III negotiations, Wilbert McCleod Chapman had left the State Department to work as a fishing industry lobbyist, but he maintained considerable influence over US fisheries policy. By selecting MSY as the optimum, rather than the lower maximum economic yield, which maximized revenues or scarcity rent, the US chose the goals of job creation and food security over profit maximization. It also set up a system in which US fisheries, most of which were exploited either at or above MSY, could be closed to foreign fishing, while fisheries near developing countries, which were "underexploited" based on MSY, would be open to US fishing fleets in the absence of sufficient domestic fishing capacity. This ingenious step had a profound effect on both domestic and international fisheries management (Finley 2011).

Although states reached agreement on the Law of the Sea in 1982, it remained inactive for many years due to a key structural feature of international governance. Under long-standing international law, once a treaty is signed it must be ratified by a certain number of states before it enters into force. The number of ratifications is usually stipulated within an individual treaty and is negotiated as part of the regime creation process. For UNCLOS III, entry into force required ratification by 60 states. This took 12 years and so, even though it was signed in 1982, the UN Convention on the Law of the Sea did not enter into force until 1994. Industrialized countries chose not to ratify the agreement until certain issues were resolved regarding deep-seabed mining. However, most states went ahead and implemented other portions of the agreement, including the extension of the territorial seas and the legal institution of EEZs (Juda 1996). Borders between adjacent EEZs were often contested, but by 1996 most boundaries had hardened and many coastal states had developed the capacity to enforce their fisheries management measures (Peterson 1995, 257).

Since more than 90% of commercial fishing still takes place within 200 miles of land, the establishment of EEZs had a significant impact on the expansion of the global fishing industry. As noted in Part I, the 1990s was a period of reduced production by fleets from historically dominant fishing countries and increased production by fleets from developing countries. After the first unilateral declarations of expanded jurisdiction, coastal distant-water states started intensive bilateral and multilateral negotiations over rights of access. Many states that did not have much fishing capacity either charged distant-water fleets for access or encouraged development of their own fleets via subsidies. In this redistribution of fishing effort, states that were heavily dependent on distant-water fishing, like Japan, struggled to purchase sufficient access rights to maintain their fleets. On the other hand, states like Canada and Peru were able to expand their domestic fleets to replace excluded foreign fishers. Hypothetically, management should have improved within the EEZs, since giving control of fisheries to a single country eliminated the problem of open access at the international level. However, the overall effect on fisheries within the EEZ was not always positive, and in many cases fisheries depletion continued in spite of the institutional change (see chapters 6 and 7).

5.6 Negotiating Chaos

In addition to negotiating the new Law of the Sea, from the 1930s through the 1970s states worked to exclude outsiders or protect their own access rights by establishing new RFMOs. Indeed, 1937 was a seminal year for international fisheries management. Many fisheries management agreements were negotiated, but most either were not ratified or did not provide for the creation of a management organization. Nevertheless, these treaties laid the groundwork for the new generation of RFMOs that was created after World War II. Fisheries were seen as an important source of food security and economic development after the war, and many countries subsidized the building or rebuilding of modern, industrialized fleets. This increased fishing pressure on coastal and international stocks and altered the economics of fishing, so that historically dominant fleets were vulnerable to competition from new entrants. In response to the resulting political problem signal from their domestic fishers, historically dominant fishing states negotiated quite a few new RFMOs to protect the economic status quo. At the same time,distant-water fishing states were forced to negotiate access to important fisheries in areas claimed by coastal developing states. This lead to the creation of RFMOs for stocks that were not yet overexploited but were the subject of international conflict.

5.6.1 Whaling

The International Whaling Commission (IWC), one of the first post–World War II RFMOs. It was built upon previous attempts to regulate pelagic whaling. The most notable of these was the Convention for the Regulation of Whaling, which was negotiated in 1931, but did not enter into force until ratified by Great Britain in 1935 (Oda 1989). Interestingly, this was one year after the Japanese started whaling in the Antarctic, which had previously been the exclusive hunting ground of British and Norwegian whalers and was regulated by the government of the (British) Falkland Islands with cooperation from Norway (Clark and Lamberson 1982; O.J.B. 1937). This suggests that exclusion played an important role in British policy making but, as in the case of halibut, economic and conservation concerns were also important in this period. Rapid technological improvement and massive entry into the fishery created a glut in the market in the 1930s, making reduction of total catch desirable to bring prices back up. The 1931

Convention for the Regulation of Whaling encouraged countries to cooperate, mainly by helping their industry representatives reach compromises about limitations on effort and also by putting in place conservation measures for calves and pregnant females as well as particularly overexploited species of baleen whales (Johnston 1965; O.J.B. 1937). Industry cooperation quickly unraveled and "pirate" whaling undermined other measures, so negotiations opened again in 1937. A new agreement entered into force in 1938, and was amended several times, but no international commission was established. By 1939, the commencement of hostilities in World War II ended all cooperation on whaling (Oda 1989).

After the war, whaling countries subsidized the rapid rebuilding of their fleets, including the proliferation of large mother ships that facilitated exploitation of whales on the high seas. Concerns about the conservation and allocation of access rights to whale stocks returned to international prominence. New negotiations on whaling agreements began in 1944, and in 1946 the United States took the lead in calling for a new whaling convention in anticipation of a return to heavy overexploitation of whale stocks and greater conflicts among whalers (Johnston 1965, 401). The resulting agreement created the International Whaling Commission and gave it considerable power to reduce whaling effort. At the same time, the agreement required that the IWC should consider the needs and interests of the whaling industry and consumers of whale products, and also forbade management measures based on the country of origin. Scientific foundations and optimum utilization were other primary obligations found in the new agreement (Oda 1989). As expected by the AC/SC framework, these contradictory goals are politically expedient responses to the discordant views of competing interest groups. They hobbled the new whaling commission and allowed the core problems of overexploitation and overcapitalization to worsen.

Although the IWC took a number of steps to restrict whaling effort and protect declining stocks, continued conflict over allocation of access rights and lack of enforcement consistently undermined regulations and generated the typical pattern of responsive management. The primary limits on effort were quotas set for the total catch of whales in "blue whale units," in which one unit equaled one blue whale, two fin whales, two and a half humpback whales, etc. The first total allowable catch was set at 16,000 blue whale units. Whaling fleets from various countries were left to scramble to catch as many whales as they could before this limit was reached and the fishery was closed for the year. This altered the economics of the fishery, reducing profits for historically dominant UK and Norwegian fishers and opening up room for entry by subsidized fleets from Japan and the Soviet Union. The result was considerable conflict at the IWC, with historically dominant countries threatening to exit the agreement if new entrants did not curb their fleets. In the meantime, pirate whaling continued and stocks declined even more (Johnston 1965, 400–401).

Like the Fur Seal Commission, the IWC was eventually transformed by exogenous forces. Public interest in whales increased in the 1970s and public protest against whaling escalated due to the IWC's failure to prevent the collapse of many stocks under its jurisdiction, including the near extinction of blue whales in the mid-1960s (Clark and Lamberson 1982, 103). While the existence value of whales increased greatly with changes in public perceptions, the commercial value of these species declined as a result of the rise of synthetic substitutes for whale oil and related products. This drove large numbers of whalers to exit the fishery and reduced objections to stricter limits on whaling from countries like the United States, where the whaling industry completely disappeared within a decade. By the late 1970s, the IWC faced considerable pressure to alter its approach to whale management, largely because of an influx of members whose only interest was the conservation of whales. As a result of this shift, in 1982 the commission placed a moratorium on commercial whaling that remains in effect today, though it is highly contested. "Scientific" harvests by Japan and commercial harvests by Norway continued during the ban. In 1994, the IWC adopted Revised Management Procedures that should hypothetically allow a return to commercial whaling of healthy stocks, but this has not yet occurred (Kock 2007, 235–236). Japanese attempts to restart their whale fishery were stymied by radical conservationist tactics in 2011 (see chapter 8).

5.6.2 Protecting the Rights of Coastal States

One of first areas exploited by steam-powered trawlers, the North Atlantic was also the long-standing subject of international fisheries negotiations. The first bilateral fisheries treaty for the area was signed in 1839 between Britain and France. The first multilateral agreement was the North Sea Convention of 1882. Neither agreement listed conservation as a major concern, and both were designed to reduce disputes among fishers from different countries. At the time, people still believed that sea fisheries were inexhaustible. This view started to change around the turn of the century and in 1902 European countries formed the International Council for the Exploration of the Seas (ICES), which fostered international scientific collaboration but served only as an advisory body (Johnston 1965, 91). Multiple small conventions were concluded in the early 20th century, but no comprehensive treaty was signed until 1937, when nine European states agreed to implement substantive conservation measures and work to reduce conflicts associated with trawling throughout the North Atlantic. However, the agreement did not enter into force before World War II, so new negotiations were required once the conflict subsided (360).

After another unsuccessful attempt at a sweeping convention in 1943, states with fisheries in the North Atlantic negotiated area-specific treaties, rather than one broad agreement for the entire region. Negotiations on the northeast Atlantic started with the International Overfishing Conference in 1946. Ongoing talks lead to an agreement to create the North Sea Commission in 1954. It was supposed to implement regulations recommended by ICES as unanimously agreed to by member states, but it had limited management and enforcement capacity. Continued stock declines and related economic signals prompted European states to negotiate a stronger agreement in 1959. The resulting North East Atlantic Fisheries Commission (NEAFC) did not enter into force until 1964 (Johnston 1965; Peterson 1995). By this time, many distant-water fleets no longer operated in the region, so negotiations were shaped largely by pressures from more vulnerable coastal fishers. Nevertheless, negotiation of management measures was slow and little was accomplished prior to the post-EEZ renegotiation of the convention in 1978.

On the other side of the North Atlantic, the International Commission for Northwest Atlantic Fisheries (ICNAF) was negotiated in 1949 and took effect in 1950. Conservation was a primary driver of cooperation, but concerns regarding the entry of developing distant-water fleets were also important (Johnston 1965). Initially, the primary management measures adopted by ICNAF were mesh size restrictions designed to protect juvenile fish. As explained in Chapter 7, this was a popular approach to the problem of overfishing from the early 1900s on. The commission did not established a total allowable catch until 1969, after a spike in production by Soviet fleets severely reduced the biomass of haddock in the region and caused considerable concern about the activities of this relatively new entrant (E. D. Anderson 1998, 82). As noted in chapter 6, the USSR started heavily subsidizing a large fleet of factory trawlers in the early 1960s, with the intention of exploiting distant-water fisheries like those of the northwest Atlantic.

In 1972, ICNAF formalized allocation of quotas as a means to restrict access to fisheries in the region. According to the "40-40-10-10" formula, national quotas were set in proportion to the average catch for the preceding

10 years (40%), the average catch in the past three years (40%), coastal state status (10%), and the flexible "special needs" (10%). This scheme ensured that larger quotas would be granted to historically dominant coastal states, then to historically dominant distant-water states, then to new coastal states, and finally to new distant-water states. It was clearly an attempt at exclusion as well as conservation, but failed on both counts due to lack of enforcement (E. D. Anderson 1998, 82–83).

The widespread adoption of exclusive economic zones in 1977 had a profound effect on management across the North Atlantic. Regulation of most species could now be carried out by coastal states, often unilaterally; only a few straddling stocks remained open to international exploitation. Therefore, NEAFC was dissolved in 1978 and ICNAF was terminated in 1979. Both were renegotiated and reestablished in the early 1980s. NEAFC kept the same name, but ICNAF was rechristened the North Atlantic Fisheries Organization (NAFO). Under the new terms, both organizations only managed fishing outside of the national EEZs, which further increased polarization between coastal and distant-water states. At NAFO, Canada tried to prevent European fishing of cod in two small areas outside of its EEZ, respectively known as the "nose" and "tail." They were not successful until the early 1990s, when an agreement was reached with EC member states (see chapter 7). Similar problems delayed effective conservation of NEAFC fisheries as well (Bjørndal 2009).

5.6.3 Establishing the Rights of Distant-Water States

All of the RFMOs described above were negotiated when profit signals aligned with political problem signals to create strong political will for exclusion. In contrast, the RFMOs created to manage highly migratory tuna and tuna-like species were usually negotiated before there was substantial overexploitation of the stocks, when profit signals were weak, and were primarily a response to international conflict. Power was also distributed differently in negotiations for the tuna RFMOs. The coastal states in fear of encroaching distant-water fishers were developing countries in Latin America, Africa, and Asia, rather than developed states in Europe and North America. The distant-water fleets came from developed countries, including Japan, the United States, and various EU member states. Politically powerful, historically dominant distant-water states were determined to maintain access to rich fisheries in spite of coastal states' expanding claims to exclusive economic zones. They therefore negotiated RFMOs that differed substantially from their northern counterparts; international management applied to both the high seas and the EEZs, and fleets from any member country could operate within the entire convention area. This coordinated strategy, intended to satisfy US coastal and distant-water interests, was championed by Wilbert McCleod Chapman, the US State Department's first Secretary for Fisheries and Wildlife (Hollick 1981).

Chapman was instrumental in the creation of the first RFMO dealing with highly migratory species of fish, the Inter-American Tropical Tuna Commission (IATTC), which covers tuna and tuna-like species in the eastern Pacific Ocean (IATTC 1961). The IATTC had a tumultuous start in the 1940s and the 1950s, a period of heightened conflict between the United States and Latin American countries over access rights for US fishers in newly claimed adjacent zones. Like many other Latin American countries, Costa Rica claimed a large zone of national sovereignty over marine resources in 1947. However, the agreement that the United States reached with Costa Rica to ensure continued access to this zone was exceptional. Rather than simply ensure US access to waters claimed by Costa Rica, the agreement also created the IATTC as a scientific and management body open to any state with an interest in fishing for highly migratory species in the entire eastern Pacific. The 1949 agreement took effect in 1950, and the IATTC met to develop a scientific program to study tuna and tuna-like fishes and bait fishes harvested by the tuna industry, which used the pole and line method at the time (Hollick 1981).

With the Costa Rica agreement in place, the United States began pushing other countries in Latin America to join the IATTC. Successes were few and far between, however. After quiet negotiations by the US State Department and the private American Tunaboat Association, Panama joined the IATTC in 1953, resolving the issue of access to Panamanian waters. No other states joined the commission until the 1960s, by which time the US had bilateral access agreements with several states claiming jurisdiction over adjacent zones. Conflict was reduced elsewhere through unofficial agreements in which US vessels purchased fishing licenses from coastal states (Hollick 1981). Many Latin American states attended IATTC meetings as observers in the 1950s, even though they were not members; in the meantime, the US kept up pressure on countries like Colombia, Mexico, and Ecuador (IATTC 1951, 1955, 1956, 1957, 1958, 1959, 1960).

In 1961, IATTC scientists found that fishing effort targeting yellowfin tuna in the eastern Pacific Ocean was at or above their management benchmark, maximum sustainable yield (IATTC 1961). With the specter of effort restrictions looming, the United States pushed harder for other countries with fisheries in the region to join the IATTC to ensure that measures applied to US fishers would also apply to fleets from other countries.

Using access to US markets as a key bargaining chip and unofficially softening their opposition to EEZs, US negotiators convinced Ecuador (1961) and Mexico (1964) to join the commission (Hollick 1981; US Congress 1962). Here, power favored US fishers, who lobbied hard to ensure that regulations would affect all fleets. This is not exclusion exactly, but was designed to limit competition for the tuna resources in the region.

The IATTC discussed quotas for yellowfin tuna throughout the 1960s, but even when a total allowable catch was set, it was always higher than that recommended by IATTC scientists and was often higher than the actual catch for a given year (IATTC 1969, 7). Profits were high for most fishers during this period, so there was considerable pressure against strict regulation in most member states. Late in the decade, Canada (1968) and Japan (1970) joined the commission as distant-water fishing states, followed by France and Nicaragua in 1973. Japan was targeting bigeye tuna rather than yellowfin and so was not seen as an economic threat at the time. In the late 1970s and the 1980s, IATTC negotiations were disrupted by further disputes between coastal and distant-water states. These conflicts were exacerbated in the 1990s by what is now known as the tuna-dolphin controversy, but rapprochement was achieved through economic and technological innovation. The IATTC is now viewed as the most successful of the five RFMOs governing highly migratory species.

In contrast, the International Commission for the Conservation of Atlantic Tunas (ICCAT) is thought to be fairly ineffective. Negotiations to establish an Atlantic tuna commission started in 1960 and apprehension about increased harvests was the official rationale for the creation of the commission (ICCAT 1966). However, conservation was not the only impetus for creating the new regime and, in fact, by the time the convention entered into force in 1969, scientists expressed little concern about tuna stocks in the Atlantic (ICCAT 2008, 3; Webster 2009). Instead, the timing of the negotiations reflects important changes in the international system. As described previously, the 1960s were a turbulent time in international fisheries. At the start of the decade, the second UN Conference on the Law of the Sea failed to produce agreement on the size of either territorial seas or EEZs. Subsequently, tensions again flared between Latin American countries and distant-water fleets operating in the eastern Pacific and in the Gulf of Mexico. In West Africa, where most of the tuna fishing in the Atlantic occurred, many countries gained independence from colonial powers and began looking to fisheries as a source of economic development. These states joined the Latin American states and others from the "developing world" or "global south" to form the Group of 77 (G77) at the United Nations. The

G77 worked to change the tenor of discussions in international relations to focus on North-South issues rather than on the conflict between US and Soviet ideals (Hollick 1981).

As a result of this turmoil, and its experience with Pacific tuna fisheries, the United States decided to engage in a multilateral process that might forestall substantive conflict over tuna and tuna-like species in the Atlantic (US Senate 1967). US fleets were struggling at the time, largely because of increased competition from newly built Soviet and Japanese fleets, so protection of their fishers' ability to access Atlantic stocks was also a driver (Hollick 1981). On the processing side, US canners faced an undersupply problem due to the growing demand for tuna in the United States. They therefore lobbied for an Atlantic Commission to ensure an increasing supply of fish (US Senate 1967, 1:21). At a 1963 FAO working group on Atlantic tuna management, Wilbert McCleod Chapman of the United States pushed strenuously for the creation of a commission similar to the IATTC. The US continued these diplomatic efforts, and by 1966 overcame resistance from France, Spain, and Japan, which believed that existing bodies (the UN Food and Agriculture Organization and the International Council for the Exploration of the Seas) could manage Atlantic tunas more cheaply than a new commission. The US had support from countries like Portugal, Brazil, and Nigeria, but even these states expressed concerns about the size of the commission's budget (ICCAT 2008, 7–9).

Conflict between historical fishing states and coastal developing states continued after the creation of the Atlantic tuna commission. Indeed, most negotiations over total allowable catch and other regulations were marked by constant wrangling over the criteria for allocation of quotas or similar types of unequal impacts. As Webster (2009) describes in detail, these conflicts resulted in responsive governance by ICCAT. In some cases (e.g., yellowfin tuna), exogenous economic shifts reduced pressure on stocks without effective management, much as in the Pacific Fur Seal and International Whaling regimes. In other cases, particularly management of northern swordfish and bigeye tuna, the threat of entry by developing distant-water fleets increased the willingness to pay of historically dominant distant-water states and they negotiated stronger management measures for these stocks. For bluefin tuna, however, exclusion either happened too early (as with the western stock, similar to NAFO), or exogenous technological changes reduced costs of production and thereby reduced pressure for better governance (for the eastern stock). In all cases, there was prolonged failure to implement regulations that conformed to scientific advice or to enforce regulations that coincided with advice from scientists until economic and political problem signals were strongly felt by powerful member states.

Several other international fisheries commissions were created in the late 1960s and in the 1970s, but most were advisory bodies without management powers. They include the Baltic Fisheries Commission (1974), the International Commission for Southeast Atlantic Fisheries (1971), the Committee for the East Central Atlantic Fisheries (1969), the Western Central Atlantic Fishery Commission (1973), and the Indian Ocean Fisheries Commission (1967; Peterson 1995, 253–255). These organizations followed the pattern of the International Council for the Exploration of the Seas (1902) and the General Fisheries Council for the Mediterranean (1952), which coordinated science on straddling and transboundary stocks and made recommendations to member states but had no power to set management measures. Many such agreements were fostered by the UN Food and Agriculture Organization, which generally supported "rational" and "optimal" management of fisheries resources as an important source of food security and a path to economic development (Royce 1987).

Only two management RFMOs were newly created in the 1980s (as opposed to renegotiated to replace existing organizations after *de facto* establishment of EEZs). First, the North Atlantic Salmon Conservation Organization (NASCO) was signed in 1982 and entered into force in 1983. It was necessary due to changes in the governance context; the renegotiated NEAFC and NAFO had no power to regulate fishing within the new EEZs of their member states. Since salmon are transboundary species that passed through multiple EEZs, cooperative management within 200 miles of the coast was necessary to rebuild severely depleted populations in the Atlantic (Windsor and Hutchinson 1990). Exclusion was clearly a primary driver in these negotiations as one of the commission's first acts was to close almost all areas outside of the 12-mile territorial seas to fishing for Atlantic salmon (7). In spite of this action, Atlantic salmon stocks continued to decline until drastically low levels drove NASCO members to adopt a precautionary approach to management—setting total allowable catches lower than those that would support management at maximum sustainable yield. These measures did facilitate some stock rebuilding, but harvests of wild salmon remain low relative to cultured salmon (Potter et al. 2003).

Second, the Commission for the Conservation of Antarctic Marine Living Resources (CCAMLR) was founded under the Antarctic Treaty System rather than under the rubric of the UN Law of the Sea and so exists within a different governance context than other fishery RFMOs. One important variation was that the commission was created initially to protect krill as a keystone species in the Antarctic ecosystem, rather than to manage krill at MSY or otherwise optimize harvests for human use. In addition, members of the commission are both fishing states and non-fishing states. This is not true for other RFMOS, except the International Whaling Commission. At the IWC, conservation interests focused on charismatic species (whales), but at CCAMLR non-fishing states are motivated by a general ideal of Antarctic governance that includes preserving the biota and the ecosystem for the common heritage of mankind (Kock 2007). In spite of these governance factors, which favor early response, switching to effective management was still delayed. Specifically, CCAMLR frequently set precautionary catch limits but did not dedicate sufficient resources to enforcement until stocks of high-priced species were heavily overexploited. Osterblom and Sumaila (2011) document the various crises that led to a strengthening of compliance mechanisms by the commission. Illegal, unreported, and unregulated fishing remains problematic, particularly for high-priced species such as Antarctic and Patagonian toothfish, but is in decline (K. W. Riddle 2006).

5.7 Closing Loopholes

After the chaos of the pre-EEZ system, there was a period of relative stability in the 1980s but this was followed by an upsurge in the creation of RFMOs in the 1990s. There were several reasons for this new trend. First, historical fishing fleets were in trouble. The industry could no longer fill the capacity it had created, profits were down, and many vessels were operating at a loss even with assistance from their governments. Second, the boundaries EEZs were firmly set, and there was little more to be gained for domestic fleets, which had been encouraged to expand to fill the void left by exclusion of foreign vessels. Third, landings of international stocks by growing fleets from developing distant-water states like China, Taiwan, and the Philippines increased substantially. Faced with these new entrants, coastal states and historically dominant distant-water fishing states put their differences aside to adopt management measures and thereby have legal cause to curb effort based on the voluntary abstention principle described in section 5.4. That is, by creating RFMOs they hoped to limit the growth of fishing effort by developing distant-water fleets. With modifications respecting the rights of coastal developing states, the voluntary abstention principle was codified into law in the 1995 UN Fish Stocks Agreement, though it was generally accepted much earlier. More important, RFMOs began using sanctions and other methods to monitor and enforce their agreements, thereby improving their exclusionary capacity. If countries wanted markets for the fish their fleets caught, they would have to join the relevant RFMO (Webster 2013).

5.7.1 Monitoring, Enforcement, and Participation

Table 5.1 lists the 17 extant marine RFMOs with management authority by date of signature and entry into force. It also shows the number of initial members, members in 1990, and current membership. Eight new RFMOs were created from 1990 to 2013, almost half of those currently in place. Participation in RFMOs with open membership that allows access by new entrants also increased substantially. However, membership in some regimes is either closed, like the Pacific Salmon Commission (only coastal states can be members), or limited, as with the North East Atlantic Fisheries Commission (membership is open but all stocks are fully managed so new entrants would not be allowed any access rights unless new fisheries were created). As might be expected, membership in closed and limited regimes stayed relatively stable.

Given the scope and scale of the list, a majority of the known international fisheries are officially managed by existing RFMOs (see figure 5.1). Remaining unmanaged areas are in the Arctic Ocean, western central Atlantic, southwest Atlantic, and a number of relatively small areas in the Pacific (Takei 2013). The Arctic is of particular interest due to the effects of climate change. Both the thickness and the extent of sea ice are already declining there, leading many to believe that fishing in the region will increase substantially in the near future. In addition, as ocean temperatures warm, valuable species like cod and haddock are expected to move into the Arctic Ocean, further increasing incentives to fish in the region (Cheung et al. 2010). The Arctic Council has authority to manage fisheries in the area but has chosen not to use it (Molenaar 2012). Some countries, notably the US and EU members, along with Denmark on behalf of Greenland, support the development of a new RFMO for the Arctic but others, like Russia and Norway, are opposed to it, probably because they intend to increase fish production in the area when conditions are more favorable (73).

Even if all fishing areas and species were covered by RFMOs, the entire system still contains a major loophole given the goal of exclusion: free riding. Many vessels continue to (1) fish in contravention of RFMO regulations (illegal), (2) misreport or fail to report their catches (unreported), or (3) fish under the flags of non-member states to avoid regulation (unregulated). These vessels are collectively known as illegal, unreported, and unregulated (IUU) fleets (Stokke 2009, 339). Most IUU vessels fish under "flags of convenience," using documentation purchased from states with open registration

Table 5.1

Current marine RFMOs with management capacity

	Full name	Signed	In force	Number of founding members	Number of members in 1990	Number of current members	Membership procedures
IPHC (IFC)	International Pacific Halibut Commission (was International Fisheries Commission until 1953)	1923		2	2	2	Closed
PSC (IPFSC)	Pacific Salmon Commission (was International Pacific Salmon Fisheries Commission until 1985)	1937		2	2	2	Closed
NEAFC	North East Atlantic Fisheries Commission (was North Sea Commission until 1959; renegotiated in 1980, in force by 1982)	1946		5	5	5	Limited
NAFO (ICNAF)	Northwest Atlantic Fisheries Organization (was International Commission for Northwest Atlantic Fisheries until 1978)	1949		4	8	12	Open
IATTC	Inter-American Tropical Tuna Commission	1949	1950	2	10	21	Open
NPAFC	North Pacific Anadromous Fish Commission	1952	1953	4	4	5	Limited
ICCAT	International Commission for the Conservation of Atlantic Tunas	1966	1969	3	18	48	Open
CCAMLR	Commission on the Conservation of Antarctic Marine Living Resources	1980	1982	14	23	25	Limited
NASCO	North Atlantic Salmon Conservation Organization	1982	1983	6	9	6	Limited
CCSBT	Commission for the Conservation of Southern Bluefin Tuna	1993	1994	3	n.a.	5	Open
IOTC OFC	Indian Ocean Tuna Commission (started as advisory body called Indian Ocean Fisheries Commission in 1982)	1993	1996	3	n.a.	30	Open
CCBSP	Convention on the Conservation and Management of Pollock Resources in the Central Bering Sea	1994	1995	6	n.a.	6	Limited
GFCM	General Fisheries Commission for the Mediterranean (started as advisory body called the General Fisheries Council for the Mediterranean in 1949)	1997	1952	2	18	24	Open
RECOFI	Regional Commission for Fisheries (RECOFI)	1999	2001	6	n.a.	8	Open
WCPFC	Western and Central Pacific Fisheries Commission	2000	2004	15	n.a.	33	Open
SEAFO (ICSEAF)	Southeast Atlantic Fisheries Organization (preceeded by the International Commission for the South-East Atlantic Fisheries created in 1971)	2001	2003	8	n.a.	7	
SIOFA	South Indian Ocean Fisheries Agreement	2006	2012	10	n.a.	5	Open
SPRFMO	South Pacific Regional Fisheries Management Organization	2010	2012	12	n.a.	12	Open

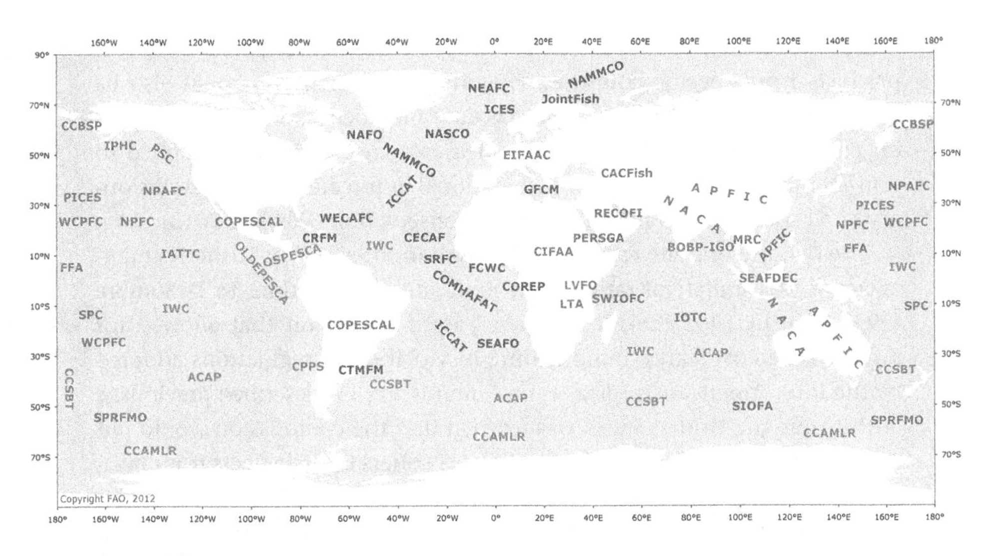

Figure 5.1
Map of RFMO coverage. Source: FAO 2013.

(that is, states in which citizenship is not required to register a vessel), lax regulations, and little enforcement infrastructure (DeSombre 2005). Some use counterfeit flags, and often the names of vessels are changed to avoid enforcement. High-priced fish that can be caught on the high seas, far from enforcement bases, are most often targeted by IUU fleets, which at once maximize the benefits and minimize the potential costs of free riding (K. W. Riddle 2006).

The practice of switching flags to avoid regulation goes back at least to the late 19th century, when some British owners registered their vessels with Norway to avoid unilateral closures on the Moray Firth fishing area in Scotland (Fulton 1911). However, the use of flags of convenience grew during the decades-long struggle over EEZs, and increased substantially again as RFMOs started regulating international fisheries. Free riding by IUU fleets and by flag of convenience states was a major driver of the negotiation of regulations that limited entry into international fisheries and the development of international methods used to monitor and enforce those rules. In this way, a non-member state could be punished if a large number of vessels flying its flag fished in contravention of RFMO rules. Monitoring methods included onboard observers, positive and negative vessel lists, and trade documentation schemes (in which for legally harvested landings a certificate is issued by the flag state and travels with the fish through each point of sale). International enforcement usually takes the form of trade restrictive measures, or bans on imports of specific species or of fish products from specific countries. Hypothetically, members could also be punished for non-compliance but, because members can block consensus on proposals in most RFMOs, trade measures are usually only applied to non-members. Because of this, international enforcement is an exclusionary tactic as much as a means to ensure conservation (Webster 2013).

The first use of trade restrictive measures in international fisheries management was unilateral rather than multilateral. According to DeSombre (1995, 56), in 1962 the US Congress passed legislation that allowed for sanctions against states found fishing in violation of regulations adopted by the Inter-American Tropical Tuna Commission. As described previously, at this time the United States was worried that the commission would put in place conservation measures binding US fishers but not fleets from non-member countries. Thus, it wanted to ensure participation in the IATTC by all states with fleets fishing in the area. Membership in the IATTC did not settle all disputes between the United States and all other coastal states in the eastern Pacific Ocean, so in 1976 the US Congress approved use of sanctions against countries that seized US fishing vessels within their EEZs as part of the wide-reaching Magnuson Fishery Conservation and Management Act. In the 1980s, the US boycotted tuna from several countries to punish them for seizing US tuna vessels within their EEZs. These ongoing conflicts between distant-water and coastal fleets in the eastern Pacific Ocean resulted in the breakdown of cooperation at the IATTC (DeSombre 2000; D. D. Murphy 2006).

Later, the United States again imposed sanctions on South American tuna as part of a general embargo of tuna captured in association with dolphins in the eastern Pacific Ocean.[1] The Marine Mammal Protection Act of 1972 authorized the use of sanctions on all yellowfin tuna imported from countries with fleets fishing in the region unless their plans to limit dolphin mortality were cleared by the US National Marine Fisheries Service (DeSombre 1995, 57). However, this part of the law was not implemented until 1990, when the US government instituted an official boycott of tuna that was not labeled "dolphin-safe." Implementation of sanctions was instigated by pressure from the Earth Island Institute and other conservation interest groups in cooperation with the US tuna canner StarKist (or its parent company, Heinz). In 1991, Mexico took the US to arbitration through the General Agreement on Tariffs and Trade (GATT, the predecessor to the World Trade Organization), and the boycott was found to violate international law. Together with the EU, Mexico brought suit against the US again in 1992, and again won its case but did not take retaliatory measures. The

US continued the boycott for several years until technological solutions to the problem of dolphin by-catch were found (D. D. Murphy 2006).

Around the same time, members of ICCAT adopted trade restrictive measures on bluefin tuna imports from countries whose fleets were found fishing in violation of ICCAT regulations. Trade measures are inherently exclusionary, but, unlike the dolphin boycott passed by the US Congress, ICCAT's bluefin measures were acceptable under Article XX of GATT because they were adopted by a multilateral body in support of conservation goals, were applied to specific countries and to single species, and did not ban general trade with a country (Le Gallic 2008, 862). In this case, the US supported trade measures but Japan was the most ardent advocate for trade-based enforcement of conservation regulations for bluefin tuna and bigeye tuna in all five tuna RFMOs. Japan is by far the largest importer of sushi-quality fish, of which bigeye and bluefin are two important species, and so it has both the incentive and the capacity to curb IUU fishing through trade-based monitoring and enforcement. Furthermore, the Japanese quickly realized that IUU fleets moved from ocean to ocean to avoid trade restrictive measures, so they initiated the Kobe Process for harmonization of tuna management across all five tuna RFMOs (DeSombre 2005; Scheiber, Mengerink, and Song 2008). Barkin and DeSombre (2013) refer to this as "the balloon problem," in global fisheries, but it is really an extension of Mancur Olson's (1971) concept of roving bandits, defined as individuals who have no strong ties or dependence on a resource, but rather use their mobility to deplete one open access resource after another.

In the 1990s, as prices for bluefin tuna, Patagonian toothfish (Chilean sea bass), orange roughy, and other popular species increased, the problem of IUU fishing grew apace and more RFMOs began to use trade measures to monitor harvests and to sanction interlopers. However, RFMOs have a *de facto* norm of governance based on consensus, so sanctions that could easily be adopted to punish non-members could not be applied to member states. This was one of the primary drivers of the growth in membership observed in open RFMOs in the 1990s. Early on, a few states like Belize and Panama were among the first to be sanctioned as IUU flag states. Trade restrictive measures applied by ICCAT and other RFMOs caused large economic costs in these countries but IUU states were not allowed to join the commission until they eliminated IUU vessels from their registries, went through a lengthy process to reform their vessel registration programs to prevent future IUU activities, and improved their fisheries management institutions. In this way, trade restrictive measures were kept in place for years until states could prove that they were fully in compliance with RFMO rules

(Webster 2009). Learning from the experiences of these first IUU fishing states, non-member states with fleets targeting an unregulated stock often chose to join an RFMO as soon as management measures were put into place, rather than risk being labeled as an IUU state (Webster 2009). This was primarily an issue at open RFMOs. In closed or limited RFMOs, IUU states could not join or, if they did join could not expect any quota allocation, so trade measures are essentially permanent exclusionary mechanisms in closed regimes.

As with EEZs, codification of trade-based measures occurred after widespread *de facto* application in multiple RFMOs. Negotiations for the *United Nations Agreement for the Implementation of the Provisions of the United Nations Convention on the Law of the Sea of 10 December 1982 relating to the Conservation and Management of Straddling Fish Stocks and Highly Migratory Fish Stocks* (also called the Fish Stocks Agreement, or FSA) began in 1993 and ended in 1995. The FSA establishes detailed minimum standards for management of straddling and highly migratory stocks, including requirements for harmonization of regulations between EEZs and the high seas, necessary monitoring and enforcement mechanisms for the high seas, and recognition of the needs of developing states in policy formulation (United Nations 2013a). In essence, it legitimizes the exclusionary strategies described above, including the use of trade documentation and sanctions as well as recognition of access rights for coastal and developing states. The FSA also called for ecosystem-based management and use of the precautionary approach (see chapter 7). Fifty-nine countries signed the FSA in 1996. It entered into force in 2001, after 30 states ratified it. At the most recent count, there are 81 ratifications, representing a majority of all eligible states and fishing entities (United Nations 2014).

The fight against IUU fishing was also bolstered by the *1993 FAO Agreement to Promote Compliance with International Conservation and Management Measures by Fishing Vessels on the High Seas and the 1995 Code of Conduct for Responsible Fishing (FAO 1995). Implementation of the Code of Conduct was further codified in the International Plan of Action for the Management of Fishing Capacity (IPOA-FC; FAO 1999) and the International Plan of Action to Prevent, Deter, and Eliminate Illegal, Unreported, and Unregulated Fishing* (IPOA-IUU; FAO 2001). These are all voluntary instruments that set standards and provide advice to countries, fishing entities, and the fishing industry. The IPOA-FC set a goal of "equitable, efficient, and transparent" management of global fishing capacity by 2005. Although several countries, including Japan, China, and some EU members have reduced the number of fishing vessels in their fleets in the past decade, this trend is offset by increased capacity in developing countries. Total marine fishing capacity was 3.23

million vessels in 2010. This is similar to previous years, so overall capacity is relatively stable, but fishing power is increasing due to the technological improvements described in chapter 3 (FAO 2010b, 10–11).

Managing fishing capacity is an important aspect of regulation on the high seas because overfishing within the EEZs drives vessels onto the high seas, contributing to overcapacity in international fisheries (see chapters 2 and 7). This, in turn, creates greater internal and external pressures on RFMOs. Internal pressures include conflicts among member states over allocation of the resource (see below), while external pressures come primarily from the IUU fleets. The IPOA-IUU further recognizes the connection between domestic and international management by reiterating the obligations of coastal states to regulate, monitor, and control vessels flying their flag in keeping with regulations adopted by the appropriate international bodies (i.e., RFMOs). In addition, it calls on coastal and port states to take any actions necessary to prevent or deter IUU fishing and provides ground rules for use of trade-based monitoring and enforcement mechanisms (K. W. Riddle 2006). In spite of the many measures to prevent, deter, and eliminate IUU fishing provided by the IPOA-IUU, there is still some confusion regarding the rights and responsibilities of states in this context, as evidenced by appeals for clarification from the International Tribunal for the Law of the Sea, which was established with the enactment of UNCLOS (ITLOS 2013).

With validation from established international law, RFMOs continue to improve their ability to monitor and enforce regulations. As figure 5.2 shows, trade-based measures are currently in use by nine of the seventeen existing management RFMOs, and eight of those RFMOs also are tracking landings through statistical document programs. Twelve RFMOs currently keep "positive lists" of vessels that are permitted to fish in specific regions at specific times and/or "negative lists" of vessels caught fishing in contravention of management measures. Non-member states whose vessels are frequently found on negative lists may be labeled IUU states and therefore subject to trade restrictive measures. More direct methods of monitoring fishing activities are used by ten of the seventeen RFMOs, including vessel-monitoring systems (VMS) that use global positioning systems to track the location of fishing vessels and onboard observer programs. Employed by the secretariat or by the flag state, onboard observers can verify catch size and location reported by fishers. Finally, NEAFC pioneered the use of port inspection schemes in 1997 and now eight RFMOs allow for search and detention of fishing vessels by the port authorities of non-flag states to improve the implementation of their management measures (DeSombre 2005).

In spite of the proliferation of new monitoring and enforcement mechanisms, IUU fleets are still actively harvesting high-value international fish

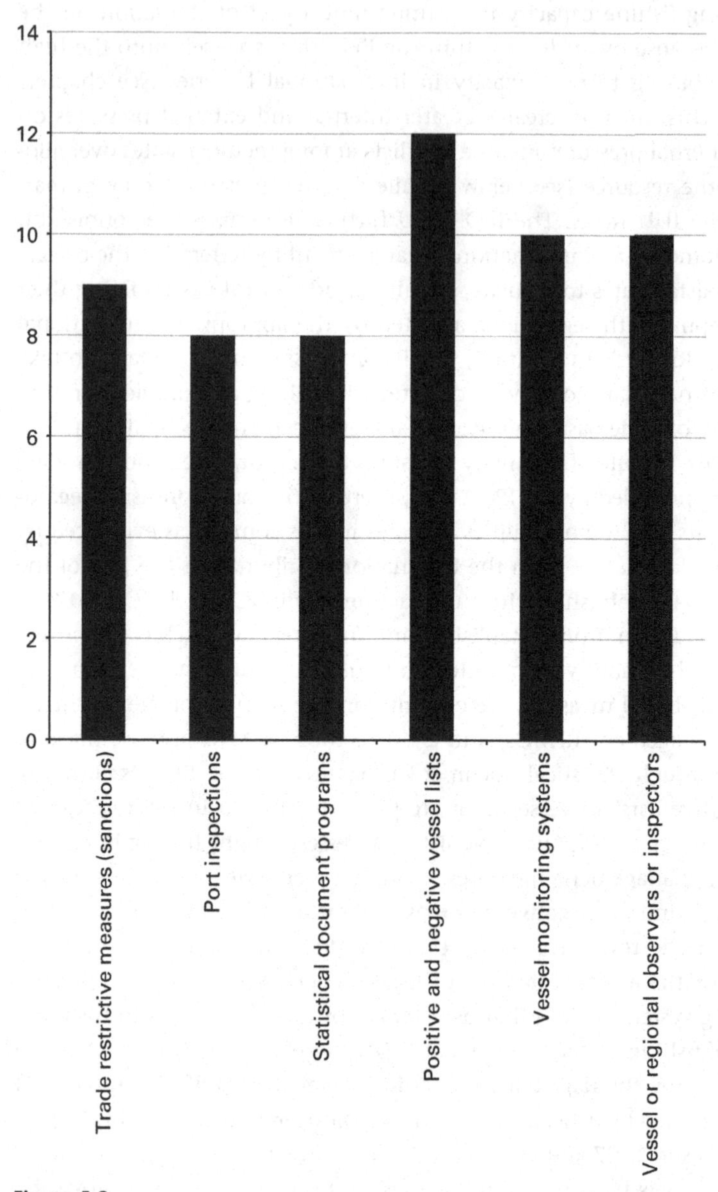

Figure 5.2
Numbers of marine RFMOs employing each monitoring or enforcement mechanism for at least one stock.

stocks. Indeed, these "fish pirates" are highly innovative and find many ways of circumventing new rules and regulations, which is why IUU fishing is still a primary concern for many RFMOs (Lugten 2010; Osterblom et al. 2010). Much like the pirates of the 17th and 18th centuries, IUU fleets thrive by switching their fishing grounds and relying on close connections to land-based organizations in states where management capacity is relatively low. In fact, IUU fleets depend on the benevolence of their respective flag-of-convenience states. Sanctions can be used to drive these fleets to switch flags or to use counterfeit flags, but the only lasting solution to IUU fishing is enforcement on land. Port state inspections are a first step, but, given modern international norms, coastal states around the world will have to choose to roust out IUU fishing interests. Unfortunately, this is not likely to happen in the near future, as economic benefits from IUU fishing increase with growth in demand for specific species, and vested interests are well entrenched in the global political economy.

5.7.2 Defending the Rights of Developing Coastal States

In addition to international monitoring and enforcement mechanisms, increased recognition of the rights of developing coastal states also amplified the willingness of non-members to join RFMOs. Prior to the 1995 Fish Stocks Agreement, many developing coastal states were skeptical of RFMOs, seeing them as a forum for regulatory control by historically dominant coastal or distant-water fishing states. Therefore, they avoided either joining or creating RFMOs prior to the formal protections of the rights of developing coastal states codified in the FSA (Sydnes 2002). The Western and Central Pacific Fisheries Commission, which oversees management of highly migratory species in a vast area, is a good example of this behavior. Much of the WCPFC's territory is within the EEZs of developing island states, which work together on fisheries and many other issues through organizations such as the South Pacific Community (SPC) and the Forum Fisheries Agency (FFA). Before the FSA, SPC member states disagreed with distant-water fishing states over the extent of management by any new RFMO in the region. They wished to retain control of fisheries within their own EEZs, while distant-water states preferred an RFMO that would manage tuna throughout its range. By reinforcing developing coastal state claims, the FSA provided impetus for the creation of the WCPFC and also reassured SPC states that their needs and preferences as coastal developing countries would be respected (Aqorau and Bergin 1998).

With the FSA in place, the trend toward increasing both the number of RFMOs and level of participation in existing regional fisheries bodies escalated further. Coastal and developing states felt that their interests would be respected in fisheries organizations and even began pushing for the creation of new RFMOs, like the Southeast Atlantic Fisheries Organization (SEAFO), to protect straddling stocks from exploitation by new distant-water fleets (Sydnes 2001). Nevertheless, the FSA codified rights of access for historical as well as developing and coastal states, so struggles among these groups continued, with each trying to exclude the other. In the SEAFO process, coastal states sought direct application of the FSA mandate, including both conservation and exclusionary measures, but distant-water states like Japan and the EU opposed the more restrictive policies, such as limited use of the objection procedure. Ultimately the SEAFO agreement, which was the first to be negotiated after the implementation of the FSA, was a compromise between coastal and distant-water states that reflected the practice of existing RFMOs more than the aspirational ideas put forth by the FSA and related texts.

Conflicts over allocation are still one of the largest problems within most RFMOs. As described above, the encroachment of new entrants and IUU fleets—driven largely by the economic responses described in Part I of this book—can alter the incentives to cooperate, thereby improving management, but such responsive governance is not always successful. ICCAT's failure to prevent steep declines in stocks of bluefin tuna is just one prominent example of a common problem (Webster 2011). Either exclusion occurs too early, preventing external pressure for cooperation, or exogenous changes in prices or production technologies widen the profit disconnect, which in turn reduces demand for management at the international level. Thus, responsive governance is a double-edged sword. It can generate conservation when the profit and power disconnects are narrow, but it can also prevent conservation when problem signals are suppressed (Webster 2009, chapter 11).

From the evidence described in this chapter, it is clear that exclusion is a major driver of fisheries policy globally. Defense of fishing territories is an ancient practice that evolved into the modern system of domestic management within 200-mile exclusive economic zones and cooperative international management of transboundary, straddling, and highly migratory species. Though such exclusion can reduce the tragedy of the commons in fisheries by transforming open access resources into club goods, it is not sufficient to prevent either overexploitation or overcapitalization as long as the economic drivers of expansion continue to widen the profit disconnect in fisheries around the world. Furthermore, the structural context is a critical determinant of the nature of exclusion. Resources available, both in terms of the size and composition of fishing fleets (coastal versus distant water) and the power and expense of military action shaped multiple changes in fisheries governance. International norms regarding the use of violence were also critical, including the public monopolization of violence on the high seas and the international norms against violent conflict between states that arose after World War II. Ultimately, action was taken most often when political and economic incentives aligned in favor of exclusion.

6 Expansionary Measures

Like exclusion, expansionary management measures are another expedient response to the tragedy of the commons in fisheries. These measures provide funds, access to capital, and other supports for fishers and the fishing industry. As such, they exacerbate the economics of expansion described in Part I. Generally, expansionary measures are preferred by fishers and related interest groups over restrictive measures that require reductions in catch or fishing effort. Subsidies, or direct and indirect payments to fishers and other industry actors, are the most commonly recognized form of expansionary management; others include government spending for research on and implementation of the improved propagation of species, usually through establishment of hatcheries, the development of new gears, processing methods, marketing techniques, and related technological improvements, and public investment in port infrastructure and shipping facilities. All of these measures increase the profit disconnect by reducing cost of production for fishers or increasing their access to new markets. However, direct subsidies and related institutions also widen the power disconnect by aligning government goals with industry goals and giving decision makers a vested interest in the success of fishing fleets. In fact, direct subsidies create a positive feedback effect known as Ludwig's Ratchet, which amplifies the cycles of expansion described in Part I (Hennessey and Healey 2000). This effect is self-perpetuating and difficult to slow down but it eventually places greater and greater burdens on public coffers, increasing political opposition to expansionary programs.

Nevertheless, governments invest in expansionary management measures because they are often politically expedient. Lobbying for *supportive subsidies* is a common fisher response when profit signals are strong. These expansionary measures allow fishers and other industry interests to continue fishing even when costs of production are high relative to revenues and economic responses are insufficient to counter the core problems in the action cycle. Fisheries entrepreneurs and secondary actors may also lobby for *developmental subsidies*, which are designed to build up new fleets or to expand into new regions or markets. Unlike supportive subsidies, developmental subsidies can precede economic problem signals, although the development of new fisheries may be seen as a way to support overcapitalized fleets. For both types of subsidies, it is important to note that decision makers also use expansionary measures to pursue exogenous goals, including food security, employment, and economic development. As described in chapter 5, states compete over access to shared resources, and provision of subsidies or other supports can help domestic fleets capture more fish than foreign fishers. Thus, the short-run benefits are weighted more heavily in the decision process than the long-run political and economic costs.

This chapter shows how expansionary management measures escalated from the 17th century on as a result of endogenous and exogenous drivers in the global fisheries action cycle. Industrialization in particular drove growth in expansionary management programs, both because the high costs of large vessels necessitated government assistance and because of the increasing economic power of large-scale commercial fishing and processing corporations. As will be discussed in section 6.1, these interactions created feedback loops between the action cycle and governance in the structural context that reinforced the economics of expansion. In other words, responses to problem signals in the action cycle caused changes in the governance context that reinforced power structures in an iterative, amplifying process. Section 6.2 covers the popularity of government-funded hatchery programs in the late 18th and early 19th century. Although hatcheries are no longer considered viable in most marine fisheries, this form of expansionary management was an expedient response at the time and a crucial factor in the growth of fisheries management institutions in all major fishing countries of the period. By the 20th century, issues of food security and economic development were globally significant, driving the increasing use of developmental subsidies. This began in historically dominant fishing states and then spread to coastal developing states. Section 6.3 focuses on the effects of Ludwig's Ratchet on modern global fisheries management, including its wide-ranging impact on both the profit disconnect and the power disconnect. Section 6.4 explores recent transitions away from expansionary management in response to the high cost of supporting an increasing array of unprofitable fleets and to exogenous shifts in international institutions related to free trade and monetary policy.

6.1 Early Subsidies

Early examples of both supportive and developing subsidies show that the economic context was a major determinant of the impacts of expansionary management measures. When the economic context was not favorable, both types of subsidies had limited effects and increases in fishing effort lasted only as long as the subsidy itself, as shown in section 6.1.1. However, when fleets were already competitive, subsidies could spark rapid increases in fishing effort and generate positive feedbacks between the economic growth of fleets and the political power of the fishing industry, as described in section 6.1.2. Infrastructure to accommodate industrialized fleets was also an important support for fishing opperations—or limit in cases where infrastructure investment was not made—as explained in section 6.1.3. Lastly, governments played an important role in the establishment of domestic and international markets for the increasing array of fish products produced in industrializing fisheries, as laid out in section 6.1.4.

6.1.1 Unfavorable Economic Conditions

British supports for the herring industry in Scotland provide an interesting example in which subsidies largely failed until exogenous shifts improved the economic competitiveness of their fleets. The gap between the British and Dutch fleets was quite wide at the beginning of the 17th century. Scottish herring fishing was carried out by small sailboats or rowboats called dories. Fishers often made their own boats and nets and had few other expenses. They also had little savings or access to capital. In contrast, the Dutch fishery engaged large vessels called busses that could catch and process 412 barrels of pickled herring in about two months. A Dutch herring buss cost £500 (more than US$124,000 in 2010 dollars), and operating costs and depreciation for a two-month voyage would require an additional £255 (more than US$61,000 in 2010 dollars). Even though profits in the Dutch fishery were high (£145/2 month trip; over US$36,000 in 2010 dollars), Scottish fishers could not afford this level of capital outlay (J. T. Jenkins1920, 106–108).[1] The Crown created Royal Fishing Companies to help raise the needed funds. Usually a company started out with some contributions from the Crown and other wealthy investors, or "society members." In addition to investing in larger vessels and bigger gear, the Royal Fishing Companies hired Danish skippers and crew to provide expertise on the Dutch style of fishing, since no Dutch skippers or crew were willing to work for the British (109).

Note that the entrepreneurs in this case were not Scottish fishers but rather British businessmen and gentry who saw the herring fishery as a sound investment backed by the king. As such, they were insulated from economic problem signals and yet had considerable influence over government policy, so this subsidy program generated a large power disconnect for the British fleets of the time. In fact, all of the Royal Fishing Companies failed, insofar as they quickly went bankrupt without the direct attention of—and subsidies from—the monarch; yet the practice continued for more than 100 years. Overexploitation was not a problem at the time, but company fleets incurred losses for other reasons, including predation by privateers, military conflicts with the Dutch and other European powers, mismanagement by company owners, and changing economic conditions. The last company was established in 1750. In spite of one early success, investment equal to more than a billion US dollars in modern currency, and additional subsidies described below, the "Society of the Free British Fishery" went bankrupt by 1772 (J. T. Jenkins 1920, 113). Here again, external factors were key. In this case, the outbreak of war altered the economic conditions faced by the Society, which still had to contend with substantive competition from well-established Dutch brands. At the same time, the power disconnect was also at play: Society members clearly treated this project as a straightforward investment, with little appreciation for the economic and ecological vagaries of fishing as a business.

Bounties were the second major type of subsidy provided to the Scottish herring fishery by the British government. They started in 1718 and continued until about 1830, with a few periods of suspension and much fluctuation in size. The first bounties were tied directly to the Scottish brand for herring, which had been established by the government to control the quality of exported pickled herring so that it could rival the Dutch brand in lucrative markets in Germany and Russia (see chapter 4). Bounties were set at about 2 shillings and 8 pence (about US$16 in 2010 dollars) per barrel of herring exported. From 1750 on the bounty was changed frequently. The effect on the scale of the Scottish herring fleet was clear. When the bounty was applied, the number of vessels increased. When it was removed, the number of vessels declined. For instance, in 1757 the bounty was increased to 50 shillings per ton (over US$400 in 2010 dollars). Subsiquently, the number of herring busses in the Scottish fleet increased from 13 in 1760 to 261 in 1766. Subsidies were then suspended and the number of vessels declined to 19 by 1770. Such cycles occurred regularly until 1776, when wars with the Netherlands and the United States again caused shortages in labor and other inputs in Britain. Privateering and losses from the military conflicts also rose again in this period (J. T. Jenkins 1920, 117–118).

These subsidies cycles are important because they show that developmental subsidies are successful only when economic conditions favor growth. When subsidy programs resumed in the early 19th century, the effect was quite different. The Scottish fleet expanded production prodigiously, from 90,000 barrels in 1809 to 444,000 barrels in 1830. There was a brief decline in production once the subsidy was removed, but it was back up again by 1834. The Scottish government continued the branding program without direct bounties because the brand had a good reputation that allowed Scottish producers to charge more for their product. Jenkins (1920) argues that Adam Smith was wrong and that subsidies and other government supports did help to jump start the Scottish herring industry. However, he ignores the impact of the Industrial Revolution and the shift in herring markets from quality competition to price competition (Duff 1884; Green 1884; Phelps 1818). These exogenous economic changes greatly altered the effect of expansionary management measures on British fleets.

6.1.2 Favorable Economic Conditions

In contrast to the British case described above, subsidies were not important in early development of the Dutch herring fleets. Dutch processors and merchants were able to pool their capital to invest in fleet improvements without government assistance. Their fleets were also highly competitive in the period, and so did not need continued support, although some would argue that Dutch actions in defense of their monopoly control over trade in Europe secured the dominance of their fleet. All this changed in the late 1700s, when the industry suffered large losses due to warfare with Britain, France, and Spain, high costs for protection from enemy fleets and privateers, and the loss of their monopoly over European markets. In response to industry lobbying, the Dutch government started to provide supportive subsidies to keep the industry afloat in 1775. Nevertheless, continued security issues and loss of market share negated the expansionary impact of the subsidy and the Dutch fleet continued to decline into the early 1800s (J. T. Jenkins 1920, 138). Beaujon (1884) argues that overly restrictive regulation of the fishery hobbled the Dutch herring industry during the Industrial Revolution, forcing it to continue to make high-priced, high-quality product even though the markets were shifting to low-priced, moderate-quality product. He points out that the industry survived and thrived in many earlier periods of maritime conflict, even subsidizing the Dutch government during some of its wars.

Many other European countries also subsidized their fisheries in the late 1800s, with varied success. In 1872, the German government tried to develop their own herring fishery using subsidies and protective tariffs, but landings were sporadic and growth was slow until the introduction of steam-powered vessels in 1898 (J. T. Jenkins 1920, 141). In the early 1900s, the Norwegian government began to raise funds to develop more advanced fleets. By 1914 Norway had a large capital stock of £118,125 (more than US$630,000 in 2010 dollars), which it used to purchase modern steam-powered vessels that greatly expanded the capacity and improved the competitiveness of its fleets (265). In Portugal, where incursions by foreign fleets were widely viewed as detrimental to domestic fishers, the government created a bifurcated tax structure to discourage use of its port facilities by foreign fleets (266). Similarly, in Britain fishers were often spared from paying duties on salt, a necessity for preserving fish for export to foreign markets (Phelps 1818, 34).[2] These practices were widespread at the time and continued well into the 20th century. Indeed, many of these early subsidies were a just one part of the larger mercantilist approach to government policy, which favored protection of domestic industries over free trade. This is another example of how governance, as part of the structural context, shapes the action cycle.

In what is now the United States, subsidies predate independence. Individual colonies started paying bounties to fishers in the cod and whale fisheries as early as 1731 (Starbuck 1878). The first record of federal subsidy programs for fisheries dates to 1789, when a bounty of five cents (US$1.25 in 2010 dollars) per standard unit[3] was placed on all dried, salted, and pickled fish exported to any foreign country. Officially, this bounty was designed to mitigate the effects of a tax on imported salt. The same act also placed duties on imported fish products to protect US producers at home (US Senate 1840, 3; Williamson 2014[4]). Two key forces shaped this act and subsequent subsidies. First, the US government needed money and preferred to tax goods rather than individuals. Second, fishers formed powerful lobbying organizations to convince decision makers to redress their "grievances and burdens." Duties on salt and the perceived unfairness of subsidies to European fleets were at the top of this lobbying agenda (10). Both of these grievances were linked to profit signals, but problems were generated by the exogenous political system rather than the core common pool resource dynamic in the action cycle.

These matters were so important in the late 1700s that Thomas Jefferson, the first Secretary of State for the United States, investigated the fishers' complaints and compiled an extensive report on the cod and whale fisheries of the United States. He concluded that the federal government should continue to offset domestic taxes on salt and other inputs in the production of fishery products, should effectively close off US markets to foreign fish producers through imposition of duties, and should work to remove barriers to entry for US fish products in foreign markets (Jefferson and Sabine [1791] 2011, 2). Jefferson noted that US fishers had many competitive advantages over the Europeans, including proximity to the fishing grounds, which lowered both fixed costs and risks of wreckage, the lower costs of vessels, provisions, and casks in the US, and the superior skills of US fishers. Thus, underlying economic conditions favored US fishers but the prevalence of protectionism on all sides prevented the growth of US fleets (4).

In 1792 Congress acted on Jefferson's advice, paying fishers a bounty to offset the cost of the salt tax and closing US markets to foreign competition. In 1799, lobbying by fishers prevailed and bounties on fishery products were increased even though salt taxes had not gone up in years (US Senate 1840, 12). Except for a short period of repeal from 1807 to 1813, the salt tax and related bounties on fisheries remained in effect throughout the early years of the 19th century (18–19). It is interesting to note that during this time the fisheries of the United States were economically successful. For instance, in 1837 the fisheries for whale, cod, and mackerel in Massachusetts alone had a net export value of more than US$7.5 million (US$176 million in 2010 dollars) and employed more than 20,000 individuals. The capital invested in these fisheries exceeded US$12.5 million (US$258 million in 2010 dollars; 3). Returns for major fishing towns were even higher than the state average, and the fisheries of the United States as a whole were prosperous in this period (54). As Jefferson predicted, highly competitive US fleets grew substantially with ready access to markets for their products. Thus, in the US as elsewhere, favorable economic conditions amplified the expansionary effects of subsidies while building up the political power of industrialized fisheries.

6.1.3 Infrastructure and Industrialization

Government subsidies for infrastructure improvements were also important for the success of fleets in the United States and other historically dominant fishing countries. Because there was no concomitant expansion of access to capital, US bounties increased the number of fishing vessels but did not incentivize much innovation at first. As Goode (1884) points out, American fishing methods did not change much until the second half of the 19th century. As described in detail in Part I, US cod fishers used the old schooner technology well into the late 1800s, and US whalers continued to use sailing vessels rather than steam-powered vessels into the early 1900s. US fleets changed around the turn of the century for several reasons. Some factors were non-governmental, such as the development of new processing technologies, changes in consumer preferences, and the rise of the United States as a global economic power. However, Goode also explains that these economic and technological changes would not have had the same impact without accompanying government investment in infrastructure, including development of modern port facilities, storm warnings, railroad lines, and general transportation infrastructure. These exogenous shifts expanded and modernized the resources available in the structural context. In response to demands from within the industry, the US Fish Commission also actively worked to expand available resources, finding new fishing grounds, developing fisheries for new species, introducing new fishing technologies, and marketing new fish products (Allard 1978, 302–316). Whether endogenous or exogenous, these indirect subsidies increased the supply of ports that could accommodate large-scale fishing vessels, reduced many of the risks associated with fishing operations, and opened up new markets for fisheries products, increasing overall demand.

Governments in most of the other historically dominant fishing countries also provided indirect subsidies to fishers through investment in infrastructure. Of course, entire economies benefited from these developments, since many other industries also relied on major land and maritime transportation infrastructure. As in the United States, most of these subsidies were exogenous at the national level. However, in some places, such as the town of Grimsby in Britain, local governments heavily subsidized port improvements to encourage the growth of industrial fishing fleets. At times, local organizations representing sectors of the fishing industry also successfully lobbied state or national governments for additional assistance (Robinson 1996). On the other hand, fishery expansion was stymied in areas where governments did not pursue infrastructure modernization. For instance, in the 19th century the Dutch government was not willing or able to pay for the infrastructure needed to accommodate large-scale, steam-powered trawlers in Dutch ports. This absence of a key indirect subsidy limited the resources available in the structural context and contributed to the decline of the Dutch herring fishery (Beaujon 1884).

China provides another contrasting example. In this case, exogenous port improvements and imports of Western science and technology began in the late 1800s, but were not available to fishers until the early 1900s. The Chinese government funded the construction of hundreds of lighthouses and improvements in many port facilities to reduce risks to trade and to facilitate use of new technologies. The government also subsidized the importation of Western vessels starting in the 1870s, helping to build a fleet of 29 steam-powered ships with a combined capacity of more than 20,000 tons by 1882. However, these vessels were owned by the Chinese Merchant's Steam Navigation Company and were used for trade rather than for fishing (J. D. Campbell 1884, 8). In fact, Chinese fishers were not allowed to build large fishing vessels until early in the 20th century, due to concerns about links with pirates (Muscolino 2009, 18). Thus, while the construction of lighthouses may have made fishing safer, many of the other general funds spent to develop Chinese ports did little to benefit fishers directly.

Because of its policy of national seclusion, Japan started modernizing much later than other historically dominant fishing countries. When it did open up to the outside world, the Japanese government invested heavily in the development of industrial fishing fleets. In the late 19th century, the Meiji government actively supported the westernization of national infrastructure and domestic industries, including capture and processing of fish and other marine species for domestic consumption and export (Makino 2011, 24). Subsidies were a major part of this plan to "catch up" to industrialized countries like the United States and Britain, and fisheries subsidies were designed specifically to reduce crowding and increase profitability for coastal fisheries while also increasing profits for distant-water fleets and contributing to domestic food security and international trade with Western countries. Given that all actors benefited either politically or economically, government-supported modernization was very popular. However, the benefits of the first subsidy program, initiated in 1897, were limited. Only steam-powered vessels greater than 50 tons and sailing vessels greater than 30 tons received subsidies. As it turned out, this measure mainly encouraged growth in the sealing industry, which was a major cause of tension between Japan, the United States, and Russia at the time. Seeing that the expected economic expansion had not materialized, in 1905 the Diet extended subsidies to sailing vessels of 20 tons or more. Combined with the spectacular and widely publicized catch by Japan's first gasoline-powered 25-ton sailing vessel in 1906, this catalyzed the growth of industrialized pelagic fishing fleets (Kitahara 1910; Makino 2011, 91).

As the British had done centuries earlier, the Japanese government imported expertise from abroad during this period and invested in several "training" or "experimental" vessels. For instance, the government-owned *Unyo-maru* was a 448-ton, steel-built, 130-foot-long fishing vessel with triple expansion engines of 300 horsepower each. It also had a cold storage room, mechanized gear deployment and retrieval systems, and many other amenities of the day. The *Unyo-maru* was used to experiment with several different types of fishing gear and to train Japanese fishers to use these new technologies. The smaller but more powerful *Hayatori-maru* was used to explore new fishing grounds and to inspect or guard fishing vessels outside of the Japanese territorial sea. It was equipped with a searchlight, a wireless telegraph system with a range of 200 miles, and a 1,000-fathom trawl winch, which allowed trawling to expand beyond Japan's exceptionally short continental shelf (Suisankyoku 1915, 6–10). In addition, the Japanese government encouraged the importation and then the domestic production of European-style trawlers and Norwegian-style whaling vessels. It also brought in Norwegian harpoon gunners to train Japanese whalers in that method (11, 13).

In the 1880s, Japan was home to a very large fleet of small artisanal fishing vessels and a few large-scale, labor-intensive commercial fisheries (Okoshi 1884). By 1910, Japan had a growing industrial fishing fleet of 124 steam-powered vessels, 669 European-style sailing vessels, and 1,674 motor boats. They also processed much of their catch domestically and exported some of it to China, Europe, and the Americas (Kitahara 1910, 378-379). By the beginning of World War II, Japan was the largest producer of fresh and processed fish products in the world (US Congress 1947a, 12). Contemporary observers note support from both the government and the Japanese public as a primary driver of this growth because the government provided large amounts of resources for modernization and the public literally "ate up" the resulting increase in the supply of fish (Kitahara 1910). Nevertheless, as with the British herring fisheries in the 18th century, the success of Japan's developmental subsidies depended heavily on their competitiveness in the context of global markets. In particular, relatively low labor costs gave Japan a competitive edge over US and European producers, where earlier industrialization had resulted in higher per capita income in this period (Finley 2011).

6.1.4 Markets and Trade

Subsidies were also provided to reduce transaction costs in the fishing industry and to help develop and market new preservation technologies. The exogenous political goals of food security and economic development were primary drivers of these interventions. As the size of harvests increased, so did the number of fishers, processors, and consumers. This generated high transaction costs and also facilitated fraud, as discussed in chapter 4. To solve these problems, governments often worked with local entrepreneurs to establish markets that centralized the buying and selling of fresh and processed fish, simplifying transactions and improving health and safety regulation. For example, Dutch towns were granted exclusive rights to market herring by the national government in the early 14th century (Beaujon 1884). The Dutch also instituted one of the first centralized fish auction houses in 1780, to prevent corruption and collusion in the sale of fish products (Teuteberg 2009). The British grappled with similar problems around the same time, and passed many different types of laws to try to ensure equitable pricing and quality control at the Billingsgate market in London (Talfourd 1884). The Fulton Fish Market in New York, the Tsukiji Market in Tokyo, the Boston Fish Exchange, and the Hamburg Fish Market in Germany are other examples of historic fish markets established and regulated to cope with the increased scale of the fish trade (Bestor 2004; Teuteberg 2009; US Congress 1947a).

Subsidies for new processing techniques included the British bounties for branded barrels of pickled herring and relief from US salt taxes described above. Somewhat later, France strongly supported the development and dissemination of canning technologies to improve food security and feed Napoleon's great armies. In 1810 the French government awarded the inventor of canning, Nicolas Appert, a large and well-publicized cash prize of 12,000 francs (about US$41,500 in 2010 dollars; Officer 2014). The prize came with the condition that Appert would not patent his invention and would freely share this new technology with the world (Drouard 2009, 179). The French government also supported canning through large purchases of canned products, mainly to feed the armed forces, and through public education about the safety and desirability of sardines and other canned goods (185). Here, subsidies increased the resources available in the structural context, benefiting canners directly, but also allowed for increased demand for raw materials, which mitigated price signals for fishers.

Ongoing government support via research and development was key to the growth of many fish-processing industries around the world, although the literature tends to emphasize the importance of entrepreneurship instead. Many of the scientific and management bodies established by governments in historically dominant fishing countries primarily explored methods for increasing fisheries production and demand for fish products, rather than methods for restraining fishing effort. For example, the US Fish Commission began helping the industry to develop new methods for freezing fish for transport in 1879. According to Hobart (1996), this was the first of many commission activities that focused on increasing the food supply through identifying new fisheries, developing new processing methods, and helping to market new fisheries products to consumers. In the 1930s, the US Bureau of Fisheries began publishing *Fishery Market News*, a monthly bulletin that was intended to help fishers with their marketing and promotion efforts (17–18). Milazzo (1998) and many subsequent authors generally classify this type of research and development as a "beneficial" subsidy, one that promotes fishery conservation and management. Sumaila et al. (2010b, 204) assert that fisheries R&D programs are beneficial because they are "geared towards improving methods for fish catching and processing, and other strategies that enhance the fishery resource base through scientific and technological breakthroughs." However, it is clear that some fisheries research and development can increase capacity in fisheries substantially above sustainable levels. In fact, as discussed in subsection 6.3.3, the expansionary effects of R&D increased in the 20th century as more countries imitated the Japanese approach to fishery development.

6.2 Fish Culture

In addition to helping fishers and secondary actors to expand fishing effort and keep prices high relative to costs, for a time governments also hoped to bolster fisheries by removing biophysical limits on fish production through aquaculture and the "seeding" of wild stocks. If successful, the practice of culturing fish from fertilized eggs and then depositing large numbers of juveniles back into the wild would increase stock size and growth, expanding the biological resource base and reducing the profit disconnect (all else equal). Though ultimately unsuccessful for marine fisheries, this practice was encouraged in the United States and Europe during the late 1800s and then spread to other countries in the early 1900s, significantly influencing the evolution of many new fisheries management bodies. Hatchery programs were wildly popular with fishers, secondary industry actors, and the public, which made them politically expedient and encouraged early response by decision makers. This, in turn, increased investment in management organizations during a critical period in their development. Indeed, though hatchery programs failed for the vast majority of marine fisheries, they had a lasting effect on the structural context by fostering formal fisheries management bodies at the domestic level.

6.2.1 Mariculture

Although many see aquaculture as an alternative to capture fisheries, it is not really an endogenous response in the fishery action cycle. Historically, aquaculture and similar technologies were not often used by commercial fishers but rather were pursued by private entrepreneurs who sought government support to develop large-scale cultivation strategies. Some of the entrepreneurs were also recreational fishers who wanted to restock sport fish fisheries. Others were scientists like Victor Coste, who happened to be a close friend of the French court and therefore was able to garner considerable funding for his work from the empire (Le Gal 2009). Although criticized for not sharing his techniques, Coste popularized the idea of fish culture and paved the way for substantial government investment in this expansionary management measure (Allard 1978; Maitland and Day 1884).

Oyster fisheries were one of the first to be severely overexploited in Europe, largely because of the development of powerful dredges that increased harvests substantially but also destroyed the oyster beds (Kirby 2004). By the middle of the 19th century, economic problem signals were strong in these coastal fisheries and oyster fishers, who had little recourse to any of the economic responses discussed in Part I, pressured their governments for assistance. Several governments responded with restrictive measures to reduce competition and allow oyster populations to rebuild, but these were difficult to enforce outside of the 3-mile territorial sea and could not undo the damage already done. In 1870, the government of the Netherlands experimented with a new management plan for oysters. It closed the Yerseke bed (7,720 acres) to capture production and leased plots of different sizes (12–150 acres) to private individuals for oyster culture. The experiment was economically successful. The rent charged for the entire area increased more than tenfold from the first lease in 1870 to the second in 1885 (Hubrecht 1884, 88). The experiment was also politically successful. The Dutch government rapidly expanded the program to additional oyster beds along other portions of the coast (97). It is interesting that a private entrepreneur first proposed this plan to the government of the Netherlands in 1867, and that subsequently the government invested in scientific studies by researchers at the Dutch Zoological Station to improve techniques used in oyster culture (87–89). In this case political response was triggered by problem signals from fishers but was shaped by relatively new actors in the structural context.

Similar processes influenced government involvement in oyster cultivation in Canada, Norway, and Australia during the late 1800s (Shea 1884; Adams 1884b; Ramsay 1884). Oyster culture was also practiced widely in the United States and, indeed, the oyster industry was by far the largest fishing-related industry in the US in the late 1800s (Goode 1884). Oyster culture took hold in some areas of the US, but in others harvesters of wild oyster stocks strongly opposed the leasing of oyster beds for cultivation. These individuals perceived this privatization of the resource as an encroachment on their long-standing rights of access. They feared that the resultant enclosure of the oyster beds would benefit those with sufficient wealth to purchase leases from the government at the expense of those with little financial capital. In such situations, governments (local, state, and federal) stood as arbiters of conflicts between fishers and culturists (US Congress 1947a). The result was a variegated pattern of closed and open oyster beds, depending on how much leverage each group held over local decision makers. Culturists were usually wealthy and had access to decision makers, but oyster fishers were numerous in some areas and wielded the power of their votes.

6.2.2 Hatcheries

Private entrepreneurs also drove the development of hatcheries for salmon and other fish species in the second half of the 19th century. The US case is one of the best examples of the feedback effect between fish culture and fisheries management organizations. Early US culturists of the 1850s, like Theodatus Garlick, H. A. Ackley, and J. C. Comstock, wanted to restock New England's overexploited rivers and streams with desirable food fish. Some of their work was undertaken with funding from state governments, but much was self-funded. Many of these early businesses failed, but Stephan H. Ainsworth managed to develop a market for his artificially raised brook trout in 1859. After his success, so many entrepreneurs entered the industry that prices for brook trout dropped precipitously in the early 1860s (Allard 1978, 116–117). In this way, aquaculture resembled fisheries, even though it is a private industry with closed rather than open access to the fish stocks. While the commercial success of brook trout aquaculture encouraged increasing investment in the industry, the biological success of this artificial propagation led some state-level fisheries managers to invest in artificial seeding of local rivers, which were already heavily overexploited, with few remaining commercial fisheries. This process started in Vermont, New Hampshire, and Massachusetts, but the rest of New England soon followed suit. Restocking rivers became a common alternative to other policies, such as constructing fishways around dams and limiting fishing effort. The four major species propagated in New England rivers in the 1860s were brook trout, black bass, shad, and salmon (Allard 1978, 117–118). Recreational fishing interests also lobbied for the restocking of popular sport fish species like shad and salmon (121).

Hatchery programs had a profound influence on the evolution of formal fisheries management institutions in the US and elsewhere. Because free market and US-state level production of salmon fry was insufficient to meet demand, US culturists started lobbying for the creation of federal hatcheries through the US Fish Commission in the 1870s (Allard 1978, 122–123). With support from US Fish Commissioner Spencer Fullerton Baird, several fish commissioners from states in New England, the American Fish Culturist Association, powerful recreational fishers, and a few US senators and representatives, the US Congress approved funds of US$15,000 (US$276,000 in 2010 dollars) for the establishment of hatcheries on the Atlantic and Pacific coasts in June 1872 (123–131).[5] The federal funds disbursed for hatcheries in 1872 amounted to three times the US$5,000 (> US$86,000 in 2010 dollars) that Congress approved for the creation of the US Fish Commission in 1870. While the government continued to fund the commission's basic scientific research, the work on hatcheries was much more popular with the public and continued to receive higher levels of federal and state funding. By 1877, Congress had approved US$50,000 (> US$1 million in 2010 dollars) for fish culture, and the total amount spent in various state programs was approaching US$100,000. This gap widened with the increasing popularity of culture programs, and by 1887 US$200,000 of the total US Fish Commission budget of US$268,000 was spent on fish culture (US$4.4 million of US$6.3 million in 2010 dollars; Allard 1978, 148, 262; Williamson 2014). Compared to restrictive responses (see chapter 7), this expansionary response occurred very early in the action cycle and ratcheted up quickly, largely due to the expedience of the approach and the alignment of public pressures, pressures from fishing communities, and decision makers' goals.

Political conditions were so favorable that increases in funding were approved in spite of several substantial failures in the fish culture program. Although seeding of rivers and streams with fresh-water fish was successful throughout the country, attempts to revive and redistribute stocks of anadromous fish that move between fresh water and marine areas were less effective. The Fish Commission released hundreds of millions of salmon and shad fry in eastern, southern (shad only), and western rivers, yet very few overfished populations returned and an even smaller number of stocks established themselves in new rivers. Indeed, attempts to transplant shad into southern rivers failed completely (Allard 1978). Shad catches increased in some eastern rivers, but there is no evidence to show that this was caused by the seeding itself. Salmon seeding on the East Coast was almost stopped completely due to lack of results, but in 1878 the return of 600 adults to the Connecticut and smaller rivers reinvigorated the program (265–267). Ultimately, the only successful program for the culture of anadromous fish was one conducted in western rivers where fisheries were not severely depleted and river systems were not as developed or polluted as in the east (268).

This suggests that habitat and ecosystem function are important resources in the structural context for hatcheries programs, as they are for fisheries more broadly.

In spite of these failures, or perhaps due to those limited successes, hatchery programs spread rapidly in the United States and in the rest of the world. The US Fish Commission's restocking programs were strongly supported by the American public and US sport fishers for many decades. Around the turn of the century, the Commission built hatcheries for marine species, including cod, mackerel, flounder, lobster, and scup (Allard 1978; Goode 1884; Smith 1994). The Japanese, British, Norwegian, and Canadian governments all gained public approval by building up national salmon hatcheries and "restocking" overfished salmon runs within their borders (Kitahara 1910; Fryer 1884; Hérubel 1912). Other countries engaged in cultivation of different marine and freshwater species. Italy already had a long history of aquaculture in coastal lagoons called *valle* (Hérubel 1912). In Spain, inland seas called *albuferas* were used for aquaculture in the late 1800s (Sola 1884, 360). In 1904, an International Commission of Enquiry experimented with "transplanting" young plaice and other groundfish species from British, Dutch, and Danish coastal areas to offshore banks in hopes of repopulating those once rich fishing grounds (Hérubel 1912, 131). Salmon were even transplanted from the US to New Zealand rivers in the early 1900s (Gilbert 1912, 3).

All of this government support reflected concern about declining wild fisheries, the popularity of hatcheries with the public, and the assurance of some scientists that fish culture could provide food security. In his inaugural address to the 1883 World Fisheries Exhibition, Thomas Henry Huxley, president of the Royal Society, expressed a sentiment that was common throughout the literature of the time:

> Thus, in dealing with this kind of exhaustible fishery [salmon], the principle of the measures by which we may reasonably expect to prevent exhaustion is plain enough… If the stock of a river is to be kept up, it must be treated upon just the same principles as the stock of a sheep farm. (Huxley 1884, 12)

This statement clearly reflects the scientific optimism and general positivist attitudes that shaped the structural context of the time. People had great faith in technological innovation and believed that humans could control and shape natural systems to their own ends. This belief was derived in part from recent advances in agriculture, and many authors of the period insisted that what could be accomplished on land could also be accomplished at sea. On the other hand, there was also a continuing belief that humans could not significantly impact the "major sea fisheries," such as those for herring, cod, and mackerel (Huxley 1884, 14). This view was held in spite of severe local declines in several of the major sea fisheries. More important, this optimism shaped much of the early development of fisheries management institutions and continues to this day.

6.3 Ludwig's Ratchet

In the second half of the 20th century, production by historically dominant fishing countries peaked and then declined as a result of strong, unavoidable economic signals associated with the overfishing of traditional fishing grounds, the legal enclosure of fisheries resources in Exclusive Economic Zones, and increased competition from growing fleets flagged in developing countries. Much of this transformation in global fisheries was due to economic responses as described in Part I, but there was also a strong political response favoring expansionary management measures. In fact, the growth of subsidy programs parallels the economic transformation of global fisheries, with cycles of expansion in magnitude and geographic scope during the 20th century. Furthermore, the creative destruction documented in Part I was at once fueled by developmental subsidies (on the upswing) and dampened by supportive subsidies (on the downswing). Food security and economic development remained important drivers of subsidies, but geopolitical concerns and lobbying by domestic fishers were increasingly important in this period.

Regarding supportive subsidies, Ludwig, Hilborn, and Walters (1993) describe the process of ratcheting up fishing effort in the exploitation of natural resources generally, though many of their examples are drawn from fisheries management specifically. They show how the combination of economic and political power can be used both to delay restrictive management measures and to obtain subsidies to support innovation and exploration by existing fleets as profits dissipate under open access. Hennessey and Healy (2000) dub this process "Ludwig's Ratchet" and show how it applies to the collapse in the New England groundfish fishery. In this section, I provide additional supporting evidence for Ludwig's Ratchet over a wide range of post–World War I cases. I also expand the concept to include developmental subsidies and related issues associated with economic expansion, starting with the interwar years and concluding with the growth of developing country fleets in the 1960s and the 1970s.

6.3.1 War and Subsidies

A new wave of fisheries subsidies started in historically dominant fishing countries after World War I. Many of these measures were implemented in order to provide cheap food and stabilize employment during the Great

Depression and World War II (Cox and Sumalia 2010). Some developing countries also subsidized their fisheries during the interwar years. Taiwan, or really the Japanese colonial rulers of Taiwan at the time, invested heavily in the development of large, modern fishing ports to accommodate growing fleets of industrial fishing vessels. These included deep sea trawlers, tuna longliners, and other distant-water vessels (Ministry of Economic Affairs1958; Suisankyoku 1915). Japan also subsidized the development of industrial fishing fleets in its other colonies, including Korea, Manchuria, and parts of northeast China (Muscolino 2009). By 1915, over 200,000 fishers and more than 17,000 vessels were employed in these colonial fisheries. Most of these vessels were traditional junks, but some were Japanese-owned industrial ships that used colonial bases to target fish stocks in the East China Sea. This was another facet of the Japanese government's policy to encourage distant-water fishing and ease economic pressure on domestic fisheries (Suisankyoku 1915).

Japan had a large impact on Chinese fisheries management during this period. Several Japanese-trained fisheries scientists in China lobbied for the "rational and efficient" development of domestic fishing fleets. This included calls for direct subsidies to industrialize all Chinese fishing fleets in order to expand production, reduce conflict over traditional fishing grounds, and stake China's claim to sea power in strategic regions. The Chinese government was sympathetic to these goals, even passing legislation calling for the modernization of their domestic fleets, but because funds were limited it was not possible to transform their fleets following the Japanese methods (Muscolino 2009). Indeed, some Chinese ports were dominated by Japanese fishing interests, to the detriment of local fishers and the benefit of local tax collectors. This created a major power disconnect in the region, since Japanese fleets were well insulated from problem signals by government support, low costs of production, and high mobility. Nevertheless, these ports were held up by domestic Chinese advocates as examples of great success in fisheries management, and were used to convince local leaders that industrialized fisheries could lead to greater government revenues (84).

With its limited resources, the government of the Republic of China (1912–1949) did respond to this political pressure somewhat. It established several fishery education institutes to train fishers in modern methods in the 1920s, and also set up a Fishery Experiment Station on Shengshan Island in 1930. Removal of restrictive regulations on the size of fishing vessels and related anti-piracy measures in this period allowed Chinese fishers to take advantage of more modern port facilities, the expansion of demand for fish, and the increasing availability of credit during the period (Muscolino 2009, 69). In 1931, the government passed a law exempting domestic fleets from taxes on fish and also including several measures designed to reduce direct competition from Japanese fleets in the East China Sea. It also established a central fish market in Shanghai in 1933, to improve marketing of fish and collection of taxes on fish products. This market boasted large-scale refrigeration and storage facilities, which helped to smooth out gluts and shortages in capture production. Starting in 1937, the war with Japan severely reduced fishing effort by Chinese fleets. Many boats were destroyed and many fishers were conscripted by both sides (176).

World War II virtually eliminated fishing in many countries, including China, Korea, Japan and various European states, but it reinvigorated US fisheries subsidies after a brief period of fiscal austerity toward the end of the 1930s. In particular, the US government promoted salmon production in the northern Pacific to make up for reductions in other fisheries. Large fishing vessels were often co-opted for the war effort, particularly from Southern California and New England. Labor shortages and the fear of maritime attacks also reduced production in many fisheries. Wartime fisheries subsidies were managed by the newly created Office of the Coordinator of Fisheries within the Department of the Interior, which was directed by Secretary of the Interior Harold Ickes. He firmly believed that efficient production in remaining fisheries was necessary for the war effort, so he approved the temporary removal of restrictions on effort in the salmon fishery and subsidized expansion of the salmon fleet (Finley 2011; Muscolino 2009; US Congress 1940). This sped up the action cycle for the fishery but also changed the structural context for US fisheries more broadly by creating a new department that explicitly placed food security above conservation or scientific research as a regulatory goal. This shift also widened the power disconnect in the United States, as it aligned federal policy with the interests of large fishing and processing corporations that were less sensitive to economic problem signals than small coastal fishers. The direct result was a focus on subsidies for large commercial fishers, which had the indirect effect of heightening cycles of overexploitation as explained in Chapter 7.

Fleets from other countries also benefited from US spending during the war. The Soviets took advantage of the US Lend-Lease Act to refurbish 12 of their large-scale commercial fishing vessels during the war, meaning that their fleet was subsidized by the US government (US Congress 1947a, 14). In 1945, liberated France received US$8 million (US$78.7 million in 2010 dollars) in loans for the construction of fishing vessels through the Lend-Lease program. Other segments of the fishing industry benefited indirectly from Lend-Lease because they supplied canned fish and other fish products to European countries as part of the program. In 1942, US canners exported 17.3% of their total production through Lend-Lease. That amount increased to 26.8% in 1943, then declined to around 10% in 1944 (US Congress 1944, 211; US Congress 1945, 6). Iceland supplied the United Kingdom with large amounts of fish through Lend-Lease, at least until the benefit was eliminated in 1944 (US Congress 1944, 105). During that period Iceland built up a large cache of foreign exchange, which it used to subsidize construction of a large steam-powered trawling fleet after the war (Sverrisson 2002). This revolutionized the economy of the island nation, generated a nascent power disconnect that fueled overcapitalization, fostered overexploitation, and created incentives for Iceland to exclude foreign fleets its unilaterally declared 4-mile territorial sea.

6.3.2 Food Security, Rebuilding, and Expansion

After the war, food security was at the top of the agenda for states around the world and new geopolitical issues were on the horizon. Most commercial fishing fleets were diminished due to wartime resource shortages and a large number of fishing vessels were destroyed in battle. Developmental subsidies were commonly used to rebuild fishing fleets and often came in the form of low-cost loans or other capital-enhancing actions. For instance, in the United States, arguably the dominant naval power of the time, the Office of the Coordinator of Fisheries authorized the construction or appropriation of 877 fishing vessels for private enterprise in 1945. Fifty of these vessels were decommissioned Navy clippers. Although the number of vessels purchased by the US government was calculated to restore the US fishing fleets to their pre–World War II levels, the new ships far exceeded those they replaced in size, speed, and efficiency, so US fishing power was greater after the war (Finley 2011).

In addition to building up its domestic economy, the United States also used fisheries subsidies to generate influence overseas, spending considerable sums to rebuild foreign fleets after the war. According to congressional testimony by August Felando, president of the American Tunaboat Association, in 1947 and 1948 the US government spent US$60 million (US$470 million in 2010 dollars) to rebuild the fishing fleet of Japan. An additional US$13 million (US$127 million in 2010 dollars) was used as matching funds for fisheries subsidies in the 18 countries covered by the Marshall Plan (officially the European Recovery Program) and International Cooperation Administration. Developing countries that had been affected by the war also received about US$19 million (US$149 million in 2010 dollars) from the UN Relief and Rehabilitation Administration and US$2 million (US$15.7 million in 2010 dollars) from the UN Food and Agriculture Organization for fisheries, both of which were almost entirely funded by US contributions at the time (US Congress 1963, 89; Williamson 2014). These amounts, though quite small relative to the overall spending on postwar reconstruction, were important to the fishing fleets in many countries and helped to restart the action cycle after the disruption of the war.

Most governments provided their fleets with developmental subsidies that matched or exceeded US contributions in order to assure postwar food security and economic development (Finley 2011). The rationale for developmental subsidies in these countries paralleled the US experience, insofar as geopolitical concerns and powerful domestic fishing lobbies drove government programs to unprecedented levels. For instance, Canada started subsidizing the construction of new, modern fishing vessels in 1947. By 1966, 762 new vessels were built under the program at a cost of more than US$6 million (US$47.1 million in 2010 dollars). Additional developmental subsidies were approved in 1961, specifically to expand the number of large (> 100 grt)[6] vessels in the Canadian fishing fleet. This program resulted in the construction of 217 large vessels at a cost of US$42 million (US$240 million in 2010 dollars; OECD 1970, 101; Williamson 2014). Other fishery development programs provided loans with artificially low interest rates to cover between 50% to 75% of the full price of a new fishing vessel. The rationale for these investments included bringing Canada's Atlantic fleet up to the same level as its Pacific fleet and ensuring full use of fisheries resources in both oceans (102). Like the US subsidy program, the vessels purchased with Canadian subsidies were more efficient than those they replaced, so the equilibrium level of effort shifted even farther from sustainable levels in many fisheries.

Freed from postwar constraints on the geographic expanse of its fisheries, in 1950 the Japanese government substantially increased subsidies to its distant-water fleets. These included direct subsidies for the purchase of distant-water vessels, as well as international aid packages to establish fishing bases overseas. By 1965, Japan had established 29 bases around the world, including three in Africa, three in Oceania, six in Southeast Asia, twelve in Latin America and the Caribbean, and one each in the Spanish Azores, India, British Columbia, and on the East Coast of the United States. Usually, 50% of the cost of these bases was paid for by the Japanese government. Amenities included modernization of port facilities and provision of processing and canning facilities. The Japanese government also provided training for local fishers and processors in the use of modern technologies (Christy and Scott [1965] 2011, 95).

Later in the 1960s, the Japanese government began subsidizing exploratory voyages by private companies to search for new pelagic fishing grounds. The government spent ¥250 million (about US$3.4 million in 2010 dollars) on this project in the first year alone (OECD 1970, 88; Officer 2014; Williamson 2014[7]). It also provided loans to support small-scale and medium-scale fishers in domestic waters—specifically, the government provided loans up to 70–80% of the value of any new vessel with a 6.5% annual interest rate and possible deferment of payment up to 15 years. The total value of these loans was more than ¥3 billion in 1967 and ¥6.2 billion in 1968 (US$42.3 and US$84.2 million in 2010 dollars; 89). Finally, Japan gave additional support to improve medical facilities on fishing vessels (¥5.8–6.7 million per year; US$78,800–91,000 in 2010 dollars; 90) and to the fisheries marketing sector to stabilize prices and increase food security (¥97–117 million per year; US$1.3–1.6 million in 2010 dollars; 91). While these subsidies for exploration eased pressures on coastal Japanese fisheries, they increased the profit disconnect in other coastal regions and generated multiple power disconnects as local elites favored foreign investment over domestic fishing interests.

According to a US Central Intelligence (CIA) report, the Soviet Union also invested heavily in developmental subsidies for its distant-water fishing fleets in the 1950s and the 1960s. It planned to purchase 175 factory trawlers of between 475,000 and 539,000 grt at a cost of US$560 million (US$3.6 billion in 2010 dollars). This would add substantially to its fleet, which was a total of 775,000 grt in 1955. The first set of 24 factory trawlers was completed in 1957 by West German boat builders, but there were also plans to begin production in Poland, in the USSR, and in East Germany (CIA 1959, 1–2). Armstrong (2009) asserts that the Soviets stuck to their plan and spent considerable sums on fisheries subsidies during the period. Approximately 75% of the subsidy package went to construction of new vessels. These numbers were provided by Soviet sources and may be subject to several types of distortion, but the increase in catch by Soviet fleets in this period suggests considerable investment, and it is certainly the case that all purchases of factory vessels were funded by the government rather than private sources (159). As in other countries, this expenditure affected politics as well as economics, but the effect on the power disconnect was greater. Since the fleets were owned by the USSR, the government itself had a vested interest in the continued expansion of production. Furthermore, fishers and decision makers were insulated from economic problem signals by heavy subsidies and by their distant-water capacity, which allowed them to move from fishery to fishery with little concern about the status of local stocks.

In Europe, the newly created Republic of Iceland, which became economically as well as politically independent from Denmark in 1944, used funds earned from fish sales during World War II to buy used fishing boats, negotiate low prices with foreign ship builders, and set up low-cost loans for its trawler fleet (Sverrisson 2002). The French government promoted modernization of its fleets by subsidizing the cost of freezing equipment and stern-trawling gear. The subsidy was 20% for "industrial" fishing vessels and 30% for distant-water fishing vessels. France also spent about ₣1 million (US$1.1 million in 2010 dollars) on education about new techniques and ₣1.5 million (US$1.7 million in 2010 dollars) on marketing and processing technologies in the early 1960s (159–160). The West German government was also subsiding expansion by the 1960s. They provided fishers with catch "premiums" for quality fish, payments for scrapping old vessels, fuel subsidies, and low-cost loans for the construction of new vessels. Total spending on these subsidies was DM107 million from 1962 to 1966 (about US$148 million in 2010 dollars; 196–197). Denmark, Greece, Ireland, Spain, Sweden, Norway, and Yugoslavia also spent millions in local currencies, either in low-cost loans or more direct subsidies, to develop their industrial fishing fleets. Norway and the United Kingdom also invested to rebuild their commercial whaling fleets at a cost of about US$6 million per vessel (US$33 million in 2010 dollars; Clark and Lamberson 1982; Williamson 2014).

These expansionary measures reinforced the alignment of government and industry interests and dampened economic problem signals associated with the overexploitation of many species and the rise of new substitutes. Indeed, all of these subsidies expanded both the profit disconnect and the power disconnect for European fleets. They ratcheted up effort much more than the early subsidies described in section 6.1, because the economic portion of the structural context was favorable to fleets from all of these countries. The rise of free trade in particular ended national monopolies on trade, and the proliferation of modern technologies and infrastructure leveled the playing field among developed countries. Nevertheless, economic growth also drove up costs of production, so fleets and investors started shifting their bases of operations to regions where costs were lower in the 1960s. This amplified the impact of economic development policies in existing and newly independent developing countries.

Like the developed countries described above, China used subsidies to create large-scale industrialized fleets after the civil war that led to the creation of the People's Republic of China (PRC) in 1949. However, the new government also subsidized the modernization of small-scale artisanal fleets. In fact, the PRC's expansionary program was much more effective than those of previous governments. In 1950, it established a credit relief fund to help fishers rebuild their fleets. This included encouragement of the addition of outboard motors to sailing junks and the establishment of a government-owned fleet of factory trawlers. Both subsidy programs greatly increased the fishing power of the Chinese fleet. The PRC also invested in infrastructure, particularly the construction of modern harbors in areas like Shengshan Island that were distant from economic centers but closer to fishing grounds (Muscolino 2009). In addition, it set up mutual-aid cooperatives, streamlined transport and marketing, and settled disputes through the deployment of local "cadres" of Communist Party loyalists (183). This focus on small-scale fishers is an important difference from the Soviet approach and was driven by exogenous factors in the structural context, particularly Maoist philosophy, and by the fact that China was already home to large numbers of small-scale coastal fishers. Therefore, the effect of the power disconnect was less pronounced than in European countries or the Soviet Union. China's strategy was more like that of Japan; government promoted a win-win solution by protecting coastal resources while depending on distant-water fleets for most of the growth in profits and food production.

6.3.3 Foreign Aid, Investment, and Economic Development

Although other developing countries had the same rationale for pursuing expansionary policies—economic growth, employment, and food security—their experience with fisheries expansion in the 1950s and the 1960s was different from China and the developed countries described above. In developing countries, fisheries development was driven by foreign aid and private investment, as much and sometimes more than direct government subsidies. In addition, developing countries tended to build up coastal rather than distant-water fisheries in this period. These differences reflect three characteristics of developing countries: limited access to capital, political leverage wielded by large numbers of existing coastal fishers, and the relative abundance of fish in their waters vis-à-vis the heavily overfished coastal zones around historically dominant fishing countries. Indeed, as explained in chapter 3, these rich waters attracted substantial foreign investment, as now heavily subsidized fishers from historically dominant fishing countries searched for new fishing grounds in response to rapidly increasing economic problem signals.

As in earlier periods, the expansionary effect of subsidies and other aid flows was heavily context dependent. Notably, economic conditions were favorable for some developing countries and large-scale industrial fisheries in Latin America, South America, and East Asia grew rapidly, well beyond the initial value of subsidies and aid provided by governments. In fact, expansion was so fast that coastal fisheries were quickly depleted, forcing countries to provide supportive subsidies to their new industries and/or develop distant-water fleets. On the other hand, aid-based attempts to build industrial fishing fleets in Africa failed in most cases, but subsidies often improved the efficiency of artisanal fleets, increasing production. In this region, economic conditions favored one sector of the industry over another, for the most part. Industrialized fleets succeeded only in Ghana and Senegal, largely due to the efficacy of local entrepreneurs. Similar experiences were observed in some parts of India, which focused largely on production from inland waters until the end of the 20th century (Agarwal 2007). Cases in this category are numerous, so details are provided only for a few exemplars.

Thailand is a good example of a small country that increased fisheries production substantially with the help of foreign assistance and domestic subsidies. In the 1960s, the government of West Germany funded several experimental fishing expeditions using commercial trawlers in the Gulf of Thailand. The Germans also trained Thais to use the trawlers, to construct and repair trawl nets, and to convert existing privately owned vessels to use trawling technologies. The harvests brought in by these experimental fishing vessels were strikingly large and profitable. This instigated considerable private investment in construction of new vessels and related processing technologies. Between 1960 and 1963 the number of trawlers increased from 99 to more than 2,000, and by 1977 it exceeded 5,800. The number of motorized shore craft also increased, while the number of non-motorized craft declined greatly (Royce 1987, 96–97).

Similar combinations of public investment and private development resulted in the industrialization of fleets in Korea, Taiwan, and the Philippines during this period. In these countries, investment and aid contributed to the development of new distant-water fleets in addition to the expansion of coastal fisheries, so the amount of investment was much higher than in countries such as Thailand. Because of this, industrialization increased the power disconnect and the profit disconnect within these countries, because control over resources primarily rested in the hands of wealthy domestic or foreign elites. At the international level, the entry of new distant-water fleets from developing countries increased conflicts over shared fish stocks but also added to economic pressures to improve management as described in chapter 5 (Gulland 1974).

Private investment was also much more important than subsidies in the development of industrial fleets in Central and South America. Peruvian fleets showed the strongest growth in the 1950s, largely due to the increase in international demand for low-quality fish for production of animal feed, the collapse of the Pacific sardine fishery, and the proximity of huge, unexploited stocks of Pacific anchoveta, which happened to be perfect for the reduction industry (Ludwig, Hilborn, and Waters 1993). Between 1953 and 1962, the Peruvian fleet of large vessels (> 35 feet in length) expanded from 50 to more than 1,000, half of which were greater than 65 feet in length (Christy and Scott [1965] 2011, 119). According to Royce (1987, 104), most of this growth was privately funded, starting with a single reduction plant that was capitalized as a joint venture between a US firm and a Peruvian company. Finley (2011) points out that Peru also was part of the Point Four Program, which the United States instituted in 1949 to propagate US (i.e., capitalist) values through the dissemination of technological innovation. Mexico, Peru, Colombia, El Salvador, and the Dominican Republic all participated in Point Four projects designed to transfer advanced fishing and processing technologies. Industrialized fisheries in these countries grew steadily thereafter, also as a result of foreign and domestic investment; however, none displayed the spectacular growth experienced in Peru, which was attributable to geographic and economic serendipity.

Relative to Southeast Asia and Latin America, larger influxes of aid had a smaller impact on most African fisheries during the 1950s and the 1960s. With a few exceptions, attempts to establish large-scale fleets in sub-Saharan Africa (excluding South Africa and Namibia) failed completely (Haakonsen 1992, 44). Industrial vessels languished in African ports for many reasons. Some were cultural, related to the preferences of local communities and their willingness to engage in industrial fishing activities. Others were structural, particularly the lack of technological expertise or markets for spare parts that are necessary to maintain industrial fishing fleets. There were also political problems generated by high levels of corruption, which discouraged foreign and domestic investment and also inhibited the development of transportation infrastructure. In addition, economic conditions were unfavorable. Haakonsen (1992) argues that industrial fleets could not compete with highly efficient artisanal fishers using small, motor-powered vessels and hand-operated gear. All of these factors minimized profit and power disconnects and would have resulted in early governance response by local communities if not for the incursion of subsidized industrial fleets from Europe, the United States, and Japan. These foreign fishers were well insulated from problem signals and had significant capacity to rapidly overexploit local fish populations, so local management institutions were completely undermined in most cases.

Ghana was the primary exception to the general trend for Africa in this period. As noted in chapter 2, the Japanese established a fairly large fleet of baitboats in Tema, Ghana in the late 1950s. Like Peru, Ghana had a geographic advantage, due to its location on the Gulf of Guinea, which was a rich fishing ground, particularly for tunas and other commercially valuable species. Ghana also had several well-connected entrepreneurs who understood the political economic context in the country and were able to build up successful industrial fleets. The two most well-known Ghanaian fisheries entrepreneurs were R. Ocran, founder of the Mankoadze Fisheries Limited, and E. N. Soli, founder of Soli Fisheries (Haakonsen 1992, 45). They ran the two largest canneries in Tema and also provided capital to maintain the pole and line vessels left by the Japanese and to motorize the traditional canoe-based fishery. This paralleled early development of steam power in Europe, in that small-scale coastal fleets adapted to survive increasing competition from larger industrial fleets, stratifying the industry. However, the power disconnect remained, both because local industrial fleets were managed by an elite few and because distant-water fleets also heavily fished the area. By the 1960s, major commercial fisheries in the Gulf of Guinea were managed by the International Commission for the Conservation of Atlantic tunas rather than local African governments.

Wherever developmental subsidies resulted in a substantial increase in local production, the similarities to the paths followed by historically dominant fishing countries are striking. Rapid depletion of coastal and offshore stocks would result in recession in the fisheries sector. Government would step in to provide supportive subsidies for existing fisheries and, in some cases, would also establish developmental subsidies to build up distant-water fleets. This happened in Thailand, Peru, and Ghana, as well as in many other countries in the 1960s. Nevertheless, national governments and international aid agencies continued to view fisheries as a source of long-term economic development, so programs of subsidies and aid packages continued throughout the 1970s and the 1980s. In 1981, the UN Food and Agriculture Organization recommended investment in fisheries sectors as an engine for growth in the developing world (Royce 1987, 3).

The same pattern was observed in historically dominant countries, though quite a bit earlier. The United States was already providing supportive subsidies for New England fleets in 1957 and started providing supportive subsidies for Pacific fleets in the 1960s (US Congress 1958; Finley 2011). After a long period of decline, Dutch fisheries rebounded in the early 1900s, but by the 1960s the government of the Netherlands was forced to provide supportive subsidies to Dutch fleets in the form of guaranteed low-interest loans just to maintain capacity at existing levels. At the same time, the government began to explore distant-water fisheries to increase the options available to Dutch fishers (OECD 1970, 305–306). It also provided credit to improve the processing sector in hopes of stabilizing the fishing industry as a whole (OECD 1965, 143). The United Kingdom took an even stronger stance. In consultation with the fishing industry, the government established a 10-year plan to eliminate supportive subsidies in 1962. Their goal was to make the UK fishing industry completely independent of government aid (197). Although the UK did not attain its goal (see figures 6.1 and 6.2 below), this was one of the first indications that governments could perceive supportive subsidies as an unreasonable burden on the public purse.

6.4 Opposition to Subsidies

Since the 1960s, there have been many attempts—some successful, some not—to reduce supportive fisheries subsidies around the world. Advocates of subsidy reduction hope to confound Ludwig's Ratchet and thereby eliminate increasing costs to the public. Often the domestic structural context plays an important role, since one of the main justifications for continuation of supportive subsidies is to maintain jobs and other economic benefits. In countries like Spain, where fisheries workers are unionized, government plans to reduce fisheries subsidies sometimes generate major strikes, including the closure of ports and disruption of trade. Furthermore, as shown above, subsidies widen the power disconnect by creating vested interests that are insulated from problem signals. Each turn of the ratchet causes effort to increase, which causes stocks to decline and triggers profit signals that drive political demand for more subsidies. However, as Asgeirsdottir (2008) points out, the balance of power may change over time, as fisheries become less profitable due to the maginfied CPR dynamic and because new industries become more important both economically and politically.

6.4.1 Free Trade

Because of the vertical interplay between international regimes and domestic governments, new actors and exogenous changes in the international structural context can also create opposition to expansionary management measures (Young 2002).The earliest calls for reduction of fisheries subsidies came from proponents of free trade rather than from conservation interests. The Organization for Economic Cooperation and Development (OECD) drew attention to fisheries subsidies in a 1965 report. At the time, the OECD recognized that developmental subsidies could be useful for building up an "infant" fishing industry, but it also pointed out that supportive subsidies distorted international markets for fish products. Furthermore, the report showed that subsidies in general and supportive subsidies specifically were increasing in the 1960s, and recommended that this trend should be reversed (OECD 1965). The OECD continued to argue for subsidy reductions by its member states in subsequent reports (OECD 1970; Cox and Schmidt 2002).

Eventually aid and development organizations also recognized that subsidies could have a negative impact on trade and on developing economies once overexploitation set in. In 1993, the UN Food and Agriculture Organization produced its own report on the negative effects of subsidies on fisheries management and began to work on ways to reduce such incentives to overcapitalization. The issue of subsidies is featured prominently in the FAO's International Plan of Action for the Management of Fishing Capacity (IPOA-FC). It was also the main subject of their Expert Consultation on Economic Incentives and Responsible Fisheries in 2000, and was investigated in several reports and technical documents published by the FAO (FAO 2002b, 92).

As a result of these combined problems—overexploitation of fish stocks and distortion of international trade—fisheries subsidies became an important part of the General Agreement on Tariffs and Trade (GATT) and subsequent World Trade Organization (WTO) negotiations. Through these processes, global mechanisms for the propagation and enforcement of free trade evolved slowly over several decades. In 1994 the WTO finally adopted the Agreement on Subsidies and Countervailing Measures (SCM). This agreement provides an operational definition of subsidies, groups subsidies according to whether they are "prohibited" or "actionable," and delineates the measures available to a state that is negatively affected by another state's subsidies (WTO 1994). However, by the late 1990s it was clear that the SCM was not a sufficient mechanism to achieve its purpose, because it primarily focused on direct rather than indirect subsidies and because it considered only issues of trade distortion but not environmental degradation. Starting in 1999, members of a group of countries known as the "Friends of Fish" (Australia, Chile, Ecuador, Iceland, New Zealand, Peru, the Philippines, and the United States) proposed modification of the SCM to cover fisheries subsidies. Japan, whose fleets are heavily dependent on subsidies, opposed any changes to the rules, but proponents prevailed and fisheries subsidies were added to the WTO agenda in the Doha Mandate of 2001 (Delvos 2006, 341–342).

In addition to the "Friends of Fish," the EU and Chile also came out in favor of changes to the SCM, including expansion of the so-called "red box" of banned subsidies to include all capacity-enhancing measures. In spite of their alignment on the need to reduce fisheries subsidies through the SCM, these states could not agree on the specific mechanisms that should be used. They also disagreed on the exact nature of the subsidies that should be banned. Each state sought different definitions that would skew the new rules in favor of domestic interest (Delvos 2006, 357–358). Meanwhile, China, which subsidizes its fleets heavily, proposed that certain subsidies should go in the defunct "green box" under the SCM, meaning that they would be allowed. All the while, Japan continued to resist changes to the SCM, now with support from South Korea. A group of small coastal developing states also opposed changes because payments for access to EEZs would be included in banned subsidies; this was a major economic threat to countries with few other means of earning foreign exchange (359).

In 2005, the WTO adopted the Hong Kong Ministerial Declaration, which recognized the needs of coastal developing states and provided for their protection in future negotiations on fisheries subsidies (FAO 2006, 61). Two years later, the chair of the Negotiating Group on Rules put forward a draft text for a fisheries subsidies amendment that generally satisfied conservation interest groups and "Friends of Fish" states. This eight-part plan would have enacted sweeping changes to reduce capacity-enhancing subsidies and generally harmonize international fisheries management and trade policy. It also contained general exceptions for "beneficial" subsidies and specific exceptions for the transfer of sustainable development technologies to developing countries. However, there is still no agreement on the proposal or any other amendments to the SCM for fisheries subsides. The "Group of Small and Vulnerable Economies" (70 states) opposes the plan because it does not provide enough exceptions for developing countries, and other states believe that the burden of required management is too great in the draft text. Though Japan decided to accept changes to the SCM in the 2000s, it still opposes the current version because it does not contain enough measures to curb illegal, unreported, and unregulated (IUU) fishing (Meliado 2012). In the future, given the political and legal problems that remain unsolved, it may take more time and more lobbying to achieve an enforceable agreement to reduce fisheries subsidies through the WTO.

6.4.2 Measuring Subsidies

Much of the impetus for the later WTO discussions was driven by lobbying by conservation interests like Greenpeace and the World Wildlife Fund (WWF). These groups also worked at the domestic level to convince governments to reduce fisheries subsidies unilaterally (Delvos 2006, 353). For instance, in 1997 the World Wildlife Fund generated a study of subsidies in 32 major fishing states to shed light on the problem and convince governments that they should do something about it. This study was essentially a meta-analysis of reports on fisheries subsidies from the Asia and Pacific Economic Cooperation (APEC) organization, from the OECD, and from the WTO. The WWF's estimates for specific countries are shown in figure 6.1. The average volume of assistance was calculated to be more than US$220 million in 1997. Japan was by far the largest contributor to its domestic fishing industry, with subsidies totaling almost US$3 billion in that year alone. The United States and the European Union were second and third,

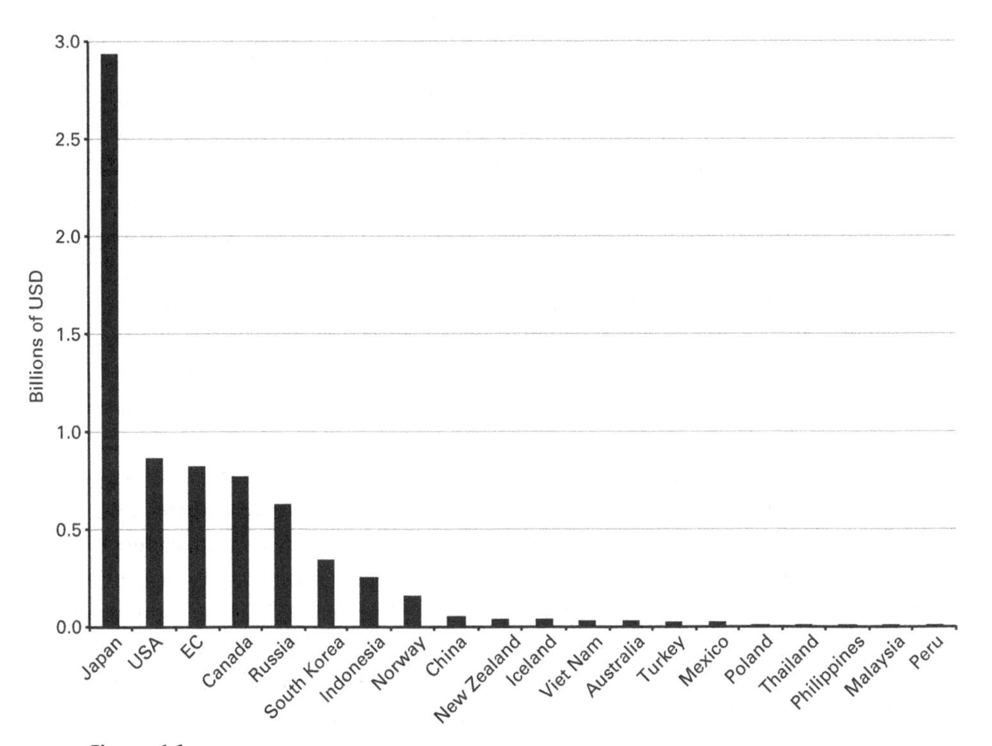

Figure 6.1
World Wildlife Federation estimates of fisheries subsidies for selected countries in 1997. Source: WWF 2001, appendix 1.

with subsidies of about US$868 and US$824 million, respectively (WWF 2001, appendix 1). The World Wildlife Fund and several other conservation groups continue to use these estimates to highlight the problem of subsidies on their websites and in public relations campaigns.

Unfortunately, unilateral attempts to reduce subsidies are problematic given the international nature of the fishing industry. No state is willing to reduce subsidies, because all states recognize that this would cripple their fleets relative to those from other states. This is a common problem in international negotiations regarding fisheries. Here it is important to note that pressure continues to mount for reducing fisheries subsidies, even as losses rise due to increasing overcapitalization and overexploitation around the world. In 2009 the World Bank and the UN Food and Agriculture Organization published a study, titled *The Sunken Billions*, in which they argued that reducing subsidies to the fishing industry was a necessary part of the strategic reorganization of global fisheries. By their estimates the industry was losing as much as US$50 billion annually (World Bank and FAO 2009, 47). Scholarship on fisheries subsidies and related impacts further increased awareness of the problem and, like the WWF study, helped to show which states were most responsible. In the most recent study, Sumaila et al. (2010a) estimated that the total amount of money spent to subsidize the fishing industry in 2003 was between US$25–29 billion. Of this, the vast majority was spent on capacity-enhancing subsidies and fuel subsidies.

Sumaila et al. (2010a) examine three types of subsidies in global fisheries: *beneficial* subsidies (improves fisheries management, e.g., spending on scientific studies, enforcement, etc.), *capacity-enhancing* subsidies (increases overcapitalization as described above), and *ambiguous* subsidies (impacts on management unclear). Country-level estimates from 2003 for each of these categories are interesting. The size of the fishing industry and resources available to each country are clearly important indicators of the magnitude of subsidies. Where subsidies are high, the fishing industry has considerable political and economic clout. Asia is by far the largest in area and population, so it is not surprising that approximately 58% of all fisheries subsidies (US$15.7 billion) are spent in this region. Japan has the highest subsidies in the region and the world (US$4.6 billion), followed closely by China (US$4.1 billion). The United States is the third-largest provider of subsidies (US$1.8 billion), followed by Russia (US$1.5 billion). Members of the European Union also subsidize heavily given their populations, with a combined total expenditure of more than US$2.4 billion. Spain, France, and the United Kingdom are the largest fishing countries and the biggest providers of subsidies in Europe. In Latin America and the Caribbean, the largest economies—Brazil (US$0.41 billion), Argentina (US$0.36 billion), and Peru (US$0.20 billion)—provide the largest subsidies. In Africa, Namibia provides the highest amount of subsidies (US$0.12 billion), followed by Morocco (US$0.09 billion). Angola, Senegal, and South Africa all spend about US$0.07 billion on fisheries subsidies. While there is substantial variation between regions (see figure 6.2) there is also substantial variation within each region.

It is also useful to look at subsidies in relation to the structural context. Part of the variation shown in figure 6.2 is due to differences in population, but, for instance, India's population is close to China's yet India spent only about US$1 billion on fisheries subsidies in 2003. The difference is likely due to governance aspects of the structural context. As a centrally planned economy, China ultimately chose to follow the Japanese model of fishery expansion, while democratic India primarily subsidized inland and coastal artisanal fleets, which provide employment for large numbers of individuals. In other cases, governments might be willing but unable to subsidize fisheries, because of a lack of funds, or, as in Africa, dependence on foreign aid and the wide-spread failure to produce industrialized fleets could negate the demand for large-scale subsidies. Perhaps most important, the countries with the highest costs of production (i.e., Japan, EU members, and the United States) also provide the highest subsidies to their fishers.

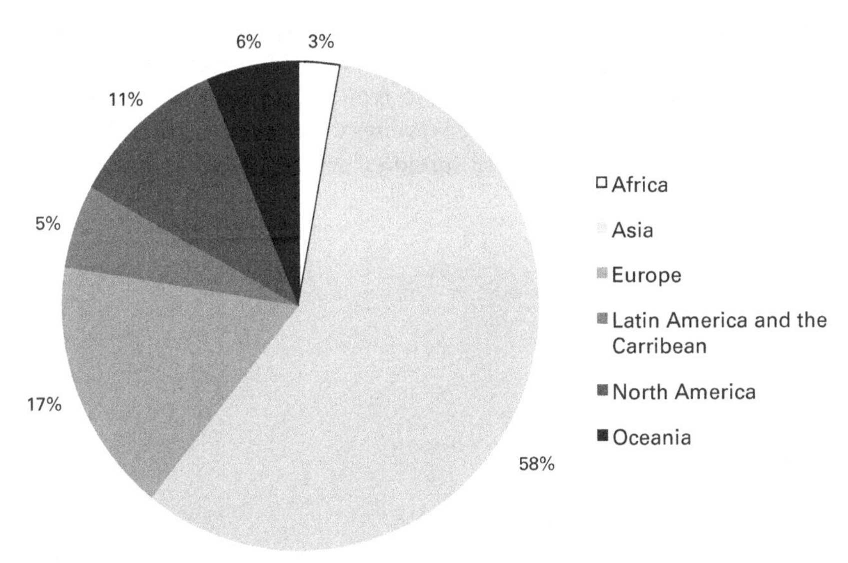

Figure 6.2
Regional distribution of fisheries subsidies. Source: Sumaila et al. 2010a.

This historical record shows that these subsidies were established to support these fleets in the face of increasing competition from developing country fleets with lower costs of production. With increasing problem signals, developed countries are more willing to forgo expansionary management measures in exchange for exclusionary regulation, but this raises issues of unfairness for developing countries, which seek to benefit from the same resource-based industries that fueled the growth of historically dominant fleets.

Expansionary subsidies played a major role in the development of fishing fleets and of fisheries governance. When economic conditions were unfavorable, the effects of subsidies and similar interventions were small. However, when economic conditions were favorable, positive feedbacks between political and economic factors in the AC/SC framework greatly amplified their expansionary effects. As long as globalization continues to drive capital from one country to another in search of lower costs of production, Ludwig's Ratchet will continue to intensify the cycles of overexploitation and overcapitalization described in Part I. On the other hand, the negative effects of fisheries subsidies can galvanize political groups with interests in preventing overexploitation, in reducing barriers to free trade, or even in reducing government expenditure generally. Thus, anti-subsidy pressure can also be expected to increase over time. Given these trends in opposing forces, it is difficult to say if or when fishery subsidy reduction will emerge from the political-economic process described above. However, one can say that measures designed to reduce overcapitalization and to rebuild fish stocks may also reduce pressures to subsidize fleets by increasing profitability. These "beneficial subsidies" will be addressed in the next two chapters.

7 Conservation Measures

Extensive use of restrictive conservation measures is a modern phenomenon. A few rules were used to protect juveniles and spawning populations in historical fisheries management regimes, but the vast majority of management measures were designed to reduce conflicts over access rights. Governments did not start using restrictive regulations to solve the core problems of overfishing, overcapitalization, and ecosystem disruption until well into the 20th century. In part, this delay can be attributed to a lack of understanding of the effects of fishing on marine ecosystems, particularly the belief that ocean fisheries were inexhaustible. However, even after considerable advancement in scientific understanding, the evidence shows that the problem signals generated by scientists were consistently discounted in the policy process due to the combined effects of the profit disconnect and the power disconnect. Indeed, the dampening of economic, political, and scientific problem signals prevented the effective implementation of restrictive management measures for conservation and permitted expanding cycles of serial depletion that followed the cycles of economic expansion described in Part I.

In this chapter, I refer to two primary types of restrictive measures: *conservation measures* and *indirect exclusionary measures*. The former are designed to conserve the stock of fish; the latter are used to restrict production by some fishers to the benefit of other fishers. Where some groups of fishers are more powerful than others and where these groups are differentiated by gear type, targeted species, fishing location, or some other technological factor, exclusionary restrictive measures are common. This is particularly the case in fisheries where open access is the *de facto* norm so direct exclusionary measures are not feasible. Indeed, with the rise of the norm of *mare liberum* and the decline in violent conflict on the high seas, fishers increasingly called for indirect exclusionary measures to reduce competition without actually conserving stocks. In contrast, conservation measures are specifically designed to address the core problems in the action cycle, usually by reducing catch or effort in order to maintain or rebuild stocks or ecosystems. Furthermore, *strong* conservation measures both conform to the best available scientific knowledge and are well enforced. This distinction is important because policy makers often implement half measures that either fail to conform to scientific advice or are not well enforced. These *weak* conservation measures are usually enacted when policy makers face conflicting signals from industry and conservation actors or when existing rules (structural context) require some conservation response but the political climate favors inaction (action cycle). In such cases, regulations that have conflicting goals, fall short of scientific advice, or are unenforced provide the appearance of conservation without the costs of restrictive management.

There is a complicated relationship between intent and effectiveness. The categories above apply to the intent of decision makers and other political actors; all three types of measures are usually described as "conservation" but some are intended to exclude, others are intended to obfuscate, and a few are intended to conserve fish stocks. As explained in chapter 1, absolute effectiveness is extremely difficult to measure, so relative effectiveness is more practical. It is measured by evaluating goal attainment alone, rather than causal connections between regulations and goals. For instance, if management at maximum sustainable yield (MSY) is the goal, then a regime is relatively effective when the stock is large enough to support MSY and fishing effort is at or below MSY, regardless of exogenous factors. When effectiveness is measured relative to the structural context, strong conservation measures are not guaranteed solutions to core problems in the action cycle and weak or indirect exclusionary measures may still be relatively effective if environmental and economic conditions are favorable. Nevertheless, it is useful to consider decision makers' intent when evaluating restrictive measures, because, all else being equal, strong conservation measures are more likely to be effective than weak or exclusionary measures.

The relationship between effectiveness and expedience is also important. Strong conservation measures are usually difficult to implement politically and so are often a last resort of policy makers. As a result, conservation response is usually delayed, although weak or indirect exclusionary measures are often adopted earlier than strong conservation measures. Historically, the period of delayed response prior to implementation of conservation measures was lengthened by both endogenous and exogenous forces. Subsidies and expansionary economic responses drastically increased the political and economic power of large industrialized fishing fleets while also insulating these actors from problem signals. Similar dynamics generated strong disconnects for secondary industry actors as well. Processors and marketers lobbied against restrictive measures so they could continue to expand globally, at the expense of local fisheries and ecosystems. Small fishers were left behind in this process. Although they lobbied for government protection as well, they were usually either ignored or given subsidies but in either case they also invested in innovation that made them less sensitive to problem signals, thereby easing political pressure in favor of catch and effort limitations. At the same time, exogenous forces underscored policy maker concern about issues such as food security, employment, and economic growth, providing incentives for the support of industry expansion. All of these factors widened the profit disconnect and power disconnect, delaying the imposition of strong conservation measures while favoring the creation of politically expedient expansionary management measures.

With the formal institutionalization of science-based management in the 20th century, many hoped that fisheries governance would become proactive—but that did not occur. Given modern methods, scientific signals of overexploitation usually occur fairly early in the action cycle; however, they tend to be overpowered by the profit disconnect and related power disconnects. Because scientists are sensitive to core problems such as overfishing and ecosystem disruption, their political marginalization contributes to the power disconnect. That said, as economic problem signals become stronger with repeated iterations of the action cycle under ineffective management, affected groups may appropriate scientific information to bolster their own policy positions, which helps to strengthen political problem signals received from scientists. In fact, most interest groups choose to appropriate scientific advice that supports their position on a given issue and to disregard or delegitimize conflicting findings. Thus, even "bad" science may be influential when it is appropriated by powerful actors. These factors generally make solid science necessary but insufficient for early and effective responsive governance.

Nevertheless, implementation of conservation measures and related switching from ineffective to effective management as illustrated in figure 1.1 was observed in multiple cases, usually after prolonged periods of overexploitation and overcapitalization. When fisheries collapsed or stagnated, industrial fleets moved on to new regions and local small-scale fishers were either forced to exit or were willing to accept conservation measures to regain their livelihoods. In coastal fisheries, these changes removed much of the political pressure against conservation-based management, at which point governments could step in to curb fishing pressure by smaller-scale fleets, allowing stocks to rebuild. In historically dominant fishing states, governments also invested heavily in fisheries science and in the development of fisheries management bureaucracies in response to an increasing number of crises. Exogenous development in mathematics, oceanography, and other sciences further catalyzed understanding of fisheries management, expanding the set of potential management measures available in the structural context. Together, better science and formalized fisheries management institutions laid the groundwork for improved fisheries governance. However, the power and profit disconnects ensured that the tendency to manage overfishing rather than prevent it remained pervasive, and was even institutionalized at the international level. Furthermore, extension of formal management frequently undermined informal governance systems and even resurrected the tragedy of the commons as outsider elites moved in to appropriate benefits as described by Buck ([1985] 2010).

This chapter starts with a brief description of the serial depletion of salmon and other coastal or anadromous fish stocks in the 19th century. The discussion shows that responsive governance dominated in the period and that decision makers were not able to learn from the mistakes of others as stock after stock declined, in spite of the known dangers of overfishing. Section 7.2 goes on to show that responsive governance continued as overexploitation set in for large-scale marine fisheries. Here, response was complicated by political demands for indirect exclusionary measures that led to the creation of formal management bodies in many countries but also undermined conservation toward the end of the 19th century. Section 7.3 shows how the declining influence of coastal fleets and the increasing power of industrialized fleets shaped modern fisheries governance into a process of managed overexploitation in the 20th century. It describes how cycles of growth and collapse led to changes in management institutions and in scientific understanding. Several different scientific views arose in response to the serial depletion of marine fisheries, including more precautionary ecosystem-focused approaches and the now well-known maximum sustainable yield approach. Although each view continues in the sciences, the latter was institutionalized around the world in the 1950s, largely as a result of pressures from the fishing industry and concerns about food security and economic development. Section 7.4 demonstrates how the profit disconnect and other forces related to the economics of expansion contributed to this process of change in the structural context for fisheries globally. Finally, Section 7.5 describes problems of compliance and shows how cycles of expansion fostered the growth of management capacity in many countries around the world, but also increased the difficulty of monitoring and enforcing regulations.

7.1 Serial Depletion

One of the most frequent patterns associated with responsive governance is the serial depletion of fisheries (Berkes et al. 2006). "Depletion" was a term commonly used in the late 1800s and the early 1900s to refer to a fishery in which the catch per unit effort was substantially reduced because the size of the stock had declined (Smith 1994, 241). It is similar to the modern concept of overexploitation, but less well defined. In this period, scientists were uncertain about the effects of fishing on marine stocks, but many anadromous and coastal stocks had already collapsed. This section documents the serial depletion of the world's salmon fisheries in the 19th and 20th centuries. It shows how managers repeatedly failed to enact early, strong regulations in spite of knowledge of previous overexploitation of stocks in other locations. Although scientific uncertainty was a problem in this case and in many others, limited management capacity and lobbying by powerful actors who were well insulated from problem signals were greater barriers to effective response.

Salmon fisheries were depleted much earlier than commercially important marine fisheries. These anadromous fish travel from their freshwater spawning grounds to the open ocean and back again. During these migrations, salmon can easily be harvested in large quantities in rivers and streams. They are also targeted at sea, usually in bays or near river mouths. Pollution and the construction of dams also can reduce salmon populations, so they face multiple anthropogenic stressors. In England overfishing of salmon was fueled by growing demand associated with the Industrial Revolution. Clear signs of depletion began in the early 1800s. In fact, salmon disappeared from the Thames in 1814, and it was obvious that many other salmon runs were severely depleted by the 1840s. However, the British government did not pass regulations for the protection of salmon until 1861, by which time the total harvest was much smaller than it had been historically. This caused significant economic costs to salmon fishers and generated demand for management actions (Fryer 1884, 4). The first regulations placed on the fishery banned fixed nets, which were in such wide use at the time that they clogged waterways and created a public nuisance in addition to passively harvesting large amounts of fish. Rebuilding was slow, however, so more stringent regulations were adopted in 1865. These included mesh size limits and length limits on movable nets so they could not span an entire river (33). Combined, these regulations helped to rebuild salmon populations, but stocks did not return to pre-19th century levels due to the proliferation of dams and the increase in pollution associated with the Industrial Revolution (41).

Some hoped that managers in North America would succeed, having learned from the mistakes of their British counterparts. Speaking in 1883, a leading expert on salmon fisheries in Europe, Charles E. Fryer, extolled North American salmon fisheries—particularly in the west, where exploitation started in the 1870s—as examples of the abundance associated with "pristine" salmon runs. Furthermore, he stated that:

> There is little fear, happily, that the Canadian and United States authorities will blindly allow such a source of wealth to be sapped, but the same fate which has befallen the fisheries of the Old World will assuredly overtake those of the New if the same obstacles are placed in their way. (Fryer 1884, 24)

Fryer's optimism was misplaced. East Coast salmon fisheries in North America were already severely depleted by the 1860s. Dams, pollution, and overfishing all contributed to the problem. With multiple stocks in crisis, Canada instituted strict regulations such as time-area closures, mesh size limits, and other gear restrictions late in that decade (Joncas 1884). These regulations were specified in the Federal Fisheries Act of 1868, which was passed in response to the depletion of many Atlantic fisheries but was not well enforced and so was an ineffective response (Augerot 2000). In the United States, restricitive regulations were few and far between while investments in hatcheries increased rapidly (Goode 1884). In the 1870s, the US Fish Commission worked with state agencies in New England to construct fishways that would allow salmon to move upstream past dams to spawn in their home locations. The commission also pressed for closures of fisheries during the spawning season, when runs were strongest. Although fishways were built, they were generally ineffective. Closures were unpopular with fishers, so they were seldom enacted and never enforced. Thus, declines in salmon populations continued due to lack of effective management intervention (Allard 1978).

On the western rivers, where salmon runs were spectacularly productive in the 1870s, landings began to decline in the 1880s. Canada responded with some regulations, including catch limits, but again these were poorly enforced (Augerot 2000). Licenses were not required until 1888 (86). Serial overexploitation of increasingly low-value salmon species continued through the early 1900s, with regulation largely driven by lobbying by different sectors of the industry seeking to exclude others (87–90). Management of salmon fisheries in the western US followed a similar pattern: industrial fishers depleted river after river, often at the expense of politically marginalized local artisanal fishers. Because of this power disconnect, management in an area was always delayed until industrial fishers moved on to new fishing grounds. Starting in 1901, Alaskan fisheries were the last to be exploited commercially, but were rapidly overexploited with little attention to management until steep declines occurred. Again, the power disconnect was a critical factor, and local management institutions were rapidly destroyed by the incursion of powerful and highly mobile industrial fishers who had little interest in maintaining the resource (Finley 2011). In California and Alaska particularly, lack of knowledge could not be blamed for the delay. Neither the US nor the Canadian government learned from the British example or from their own experiences with the domino-like collapse of salmon stocks from within their own territories.

Two other countries with numerous salmon runs displayed the same behavior. Regulation of Norwegian salmon fisheries was first officially recorded in 1687. Based on much older laws, these rules were codified to reduce conflict among fishers by establishing endowments and entitlements. The first laws aimed at salmon conservation were passed in 1848. An official salmon authority was established a few decades later. The focus of their work was on expanding salmon populations by building hatcheries, as well as enforcement of existing laws regarding salmon. However, as in the United States and elsewhere, these initial regulations were not sufficient to prevent overfishing and the further decline of salmon stocks (Ministry of the Environment 1999). Norwegian salmon production increased steeply in the 1870s and the 1880s as some fishers appropriated fixed gears like those used in England. High prices for salmon in England—due to increasing demand and decreasing local supply—drove this increase in Norwegian production. Both shifts widened the profit disconnect and gave Norwegian fishers incentives to press the government to delay any strong conservation actions as long as there was money to be made (Wallem 1884). Further delay occurred because of intense conflict between river fishers and sea fishers that was not resolved until much later. However, due to increasing economic signals in both fisheries, the Norwegian government finally started tightening salmon regulations in 1891 (Ministry of the Environment 1999).

In Japan, evidence of some salmon stock declines surfaced in the middle of the 18th century, just before the Meiji revolution. During the early Meiji period, salmon fishing increased drastically due to government subsidies, revocation of sea tenure laws, and technological improvements. Landings in domestic waters peaked in 1889 at 11 million fish and then declined sharply, sending strong economic signals to all industry actors. In response to fisher demands, the Japanese government reinstated village fishermen's unions to control local access to the stocks, began funding hatcheries to increase production, and subsidized industrial fishing of salmon in Siberia and on the open ocean (Augerot 2000). These initial interventions proved unsuccessful, so more drastic measures were taken as part of a nationwide reform of fisheries laws in 1901 (see below). In spite of their experience at home, the Japanese followed a similar though more chaotic path when exploiting Russian salmon stocks, which showed signs of rapid depletion in the early 1900s (65). It appears that "learning" did not take place as Fryer hoped, although technology transfer, particularly for hatchery propagation, did occur between Japan and the United States.

Many other commercially important anadromous and coastal fisheries exhibited the same pattern of serial depletion demonstrated for salmon, though few were as widely distributed around the world. In Europe, in the United States, and in Australia, fisheries for oysters were serially depleted in the mid- to late 1800s (Hubrecht 1884; Kirby 2004). In Maine, lobster fisheries were overexploited by the 1860s. In the 1880s, fishers targeting fresh lobsters won a significant change in regulations that effectively excluded cannery lobstermen from the fishery, but this did not prevent further increase in effort or depletion of the fishery, because growing demand caused prices to increase. This profit disconnect continued to cause delayed management until a steep drop in demand in the 1920s prompted fishers to lobby for protection. Even then, they first demanded exclusion via trade measures and government subsidies before finally agreeing to relatively low-cost regulations aimed at protecting spawning and juvenile lobster (Acheson 1997). Another anadromous fish, the North Atlantic shad, was very popular in the late 19th century. It was depleted in much of its native range in less than a decade. Susceptible to the same factors as salmon, shad was also largely unregulated in the United States except for strenuous attempts to restock rivers and expand the range of the species using hatchery programs (Allard 1978). In Europe, coastal stocks of plaice, herring, and other commercially valuable species were also heavily depleted before the implementation of strong conservation regulations. In fact, as fishing effort expanded, serial depletion of stocks also increased in scope and magnitude. Widespread and unavoidable recognition of this pattern ultimately led to important changes in the structural context for coastal fisheries but this was not transferred to other stocks.

7.2 Science, Conflict, and Management

As can be seen from the examples above, the dynamics of serial depletion changed toward the end of the 19th century. The economics of expansion ensured that stocks were depleted more quickly, speeding up the action cycle. Furthermore, related stratification of the industry created many different groups of fishers, and crowding increased interactions between groups, so conflicts among fishers and demands for exclusionary measures increased. These conflicts brought in governments at local, regional, national, and finally international levels to settle disputes and regulate fisheries. Perhaps most important, by this time fishers themselves were using conservation-based arguments to support their demands for indirect exclusionary management measures such as bans on the use of more effective gears. Most likely, this change of position was generated by the highly publicized depletion of coastal and anadromous stocks. Because of these claims, governments began investing in fisheries science and ultimately created fisheries management agencies in order to settle conflicts between disparate groups. While this institutionalization of science-based management was a critical juncture in the evolution of fisheries governance, the initial impact on regulation and the action cycle more broadly was counterintuitive. When scientists investigated the conservation-based claims of coastal fishers in the mid-1800s, they often found that the fisheries were healthy and rejected assertions that new industrial fleets were depleting commercial fish stocks. These conclusions, combined with a general trend toward *laissez faire* practices in Europe and North America, ushered in a period of deregulation toward the end of the 19th century. This section covers the rise of science-based management, the resulting deregulation, and the initial effects of both processes on serial depletion and the power disconnect.

In the 19th century, most fisheries scientists believed that coastal and anadromous fisheries could be depleted—after all, there was substantial evidence from the cycles of depletion described above—but the idea that fishing could reduce stocks of "sea fisheries" like herring or cod was still hotly debated. For instance, in 1854, John Cleghorn introduced the concept of "overfishing," which he defined as depleting a stock by harvesting too many fish. This concept was based on data showing how fishers serially depleted stocks of herring, starting near the coasts and then moving farther away as near-shore stocks declined (Smith 1994, 71). His explanation was quite controversial. Other competing explanations for fluctuations in herring catches included natural predation, changes in ocean currents, the fickleness of the fish themselves in their choice of migratory paths, and divine punishment for wasting fish that could not be consumed or processed quickly enough when catches were large (8–9). Though some of these explanations will seem peculiar to present-day readers, it is important to remember that, even in the 1850s, fishers harvested thousands of tonnes of salmon but millions of tonnes of cod, herring, and similar species (ICES 2012). Debates over overfishing were not limited to the scientific discourse. Fishers also used similar claims and counter-claims in their attempts to monopolize access to resources through indirect exclusionary measures. The marshaling of conflicting information by fishers, combined with government concerns about depletion and food security, helped to institutionalize the role of science in management.

In the mid-1800s, early government funding of fisheries science started in response to exogenous concerns about food security and to endogenous lobbying by industry actors who faced strong political and economic problems signals at the time. In this period, government-funded scientists focused on finding new fishing grounds or otherwise expanding production of fish in spite of the overexploitation of historical fishing grounds. Sweden commissioned several biological studies to increase herring production as early as the 1830s, and Russia began funding systematic studies of marine fisheries in the Black Sea and the Barents Sea in 1853. Each of these efforts helped to subsidize exploration for new fishing grounds while also identifying the importance of protecting spawning and juvenile fish in historic fisheries (Smith 1994, 35; Borisov 1964, 17–20). Furthermore, both programs were foundational to the institution of science-based management but did not immediately cause much change in regulation.

Calls for increasing indirect exclusionary measures also arose in the mid-19th century due to technological innovation, industry stratification, and escalation in conflicts among groups of fishers. This pattern began in Britain, where beam trawls were first developed in the early 1800s. By the middle of the century, fishers using driftnets and other historical gear types started lobbying the British government to ban trawling by claiming that it harmed the fry of many commercially targeted species and resulted in the severe decline of fish stocks. Since there was not sufficient evidence one way or another, the government appointed several Royal Commissions to study the problem. Professor Thomas Henry Huxley was a member of the first Royal Commission, along with other prominent marine scientists. They were able to clarify the reproductive cycles of several important species, debunking many of the coastal fishers' claims. The result was the Sea Fisheries Act of 1868, which repealed most previous regulations intended to protect various species (or, really, the fishers who targeted them), leaving in place only rules for mitigating conflicts among fishers (Shaw-Lefevre 1884, 87–88).

In France, conflicts among fishers began in the 1850s just as wealthy capitalists started to acquire beam trawlers. A combination of serial depletion of oyster stocks and increasing conflicts among fishers led Napoleon III to establish the marine biological station at Concarneau in 1859. Victor Coste, the station head, was charged with undertaking research to improve methods of fishing and fish farming to increase food production (Le Gal 2009). Deregulation began a few years later. In 1862, fisheries within the 3-mile territorial sea were legally listed as open access and effort restrictions were prohibited except for some of the existing protections for oyster fisheries, which were upheld (Fulton 1911). Nonetheless, some measures included in the Declaration on Coastal Fisheries of May 1862 were designed to protect juvenile fish and to limit conflict among fishers. Interestingly, the act stated specifically that these stipulations were made at the behest of fisher cooperatives (*purd'hommes des pecheurs*), their delegates, and fisheries managers (*les Syndics des gens de mer*). However, article IV explicitly allowed the use of gear that dragged the bottom (*des filets trainants*; e.g., trawls) within the 3-mile limit as long as they were confined to areas not used by others and therefore were not "inconvenient to anyone" (*il ne present aucun inconvenient*). The edict also established mesh size limits for various types of gear within the 3-mile limit (article III) and banned all methods of fishing except angling on inland waters (article IV). This last regulation reflected the power of recreational fishers at the time (see chapter 8).

The United States went through a similar process. It started with a controversy between traditional longline fishers who accused new, more efficient net fishers of drastically depleting important commercial stocks in Massachusetts and Rhode Island. The net fishers argued that they were providing food, jobs, and economic development and that stocks must be declining due to environmental changes rather than as a result of overfishing. The conflict came to a head in 1870, when there were record low harvests in both states. Because fishing was such an important industry in Massachusetts and Rhode Island, the economic repercussions of the crisis were widespread. Thus, problem signals were strong. In light of the shortage and the concomitant conflict, the legislatures of both states attempted to intervene but quickly found that they did not have enough information to effectively settle the dispute, even though each state had a Fish and Game Commission and each had funded coastal studies of the fishes in question (Allard 1978; Smith 1994).

This dilemma was observed by a formidable scientist and public figure, Assistant Secretary of the Smithsonian Institution, Spencer Fullerton Baird. In addition to his work at the museum, Baird conducted biological research on marine life from the port of Woods Hole, Massachusetts. Hearing of the claims of coastal fishers while conducting his studies, Baird decided that the best way to resolve the conflict in New England was through federally funded scientific research into the fisheries of the area. He therefore began to lobby Congress for funding for a large-scale survey of the fish in Massachusetts and Rhode Island waters. Here again, the structural context was important in shaping the outcome of the action cycle. Baird wanted only US$5,000 (US$78,600 in 2010 dollars) for this research, but he still needed congressional authorization for the funding. At the time, that entailed the creation of a commission that could provide the basic bureaucratic underpinnings for the expenditure. This was a period of tight fiscal conservatism, so Baird's first request for funding in 1870 was rejected. However, with persistent lobbying by Baird and his colleagues, and after the plan was revised to expand the research beyond New England, Congress approved the creation of the United States Commission of Fish and Fisheries in 1871 (Allard 1978, 76–82; Williamson 2014).

It took some time for federal officials to recognize the political nature of regulations on fishing. Baird faithfully investigated the claims of fishers in Massachusetts and Rhode Island, and found that the net fishers did have some negative effect on stocks, but he recommended only a few time-area closures during spawning periods. Because neither state wanted to take action unless the other did so first, his management recommendations were not followed. The next year, this particular conflict dissipated when fish stocks rebounded exogenously. However, there were many more areas of contention for Baird and the US Fish Commission. A similar disagreement arose between traditional line fishers and new trap fishers on the Great Lakes in 1871, with a similar result. In this case, James W. Milner, an employee of the US Fish Commission, advised that a minimum mesh size for fish traps would prevent capture of juveniles and benefit all fishers on the lakes. Again, the states bordering the lakes chose not to implement the size limit (Allard 1978, 108). This illustrates an important aspect of the governance system at the time: the Fish Commission could provide advice, but states, not the federal government, regulated domestic fisheries.

With additional experience, Baird and other US fisheries scientists uncovered the political rather than scientific basis for most US management decisions in the 19th century. As in Europe, this realization generated a high degree of skepticism regarding fishers' claims. In Baird's own words:

> It very rarely happens that the enactments for the protection and regulation of the fisheries are based upon a thorough knowledge of the habits, migrations, and general relations of the fishes themselves. ... In many cases action, when taken, is the result of unfounded clamor of jealousy of fishermen using one kind of apparatus against those employing another, or, in some instances, it results from the influence

of the wealthier classes, who wish to preserve the fishing as a sport and relaxation, as against the interest of those who depend upon it for a living. (S. F. Baird 1889, 151–152)

However, because the Fish Commission had no direct control over fisheries policy, exclusionary politicking continued as a norm in state and local regulation. With the continued success of modern gears and the industrialization of US fisheries, the power disconnect eventually widened along with the profit disconnect in most states, so restrictions on mechanized fleets were removed in many areas. In addition, as explained in chapter 6, the US Fish Commission tended to favor expansionary rather than restrictive management measures for many decades after its inception. Indeed, Baird himself referred to expansionary measures as "positive management" and to restrictive conservation measures as "negative management."

In Japan, the scientific revolution that started with the Meiji era in 1868 had a similar impact, causing deregulation, but it was filtered through a different set of governance institutions. In general, pre-Meiji rules of sea tenure kept fishing effort at sustainable levels, and there were relatively few conflicts over access rights. However, population growth and demand for both food and foreign exchange generated political demands for developmental subsidies. At the same time, Western fisheries science showed that considerable "unused" biomass was available for exploitation and that fishing effort could expand without harming stocks. Therefore, in conjunction with heavy subsidies, the Japanese government chose to actively relax effort-reducing measures. It nationalized the 3-mile territorial sea in 1875, replacing many of the older rules of sea tenure with a centralized licensing system. Fishers would purchase licenses to "rent" usage of fisheries resources. There were no limits on the number of licenses granted, so this fee was the only effort-reducing mechanism. The new system opened up fisheries to many people with no background in the industry and, since fees were low relative to potential profits, led to a dramatic increase in production, which tripled within seven years of the institution of the new policy (Makino 2011, 24–25). Clearly, the new equilibrium level of effort was much higher than previous levels, and, based on the rapid decline of stocks, this policy substantially widened the profit disconnect in many Japanese fisheries.

The process of deregulation continued through the late-1800s, driven by both endogenous and exogenous forces. Exogenously, there was a general preference for *laissez faire* economic policy during this period, which resulted in widespread reduction of regulation of economic sectors in the 19th century.[1] Endogenously, massive growth in industrial fleets widened the profit disconnect and the power disconnect. Larger-scale fleets were insulated from economic signals by their ability to explore, innovate, and open new markets. They could also provide more jobs, food security, and economic development than small-scale fishers. Together with the preference for less regulation, this capacity advantage generated exceptionally strong alignment between fisheries scientists, managers, and large-scale fishers, processors, and marketers. Indeed, many fisheries scientists and decision makers of the time believed it was better to err on the side of fishers whenever there was uncertainty about the biological effects of effort-reducing measures. Professor Huxley's inaugural address to the International Fisheries Exhibition in London in 1883 sums up the feelings of many fisheries decision makers of the time:

> Now every legislative restriction means the creation of a new offence. In the case of fishery, it means that a simple man of the people, earning a scanty livelihood by hard toil, shall be liable to fine or imprisonment for doing that which he and his fathers before have, up to that time, been free to do. If the general interest clearly requires that this burden should be put upon the fisherman—well and good. But, if it does not—if, indeed, there is any doubt about the matter—I think that the man who has made the unnecessary law deserves a heavier punishment than the man who breaks it. (Huxley 1884, 18–19)

Clearly leaders like Huxley had great faith in the human ability to manipulate marine resources for their own ends, but also great mistrust of measures that would impose costs on fishers or on society. It is also interesting that he describes fishers as a homogeneous group of poor but noble artisans—a nostalgic view often found in historically dominant fishing countries. When Huxley gave his speech, coastal fishers were struggling but fleets of large-scale vessels were flourishing in Britain. Moreover, the owners of these fleets were not from the lower classes or from ancient fishing families, but rather were wealthy capitalists and entrepreneurs. This cognitive dissonance is even more interesting given that Huxley was one of the scientists and decision makers who debunked claims made by small-scale coastal fishers against trawlers in the 1860s and argued strenuously for deregulation. Indeed, even while decision makers extolled the worth of hard-working coastal fishers, they continued to enact policies that favored industrializing fleets, even in the face of growing evidence of depletion.

The case of British trawlers is a good example. In the late-1800s, steam-powered trawling fleets grew rapidly in Britain. This resulted in serial depletion, which was documented through data collected by G. L. Alward in the early 1900s. Catches of groundfish by British fleets were fairly constant from the early 1800s to about 1890, when a steep rise in the number of steam-powered trawlers increased coastal depletion of these stocks. The resultant increase in competition caused "historical" fishers, including sail-powered trawlers that were the focus of exclusionary tactics in the 1850s and the 1860s, to organize as the National Sea Fisheries Protection Association (NSFPA). This organization lobbied for regulation of steam trawlers from about 1890 to the beginning of World War I. It demanded bans on trawling in territorial waters, closures of nursery and spawning grounds, and minimum size limits on fish sold at market, as well as the expansion of British territorial seas from three miles to ten and international negotiation of management for fisheries in the North Atlantic. NSFPA activities led to the creation of several additional Royal Commissions as well as a select committee of the House of Commons (1893) and a fisheries bill before the House of Lords (1904), but little actual management ensued (J. T. Jenkins 1920).

In spite of pressures from the NSFPA, the British government maintained existing open access regulations until about 1910. Although some of the delay was due to international concerns as described in chapter 5, these two decades of inaction show that the large, steam-powered trawling fleets were very influential; they were able to prevent regulation even though the NSFPA provided considerable scientific evidence to back up its claims of depletion. By this time, significant declines in the effectiveness of trawling effort were well documented. In only 10 years, from 1903 to 1913, the catch per unit effort for herring trawlers decreased from 18.94 to 14.31 tonnes per day (J. T. Jenkins 1920, 125–126). These data were collected by government agencies, including the respective fisheries commissions of Scotland, Ireland, England, and Wales. Declines in the catch per unit effort (CPUE) were not as clear cut for other fisheries but, for instance in the case of plaice, data showed that decreasing harvests of large fish were being offset by increased landings of small and "extra small" fish. Furthermore, research by several scientists, including D'Arcy Thompson and W. S. Masterman, showed that these excessive catches of small fish were extremely detrimental to plaice stocks throughout the North Atlantic (63–69).

Eventually, mounting scientific evidence and increasing pressure from domestic lobbyists forced British decision makers to take action. However, it should also be noted that regulations were enacted to protect the coastal zone only after the catch per unit effort for industrial fleets had decreased significantly and many of these vessels had moved to offshore and distant-water fishing grounds by their own choice. In 1910, the Fishery Board of Scotland closed an important fishing ground, the Moray Firth, to trawling. Since the closure applied only to British vessels, it was not very effective at first, due to the flag of convenience problem described in chapter 5, but effectiveness improved when the ban was extended to any vessel fishing in Scotland's territorial seas (Gibbs 1922). Steam-powered trawlers were also banned in the territorial seas around Ireland and other locations in Britain during this period, but British steam-powered trawlers were already fishing off Iceland and Morocco instead of their native shores (J. T. Jenkins 1920). In 1913, the British government created a minimum size limit to protect juvenile plaice. This was part of an international negotiation involving Britain, Denmark, Germany, and Holland. The latter three countries had already established size limits for the species—largely due to similar conflicts between traditional and modern fishers—so the negotiations allowed for harmonization of practices, even though the final result was less restrictive than the limit recommended by scientists (71). As with many of the negotiations described in chapter 5, international arguments over the allocation of costs and benefits led to a dilution of management measures.

This is just one example of a pattern of regulation that was pervasive in the 19th and 20th centuries. In fishery after fishery, governments delayed politically unpopular conservation regulations until forced to take drastic measures in response to economic crisis, just as expected from the AC/SC framework. Often, multiple crises were necessary to ratchet up regulation to a point where substantial rebuilding could occur. In some cases, these cycles of responsive governance generated what Acheson (1997) called a conservation ethic among fishers themselves. Such institutions generated prolonged periods of sustainability in specific fisheries, like those targeting lobster in the northwest Atlantic and those targeting halibut in the northeast Pacific. In other cases there was no happy ending. Many fisheries either stagnated or collapsed because effective response was delayed by the profit disconnect as amplified by the power disconnect. Ecological forces played a role as well, but usually only increased or decreased the frequency and magnitude of the action cycle. The next two sections describe these escalating cycles of serial depletion in greater detail.

7.3 Expanding Cycles of Response

The general trend toward deregulation described above was a necessary political precursor to the massive escalation in fishing effort that started with steam power and gained speed with each new technological innovation in the 20th century. Moreover, as shown by the case of British trawlers, while the action cycle sped up, responsive governance slowed down due to the expansion of both the profit disconnect and the power disconnect.

Scientific uncertainty was often blamed for inaction, but the prevalence of delayed response and the resulting serial overexploitation of fisheries were well known by the middle of the 20th century. Even industry actors recognized the costs of responsive governance in this context. For instance, when testifying before the US Congress in 1947, Henry Jackson of the industry-sponsored National Fisheries Institute described multiple fishery collapses and then stated explicitly that:

> Unfortunately, the public is not aroused. Yes, I must be frank and say that even the fishing industry itself is not aroused until a given species of fish reaches such a low population that it requires a tremendously expensive, long-time project in order to restore it. Not until it becomes unprofitable to fish the species does the public seem to wake up. This has been the history of nearly all of our fishery resources. (US Congress 1947a, 10)

This section explains how the profit disconnect and the power disconnect reinforced each other over time, causing increasing delays in effective political response that paralleled the cycles of expansion described in Part I. It shows how governments ignored mounting scientific evidence of overexploitation of sea fisheries and increasing calls for regulation by coastal fishers in the first half of the 20th century. During this period, managers and even some scientists continued to argue that fluctuations in catches were due to natural rather than anthropogenic forces. By the early 1950s, it was clear that some type of effort-reducing management was necessary, even for "sea fisheries," but the power disconnect was even wider and there was little regulatory action. Modernization of coastal fleets and geographic expansion of distant-water fleets reduced domestic conflicts among groups of fishers, although conflicts over international regulation continued, as described in Chapter 5. Thus, the industry was often united against conservation management measures at home, and scientists were the only actors who were sensitive to problem signals in the action cycle. Alarmed by the rapid overexploitation of the first half of the century, some scientists argued for a "go slow" approach to proactively control fishing effort, but most governments opted for a responsive approach to fisheries governance.

7.3.1 Necessary but Not Sufficient

Increasing evidence of serial depletion allowed scientists to prove that fishing could indeed drastically reduce stocks of marine species, including herring, cod, haddock, most commercially targeted whales, and many others. Fisheries biologists further developed and tested theoretical models to identify the "optimum catch" and then "maximum sustainable yield" in fisheries, along with tools for understanding non-human drivers of variability in populations and catch size. In spite of these many advances, governance response was increasingly delayed by growing disconnects between problem signals and decision makers. In comparing cases for all major commercial fisheries and many minor fisheries of the period, I found none in which managers implemented strong conservation measures prior to the severe depletion of a given stock. In those cases in which early laws matched scientific advice, monitoring and enforcement were neglected, and both catch and effort remained higher than recommended. This subsection provides evidence of delayed response and related power dynamics, but also shows how response to serial depletion extended management options in the structural context during the first half of the 20th century.

By the early 1900s, fisheries scientists could show conclusively that some "sea fisheries," such as the trawl fisheries targeting groundfish described above, were exhaustible. However, some scientists and many decision makers were not convinced that reducing fishing effort was a solution to the "overfishing problem." They still believed that environmental forces were the primary drivers of variation in fishery production. World War I provided a natural experiment of sorts to test the effects of reduced fishing effort, particularly around Europe, where fishing stopped almost completely during the conflict. The war lasted from July 1914 to November 1918. Before the war, landings data showed considerable depletion in commercial fisheries. The catch per unit effort was quite low in many fisheries, and fishers were catching primarily small fish, even in areas where large fish were had once been plentiful. After the war, catch per unit effort was 50–100% higher than prewar levels, depending on the area fished. Fishers also caught a much larger proportion of large and medium-size fish just after the war (Smith 1994, 158–159).

Scientists from the International Council for the Exploration of the Seas (ICES), an international body charged with providing European governments with scientific advice on North Atlantic fisheries, used the "Great Fishing Experiment" of World War I to demonstrate the economic and biological benefits of reduced catch and effort levels in fisheries. However, because fisheries were profitable again, many people entered the industry, and the gains made during the war quickly dissipated. Rather than heed the advice of ICES scientists and limit fishing effort, European governments chose to use subsidies to rapidly rebuild their domestic fleets in pursuit of economic growth and food security. Given the privations of the war, the political pressure to attain both of these goals was intense. However, these short-sighted policies accelerated overfishing in the northern Atlantic and adjacent seas. By the 1930s, scientists, policy makers, and the fishing industries of Europe again faced problems of overexploitation. Moreover, while distant-water fleets could mitigate some of the loss of revenue and food supply, they could not alleviate the economic hardships faced by coastal fishers (Smith 1994, 162).

Other regions also were plagued with overfishing in the interwar years. Fisheries in Canada and the United States were overexploited on both coasts, even as far north as Alaska, and Japanese fleets had depleted fisheries near Korea, China, and Russia, as well as along their own shores (see chapter 2). Like the Europeans, most North American fisheries management in this period focused on size limits and time-area closures to protect juvenile fish. These types of regulations were preferred by industrialized fleets because they were difficult to enforce and therefore had little impact on harvests. Size limits in particular could be circumvented by discarding fish. Many coastal fishers were dismayed by the rapid depletion of their targeted stocks but did not have sufficient influence to counter industry lobbying against stronger catch-reducing and effort-reducing measures.

After a period of frenetic growth, most North American fisheries were hit hard by the Great Depression toward the end of the 1920s. Severe economic contraction generated a strong price signal that triggered fishers to demand subsidies and trade protections across the board. When such measures failed, fishing groups would also lobby for closures of less productive grounds or other methods of reducing supply at low cost. As in the past, the most expedient alternative was to argue for conservation management measures that effectively excluded other groups of fishers. These pressures led fisheries managers in the United States and elsewhere to reconsider their past approaches and the need for management interventions. Henry O'Malley, then Fish Commissioner of the United States, summed up the problem in a statement before the Senate appropriations committee in 1930. After providing data on the economic importance of US fisheries, he stated:

> We have been prodigal in the use of this resource and so long as there was an abundant supply of fish little thought was given to the future. As rapidly as the supply of new species declined we substituted others. When the fish in one area became scarce we moved on to other areas. These things are no longer possible and we are faced with the necessity of giving much more attention to husbanding the available supply. (US Congress 1930, 2)

O'Malley went on to point out that the lack of property rights at sea and the migratory nature of fishing fleets undermined the "orderly development and growth" of fisheries and complicated fisheries management. This eloquence had a specific purpose: to convince the Senate to approve a special budget of more than US$350,000 (US$3.7 million in 2010 dollars) for a five-year science and management program at the Federal Bureau of Fisheries, which had replaced the US Fish Commission in 1903 (US Congress 1930, 1). However, his words reflect the Bureau's experience with the unchecked decline of multiple fisheries, including those for commercially important marine fisheries like mackerel, cod, and halibut (Smith 1994). Through manifold small- and large-scale fishery declines, managers learned that size limits and protections on juvenile or spawning populations were not sufficient to maintain stocks given the fishing power of modern fleets. In spite of this knowledge and the growing scientific evidence provided by fisheries scientists, other conservation measures remained a hard sell in most fisheries, and the idea that one could improve the economics of a fishery by reducing fishing effort remained counterintuitive until another great experiment was carried out—this time in the Pacific halibut fishery.

As already described in chapter 5, the International Pacific Halibut Commission (IPHC; known at the time as the International Fisheries Commission) was established in 1923, in response to price signals generated by market gluts rather than in response to overfishing per se. However, by the 1920s, many near-shore stocks were already heavily overfished. William F. Thompson and other IPHC scientists showed that serial depletion of stocks of Pacific halibut started in the early 1900s, but that the economic and biological signals of this core problem were masked by the mobility of the fishing fleets. Fishers intensively focused on stocks that were close to shore first, and then, as those stocks declined and finding the fish became more difficult, they moved farther and farther offshore, leaving depleted stocks in their wake (Thompson and Bell 1934). However, there was no regulatory response to this report. By 1930, production had declined steeply, and fishers had run out of new halibut stocks to exploit. In combination with the decrease in demand associated with the Great Depression, this finally reduced the profit disconnect and fishing effort dropped precipitously as many fishers were forced out of the industry in 1931 (14).

In 1932–1933, the IPHC reinforced this economically driven reduction by modifying and extending existing time-area closures and by setting area-specific catch limits. Taking such action required renegotiation of the terms of the original treaty that established the IPHC and almost certainly would not have been possible if the fishery were not in a crisis situation. As reported by Thompson and Bell (1934), the result of this management policy was an increase in the catch per unit of effort in all three areas due to a larger stock of bigger fish. The period of the study was only three years (1930–1933), but the fishery continued to prosper. These positive results were widely publicized in the mid-1930s and this purported success was frequently cited as justification for new management institutions like the International Pacific Salmon Fisheries Commission (US Congress 1954). Indeed, the IPHC's apparent success completely altered the structural context for fisheries management in the United States and elsewhere. It can be summed up in the words of Thompson and Bell:

> It [the IPHC] hopes to determine what is the most favorable intensity of the fishery in order to get the largest permanent yield from each stock. It seeks to continue to provide regulations which will permit the fishermen to make as large catches as can be made without endangering future supply. (Thompson and Bell 1934)

Thompson and Bell's findings further validated European science on the effects of fishing and revitalized calls for conservation management measures in Europe. Norwegian fisheries scientist and ICES member, Johan Hjort, developed methods for estimating the "optimum yield" for a fishery as part of his larger research program into the serial depletion of whale stocks (Smith 1994, 220). Hjort started his research around the turn of the century and published a detailed description of the serial depletion of northern whale stocks in 1902 (Burnett 2012, 29). At first Hjort was skeptical that southern whale stocks would be equally overexploited, because they were farther from markets and infrastructure (84–85). However, in the early 1930s he published several more papers describing how innovation and growing demand for whale products drove the serial depletion of stocks and pointing out the need for some "equilibrium between the whaling industry and the annual renewal of the stock" (Hjort 1932). Furthermore, Hjort built on existing models of human population growth, such as Pearl and Reed 1920, which used sigmoidal growth curves to create similar population models for whales. Using these tools, he determined that the optimum yield for any given whale fishery was the rate of extraction that would maintain the whale stock at half its unfished biomass. At this point, any gains in survivorship due to reduced competition over limited resources would be counteracted by the lower reproduction of the smaller stock of whales (Finley 2011; Smith 1994; Caddy and Cochrane 2001).

Hjort published his whale population models in 1933, a year before Thompson and Bell's report on North Pacific halibut. Both the modeling techniques used by Hjort and the statistical techniques developed by the IPHC scientists were quickly taken up and modified by other fisheries scientists. Michael Graham of the Lowestoft Laboratory in Britain and a member ICES, particularly advanced Hjort's work by refining his models and applying them to the groundfish fisheries of the North Sea. Graham used the results of this analysis to actively argue for reduction of harvests in these fisheries in a 1935 paper (Smith 1994, 230–231). Many other scientists worked in parallel with Hjort, Thompson, and Graham to develop models that would allow them to identify the "largest permanent yield" from a stock of fish. In talking to his protégés, Graham likened this search to the quest for the Holy Grail (Beverton and Holt 1993). This perspective contrasts strongly to that of Huxley in the 1880s and shows that increasing cycles of serial depletion triggered substantial changes in fisheries science, making scientists much more sensitive to catch and catch per unit effort as problem signals. However, while the role of science was an institutionalized aspect of management, scientists did not have much influence in the decision making processes, so the power disconnect was still considerable.

By the late 1930s, demand for effective management intervention in fisheries was high as a result of the economic problem signals created by widespread overexploitation and exogenous declines in demand. With no major technological innovations in sight or much prospect of opening new markets, fishers turned to political solutions, including conservation management as a last resort. Because the most important fisheries were targeted by fleets from multiple countries, governments intensified treaty negotiations covering North Atlantic fisheries, salmon and other fisheries in the north Pacific, and whaling fisheries all over the world, as shown in chapter 5. Few treaties were signed prior to the start of World War II, and those that were had little chance to go into effect before hostilities brought an end to cooperation. Domestically, however, the historically dominant fishing countries of Europe and North America built up considerable scientific and management capacity during the interwar years. While management remained delayed, there were some instances in which, like the IPHC, domestic governance organizations established rebuilding plans in coordination with local fishing communities that led to long periods of stability. Of course, there were also cases in which fishery collapse was not avoided and management was both too little and too late.

Before moving on to the post-World War II period, it is important to consider fisheries governance in Japan, which was the largest producer of fresh and processed fish products in the interwar years. Japan engaged in responsive governance, just like other historically dominant fishing countries. However, some substantive differences in the nature of its management response are apparent. The Meiji Fisheries Law of 1901 was the most sweeping reform of Japanese fisheries policy and remains the basis for

Japanese fisheries management today. It replaced the previous policy in which license fees were the only limit on effort with a new system of territorial use rights similar to the sea tenure practices of earlier periods. These rights were area based and could be bought or sold like property rights on land (Makino 2011, 25). In addition, the government closed near-shore areas to trawling to protect coastal fishers. This action reflects a more even distribution of power between coastal and distant-water fleets compared to other countries, but is also indicative of the government's tight control over economic policy and of its geopolitical goal of regional domination (Suisankyoku 1915). Licenses were still required for offshore and distant-water fishing, but the total number of licenses per fishery was controlled by the government. The fleet targeting a specific stock would be encouraged to grow until the point where profits started to decline. When this happened, the government started sending vessels to new fisheries, gradually reducing effort in the original fishery (Finley 2011). For these fleets, responsive governance was the explicit policy of the Japanese government, but they responded to profit signals rather than to measures of biomass or political problem signals.

The Western concept of optimum biological yield was not a major consideration in Japanese management and therefore was not appropriated by Japanese fisheries scientists, all of whom were funded through government grants. Indeed, tight government control of both the fishing industry and fisheries science was and is an important attribute of Japanese fisheries management. The government also supported a large and active scientific research program focused on exploration to identify new fishing grounds and "underutilized" species of fish. In particular, it funded large-scale expeditions to identify new fisheries, document the life histories and migration of fish stocks, and collect and analyze oceanographic data related to their fisheries. There is little record of discourse between European and North American fisheries scientists and their Japanese counterparts. For the most part, Westerners viewed Japanese science as inferior—but the record indicates that the Japanese research program was rigorous even though it focused largely on collecting information that would help the industry expand production (Finely 2011). Thus, there was considerable alignment between the interests of fishers, scientists, and decision makers in Japan, and access was effectively controlled in domestic and international waters, but the focus was on maintaining profits rather than conserving stocks.

Interestingly, the Japanese approach does not appear to have been more effective than belated attempts to set optimum yields in the West. Many fisheries targeted by Japanese fleets were heavily overexploited by the beginning of World War II. Indeed, coastal fisheries in Japan were under even greater pressure than those of Europe or the United States, and the heavy economic costs of depletion led to the creation of the Meiji Fisheries Law in 1901 and to its revision in 1910. However, with little flexibility and a tendency toward oligopoly, the new laws limited access without limiting effort, so overexploitation set in as rising demand widened the profit disconnect substantially in this period. Because the Japanese government managed for profitability, this disconnect delayed strong conservation-based management and allowed core problems to escalate in spite of tight government controls on effort (Makino 2011, 25). This finding is key because it shows that severe overexploitation can result in the absence of the tragedy of the commons. Overexploitation of Japanese fisheries cannot be blamed on open access after 1901; it is entirely driven by the profit disconnect and related government policies. The war changed many things about Japanese fisheries management, but the tight connections between the government, the fishing industry, and fisheries science remain important (26–30).

7.3.2 Maximum Sustainable Yield

As described at length in previous chapters, World War II halted fishing in many industrialized fisheries and also significantly reduced the size of fishing fleets in historically dominant fishing countries. Some scientists hoped that postwar fisheries management would finally shift from responsive to proactive governance, although they did not use those terms at the time. According to Graham's protégé Sydney J. Holt:

> Those were post-war years of optimism. It seemed that with the reduction in the number of fishing vessels available, as a direct result of the war, and the expected recovery of depleted fish stocks during it, it would be politically feasible for Europeans to engage in what [Michael] Graham called "rational fishing." (Holt [1957, 1993] 2004, ii)

Graham was one of the researchers who documented the stock rebuilding around Europe after World War I and argued unsuccessfully for limiting the growth of fisheries in the 1920s. From that experience, he was well aware of the various political pressures that could prevent timely limitations on catch and/or effort. Furthermore, he believed that the only way to ensure that decision makers could be convinced to avoid the mistakes of the past was to provide them with more specific predictions of the positive benefits of proposed management measures, particularly in terms of higher landings. Here again, advances in science were driven as much by problems in the action cycle as by scientific curiosity per se. However, determining an optimum yield for "rational fishing" was not an easy task, and several endogenous and exogenous advances in science were needed before Graham could attain this goal.

Scientific Foundations The first use of calculus in fisheries population models can be traced to a paper by Fedor Ilyich Baranov in 1918. Because it was written in Russian and because the mathematics was quite advanced, the paper was inaccessible to many fisheries biologists of the time. E. S. Russell, then director of the Lowestoft Laboratory in Britain, interpreted Baranov's work in 1938, but it gained little traction until the Canadian scientist William Ricker built upon Baranov's models and generated the first differential equations that accounted for fishing and natural mortality separately in 1940. Ricker tested this model using pre–World War II data from the Pacific halibut fishery and showed that his approach produced results that were similar to Thompson and Bell's (1934) findings (Smith 1994, 307–308). This was important because the usefulness of mathematical modeling techniques in fisheries biology was controversial and because the test showed that a simple model could provide predictions that fit well with existing datasets (301). Perhaps more important, Ricker (1946) was also the first to use the term "maximum sustainable yield" (MSY) to describe the largest amount that could be consistently removed from a stock of fish without diminishing future harvests (Finley 2011).

Even with this mathematical advance, fisheries biologists still had to grapple with the problem of selectivity in fisheries. That is, fishers using different gears in different areas would catch different sizes and therefore different ages of fish. Graham recognized that this phenomenon also had political importance, since variation occurred between fleets from different countries. Thus any given regulation would have disproportionate costs and benefits at the state level. For instance, a minimum size limit on plaice for the entire North Atlantic would have little effect on British trawlers who caught large fish, but would have a severe effect on Dutch driftnetters who caught mainly small fish. Furthermore, any regulation that increased the number of adult fish would benefit the British fleets more than the Dutch. This problem apparently preyed on Graham's mind even during the war, because he asked the mathematician H. R. Hulme, a co-worker at the British Air Ministry, to work on it. Hulme famously wrote his response on an envelope during a spare moment. It was a set of equations that incorporated age structure into a simple recursive population dynamics model. That is, rather than looking at the population as a whole, Hulme's equations accounted for changes in individual year classes, or age groups of fish (Smith 1994, 310).

After the war, Graham hired Sydney Holt and Raymond J.H. Beverton to quickly transform Hulme's insight into a practical population model. This entailed considerable work, since Hulme had assumed that the weight of a fish was a linear function of its length (an assumption not supported by evidence) and that fish could live forever as long as they had enough food and were not caught either by humans or other fish. Beverton and Holt also built on Ricker's work. Another mathematician, Ludwig von Bertalanffy, helped them remove Ricker's unlikely assumption that the weight of an individual fish increased exponentially with no limit. Surmounting mathematical difficulties, Beverton and Holt published a generic model of age-structured population dynamics with Hulme as first author in 1947, and then spent the next two years working on specific models that could be applied to the trawl fisheries of the North Atlantic. Their work was disseminated through many venues, including peer-reviewed publication and workshops organized by the British government, the International Council for the Exploration of the Seas, and the newly created North East Atlantic Fisheries Commission (NEAFC). Their compiled research was finally published in 1954 and remains one of the foundational texts for fisheries population biologists (Beverton and Holt 1993).

Although Beverton and Holt's work had a profound effect on their field, and was particularly useful for the calculation of size limits and other measures related to reproduction, it did not alter the pattern of expanding cycles of serial depletion that was observed before the war. As shown by Smith (1994, 297), plaice populations rebounded due to low fishing rates during the conflict. The same was true for other commercially targeted stocks in the North Atlantic. On average, numbers of fish doubled while the total weight or biomass tripled. Age structure also shifted, and fishermen caught more large fish after the war than before. These findings were much more certain than the post–World War I assessments showing similar effects. Data collection was much improved, as were methods of analysis (296–297). Nevertheless, decision makers chose to subsidize the expansion of fishing fleets as described in chapter 6, rather than limiting effort as Graham had hoped. Few if any limits were placed on fisheries in Europe in the postwar period, and the effectiveness of newly reconstructed fleets was much higher than it had been during the interwar years. Furthermore, coastal fishers had appropriated new technologies and were now less exposed to economic problem signals. Combined, these factors led to a third phase of even more rapid depletion of coastal and offshore fisheries, with a subsequent push to expand production in distant waters.

In the United States, conditions were somewhat different than in Europe but results were similar. Several fisheries continued through the war, often with support from the federal government; many of these were severely overexploited before the war ended. In an important hearing before the House of Representatives on the "Fish and Shellfish Problems of the United States" in 1947, experts testified at length about the unchecked biological and economic decline of commercially important domestic fisheries targeting cod, haddock, pilchard, menhaden, shad, herring, and many other species (US Congress 1947a). Henry Jackson of the industry-funded National Fisheries Institute even described the rapid rise of a shark fishery around San Francisco that catered to the tastes of wealthy Chinese immigrants for shark fin soup. Shark populations in the Bay Area were severely depleted within three or four years—before scientists and managers even realized there was a fishery in need of study (40–41). The primary argument of the domestic industry was not for proactive governance per se, but rather that government intervention needed to occur earlier, before crisis set in, and that scientists and managers should be provided with more resources to help the industry avoid the "needless waste" of overexploitation (52). They now believed that management should occur at some highest possible yield and that limits should not be set on the development of "latent fisheries" (49).

Like their European counterparts, US fisheries scientists—most of whom did not have direct managerial powers at the time—also had difficulty identifying that "largest possible yield." In this respect, the collapse of known fisheries was an important source of data on the impact of fishing on fish stocks. Of all the fisheries that collapsed after the war, the case of California sardines was most critical to the development of fisheries science and management. Its collapse was also an enormous economic loss and a potent demonstration of the need for management intervention. The California Fish and Game Commission pointed out depletion in the sardine fisheries in the 1920s, but their attempts to convince the legislature to take action to limit fishing effort were blocked by powerful fishing and processing groups until the early 1950s, when the fishery finally collapsed. Scientists employed by state and federal agencies worked over the intervening period to develop convincing analyses that would show how much fishing was too much. Although none could make an exact determination, all studies showed that current catches were unsustainable and recommended reductions in effort. Managers, who were politicians rather than bureaucrats at this time, listened to industry representatives instead. Lobbyists argued that catches kept increasing and so no limits were needed. In this they ignored the increasing size and effectiveness of the sardine fleet, which masked the catch and profit signals as expected based on the AC/SC framework (Smith 1994, 241–254).

Fishing "Up To" MSY These studies of the California sardine fishery laid the groundwork for another important tool of fisheries governance and a keystone in US management: the yield-effort function. In the 1930s, scientists at the US Bureau of Fisheries started to apply methods developed by economists to the problem of overfishing in the sardine fishery. They found that although catches were increasing, they were doing so at a decreasing rate. Thus, catch followed a logistic growth function for the fishery, even though there were cyclical oscillations around that trend (Smith 1994, 249–251). Milner Schaeffer, who had been a junior member of the team studying sardines in the 1930s, continued this line of inquiry while researching tuna in Hawaii in the 1940s and while serving as the first head of the new Inter-American Tropical Tuna Commission in the early 1950s. He published his "surplus production theory" in 1954 and fit his yield-effort models to data from yellowfin and skipjack tuna landings in the Pacific. He concluded that yellowfin was probably at MSY but that skipjack appeared to be so abundant that fishing had no discernible effect on the stock. He also argued that increasing costs of production with lower stock size would create an economic equilibrium around maximum sustainable yield, and therefore believed that little effort-reducing management was necessary in fisheries (Finley 2011). Schaeffer, too, missed the importance of the profit disconnect.

Many people, including biologists, ecologists, and economists, criticized the MSY approach in the 1950s. Graham, Beverton, Holt, and several other European scientists favored a "go slow" approach. They argued that size limits and gear restrictions should be put in place as early as possible, and that regulators should limit effort before the fishery reached MSY rather than waiting until overexploitation was obvious (Finley 2011). G. L. Kesteven and other ecologists criticized all of the mathematical approaches described above, including MSY, claiming that environmental and trophic interactions should be important management considerations (Smith 1994). These critics were proved right in several cases (see section 7.4). The economist H. Scott Gordon also took Schaeffer to task for his economic interpretation of the yield-effort curve (Beverton and Holt 1993). Gordon (1954) modeled costs and revenues and showed that the open access yield would be well above the maximum sustainable yield in most cases. He also pointed out that the maximum economic yield was usually lower than the maximum sustainable yield. The latter would provide higher profits to the industry, though in theory it would also result in a lower supply of fish and therefore a higher price for consumers. This conclusion was based solely on the assumed connection between harvests and stock size; like Schaeffer, Gordon did not consider the effect of economic responses that could dampen problem signals.

In spite of these criticisms, MSY became the dominant benchmark for fisheries management in the 1950s and the 1960s (Gulland 1974). As Finley (2011) points out in her exhaustive study of the institutionalization of this approach, MSY appealed to decision makers and to industrial actors because it appeared to balance the need for conservation—or prevention of fishery overexploitation and collapse—with the desire for economic development and food security. In particular, Schaeffer's work supported the dominant policy position of the United States, which was to allow fisheries to grow just until the first signs of overexploitation, at which point management might be used to help return the fishery to MSY (Smith 1994). In other words, responsive governance was the official policy of the United States in the post-World War II period. Wilbert McCleod Chapman established this policy during his short tenure in the State Department, and his successor, William Herrington, continued to advocate it in multiple international negotiations at regional fisheries management organizations and in talks regarding the UN Law of the Sea (Finley 2011). At home, the US Fish and Wildlife Service, which replaced the Bureau of Fisheries in 1939, worked hard to help develop new fisheries in order to maximize use of ocean resources (Hobart 1996).

The management strategy that allowed fishing "up to" MSY appealed to many other countries as well. As Gordon (1954) points out, in spite of all past experiences of overexploitation and fishery collapse in near-shore waters, the 1950s was a period of renewed optimism about the wealth of oceans. The rapid growth in landings after the war led many to believe that fisheries were only limited in the local sense, and that the global potential of the oceans was unlimited. As described previously, developed and developing countries alike looked to fisheries as a major source of economic development and food security. Thus, in negotiations on the UN Law of the Sea they could agree on the ideal of rational management at the maximum sustainable yield even when allocation of access rights remained a highly contentious issue (Juda 1996). Though some Japanese scientists, like their European counterparts, criticized MSY, Japan's government embraced the concept during US occupation and continued to use it thereafter. The Soviet Union and many other states with plans for the development of fisheries did the same (Finley 2011).

There were a few other reasons for the rapid institutionalization of MSY in global fisheries management. MSY was much easier to calculate than other benchmarks, data requirements were low relative to other approaches, and results were easy for decision makers to interpret. This made it attractive to many fisheries scientists, particularly those who provided management advice. Indeed, the unprecedented growth in the number of stocks targeted in the postwar period generated considerable pressure to develop management benchmarks that would be relatively easy to calculate from catch data. MSY clearly fit this bill. It was even appropriated by European scientists who preferred a more nuanced approach but needed to conserve their limited financial and organizational resources (Beverton and Holt 1993). The relative simplicity of MSY also made it easier to communicate to fishers and other vested interests, further increasing its attractiveness as a management tool (Gulland 1974, 108). For all parties, the progressive belief that humans could control the environment using science and technology was also important. MSY was seen by many as a panacea, much as hatcheries had been at the turn of the 20th century. It would bring rationality to fisheries management and thereby solve the core problem of overexploitation. Overcapitalization and ecosystem disruption were not yet on managers' radar, but this changed with the continued failure of responsive governance in the 1970s.

7.4 Global Disconnects

Unfortunately, reality fell well short of the ideal of rational management using MSY. During the 1950s and the 1960s, the disconnects between problems, profit, and power were all greater in both scope and scale than at any other time in history. Indeed, there was a strong feedback between profit and power that widened both disconnects. As industry actors accumulated capital, they also accumulated power, which helped garner expansionary management and postpone conservation management, further increasing the wealth of actors who could avoid core problem signals. These profit-power cycles resulted in even longer delays in governance response, the related overexploitation of many commercially valuable stocks, the tremendous overcapitalization of fishing fleets, considerable habitat destruction, and the depletion of many by-catch species. In addition, conflicts among groups of fishers increased with declining catches and declining profits. By the late 1970s, much of the optimism of the 1950s had been dispelled by collapses in the great Pacific anchoveta fishery, by closures of the historic herring fisheries of the North Atlantic, and by the near extinction of several whale species (Holt 1978). Many people attributed these declines to a lack of governance on the high seas, though the collapse of smaller, coastal fisheries should have signaled that the problem had much deeper roots (McGoodwin 1990).

Writing from his wide experience with fisheries governance as a senior scientist in the Fisheries Department of the FAO, John Gulland captured the growing concern over the pace of change in fishing effort and related governance impacts in the 1960s:

> The traditional series of events is for a fishery to develop, falling catches cause fishermen to complain, a scientific investigation is set in motion, the results of this investigation show that certain management measures are desirable, and finally these are introduced. ... [However,] fisheries can now develop and reach a state of crisis in a much shorter time than that required for a scientific assessment by classical methods. The research then becomes a post-mortem rather than a source of useful advice. (Gulland 1971, 471)

Gulland's words reflect a widespread awareness of the responsive nature of fisheries governance and the potential for long-term overexploitation due to rapid economic growth and technological development. This section documents how the power disconnect further delayed response at a time when the profit disconnect and related technological improvements had vastly increased the geographic scale and the technical effectiveness of fishing effort. Section 7.4.1 describes how the power disconnect affected governance in developing coastal states after the incursion of industrialized fleets in the 1960s. Section 7.4.2 explains how domestic pressures to exclude distant-water fleets fostered the widespread acceptance of the 200-mile Exclusive Economic Zone in the 1970s but then shows how lobbying by domestic fishers continued to undermine management of coastal fisheries well into the 1980s. It also shows collapses in large-scale fisheries continued to drive improvements in science and changes in governance institutions.

7.4.1 Coastal Developing Countries

In the postwar period, most of the fleets in historically dominant fishing countries were fairly modern. Even small-scale fishers used diesel-powered vessels and mechanical devices to enhance their efficiency, and economic niches were well established. However, the globalization of commerce enabled developing countries to skip this gradual process and import or license large-scale industrial vessels that directly competed with artisanal fishers. In some cases, governments or corporations helped artisanal fleets modernize through subsidies for small motors or other incremental improvements, but the economic gap between industrial and artisanal fleets was still unprecedented. The incursion of industrial fleets also had a profound effect on the politics of response in these countries. In region after region, local management systems that had operated sustainably for centuries were undermined or replaced as industrial fleets increased production, often with the blessing of government organizations (Berkes et al. 2006). The resulting power disconnect—combined with the lack of enforcement capacity described in section 7.5—led to the uncontrolled overexploitation of commercial stocks in these regions and, as in developed countries, the eventual collapse of stocks with associated delayed response.

As noted in section 5.1, insular fishing communities often developed their own methods for managing coastal fish stocks based on endowments and entitlements. Although largely focused on exclusion, these local institutions also incorporated what is now known as traditional ecological knowledge to protect spawning or juvenile fish and generally keep stocks healthy. Some fisheries were used solely for the subsistence of coastal communities and some to supply local markets, but demand was relatively low in both instances and technological limits prevented large-scale trade in fish products. In other words, the profit disconnect and the power disconnect were minimized for centuries and responsive governance worked quite well. However, due to population growth, economic development, and trade globalization in the post-World War II period, pressures on near-shore fisheries expanded exponentially. Therefore, it was increasingly difficult for local populations to exclude outsiders—either new domestic fishers or encroaching distant-water fleets. Furthermore, the dual incentives of higher prices for fish products and the wider availability of consumer goods undermined many local management institutions as fishers chose to ignore traditional rules in pursuit of higher incomes and consumption-oriented lifestyles. The erosion of these local institutions, combined with the profit disconnect and lack of government management capacity, resulted in the overexploitation of in-shore fisheries all around the world (McGoodwin 1990).

Fishing effort and harvests in large-scale, highly industrialized offshore fisheries also ratcheted up to unprecedented levels in the coastal zones of developing countries during the postwar period. Response was delayed even further by feedbacks between the profit disconnect and the power disconnect. The Peruvian anchoveta fishery provides a well-documented example. Before 1977, most large-scale fisheries were international in scope because a majority of countries still adhered to the 3-mile territorial sea. However, Peru claimed and defended a 200-mile conservation zone in the early 1950s, so the fishery for Peruvian anchoveta was one of the few large-scale fisheries that was managed by a single country after World War II. As described in chapter 6, the anchoveta fishery grew by several orders of magnitude from the 1950s through the 1960s. Catches skyrocketed from 1,100 tonnes in 1950 to nearly 3.5 million tonnes in 1960—placing this fishery on par with the two oldest and most productive fisheries prior to 1960, Atlantic cod and Atlantic herring. This growth was largely driven by high demand for fish meal for animal feed and the collapse of sardine fisheries around the world. It continued for another decade, and landings reached over 13 million tonnes in 1970, making the Peruvian anchoveta fishery easily the largest in the world (FAO 2012a).

Along with increased employment in fishing, the Peruvian anchoveta fishery brought economic development in shore-based facilities and foreign exchange through the export of fish meal. The number of processing plants increased from 96 in 1958 to a peak of 220 in 1967. These plants employed 14,000 people, and production of fish meal became the largest manufacturing industry in Peru (Guerra 1972). As a contributor to gross domestic product, anchoveta processing rapidly eclipsed the guano industry, which had been politically and economically dominant in previous years (Murphy 1972). Although the guano industry remained important, the political ascendency of the anchoveta industry is clear. There is a natural conflict between anchoveta fishing and guano mining because the marine birds that produce copious quantities of guano can exist in large numbers only because of the plentiful food source provided by anchoveta. Thus guano producers feared that high anchoveta harvests would destroy the foundation of their livelihoods and lobbied hard to restrict the growth of the fishing industry in Peru. However, the potential for wealth creation and economic development through exploitation of anchoveta was obvious by the late 1950s, and the government supported the growth of the fishing fleets in spite of guano industry protests (Glantz 1979).

Given its goal of maximizing production and related employment and foreign exchange generation (through exports of fishmeal), the Peruvian government espoused the US-fostered ethos of management "up to" MSY. It established a Council of Hydrobiological Research in 1954, and later appealed to international organizations for help in establishing a scientific program for the study of anchoveta populations and fishery impacts. In 1960, the UN Food and Agriculture Organization (FAO), the United Nations Development Programme, and the Peruvian government started a joint research project called the Instituto del Mar del Peru. Good data were already available from the Council of Hydrobiological Research, so scientists were able to estimate stock size, fishing mortality, and maximum sustainable yield fairly quickly (Boerma and Gulland 1973). On advice from scientists, the Peruvian government established a two-month closed season in the middle of the year, when catches were light, and a total allowable catch (TAC) of 6.35 million tonnes in 1965 (Aranda 2009, 147). Aside from the closure period, the fishery would remain open until reported landings reached the TAC, and then fishing would be stopped completely until the next year. Additional measures were enacted to protect juveniles (a two-month closure at the beginning of the year, which replaced the previous closure) and to protect fishers (a mandatory five-day work week). In spite of its good intentions, the government gradually increased the TAC, based on increased estimates of MSY (due to a decline in bird populations) and because of pressure from the industry (Boerma and Gulland 1973). Unfortunately, these regulations were not well enforced, and there were no limits on fishing effort, so overcapitalization was an exceptionally large problem for both the fleet and the processing industry.

As the Peruvian anchoveta industry got bigger, its political power increased, largely due to its scale and related importance in the Peruvian economy. This growth was caused in large part by exogenous drivers of the profit disconnect, particularly demand, but overcapitalization could have been stopped if not for the power disconnect. Indeed, the government of Peru allowed anchoveta fleets and the processing industry to expand to the exceptionally large size described above in spite of repeated recommendations for effort limits from scientists and calls for reductions from the guano industry and from other fishers targeting local stocks of fish for food. By the early 1970s, the Peruvian fleet was capable of catching one third of the total anchoveta biomass in a single month (Glantz 1979). This capacity is evident in the length of the open season, which only lasted 90 days in 1971. In other words, fleets harvested the full annual TAC of 7.5 million tonnes in a month and a half. Processing capacity at the time was substantial but the industry still could not process this catch fast enough in such a short time, so much of the harvest was wasted (Boerma and Gulland 1973, 2232). The TAC and related short season caused considerable economic hardship for fishers and for processors alike. Both groups put great pressure on decision makers to raise the total allowable catch in spite of increasing warnings of overfishing from the scientific community. With cooperation from processors, fishers also increased their out-of-season harvests, so the 11.2 million tonnes of anchoveta landings reported for 1971, though much greater than the TAC, was an underestimate of the true fishing mortality (Aranda 2009, 147).

Industry responses ran counter to scientific advice, but allowances were made for economic hardship and the government did not enforce regulations. In 1970 and 1971, the stock assessment panel for anchoveta strongly recommended reducing fishing capacity and downward adjustment of TAC to account for changes in recruitment (that is, the survival of juveniles to the adult stage; Boerma and Gulland 1973). In 1972 the stock collapsed, partly due to overfishing and partly due to changes in natural conditions. This caused a severe economic crisis in Peru, and in response the military government nationalized the industry. Repercussions also reverberated through international markets as the price of fish meal skyrocketed. Farmers found new, cheaper sources of protein, particularly in the form of soy products, which dampened demand for fish meal in the long run (Glantz 1979). Environmental factors also contributed to the sudden stock decline, but the increase in the power disconnect delayed response and amplified the negative consequences of the collapse of the fishery.

Boerma and Gulland (1973) document the events of 1972 in detail. They describe how decision makers chose not to take action after early signs of stock decline, which prompted scientists to recommend a closure of the fishery in June. By this time, researchers believed that El Niño conditions were creating a recruitment failure for the species (that is, reproduction was exceptionally low and few juveniles were surviving to adulthood). In addition, they showed that remaining adults were forced into smaller and smaller areas as warm water from the western Pacific overwhelmed the upwelling of the Peruvian current, which brought both nutrients for juveniles and cold water that was the preferred habitat of adults. Concentrated into such small areas, the fish were easier to catch even though their numbers were much lower than in previous years. However, due to pressures from the industry, managers chose not to act on this scientific advice. The heavily overcapitalized fleet—with a capacity equal to three times the maximum sustainable yield for the stock under positive environmental conditions—was given three more months to harvest anchoveta, at which point landings declined so steeply that the industry itself asked for a closure. Even so, because the southern stock was not as heavily depleted as the northern stock, they successfully lobbied to keep this portion of the fishery open, over objections by scientists.

The short-run negative effect of El Niño on populations of Peruvian anchoveta was already well understood at the time. However, fishing pressure was still very high, and estimates suggested that the adult fish, or spawning stock biomass, decreased from almost 13.6 million tonnes to 0.9–1.8 million tonnes prior to the closure of the fishery (Boerma and Gulland 1973, 2233). Thus, most scientists expected that recovery of the stock might be slower than it had been after past El Niño events. In fact, anchoveta stocks remained depressed for more than two decades. By the late 1980s, scientists found this condition was due to environmental changes associated with the Pacific Decadal Oscillation, which precipitated a shift from the dominant cold-water regime to a warm-water regime in the eastern Pacific (Chavez et al. 2003, 210). Support for this hypothesis came from many sources, including sediment core samples that showed regular if unpredictable shifts from anchoveta-dominated systems to sardine-dominated systems long before the era of commercial fishing in the eastern Pacific (Sandweiss et al. 2004). Coincident with the collapse of anchoveta, sardine stocks across the Pacific surged back from their own decades-long slump, providing further evidence that environmental conditions played a major role in the life histories of small pelagics in the region (Patterson, Zuzunaga, and Cárdenas 1992).

Scientists continue to debate the relative importance of fishing versus environmental pressures in many fisheries, but several recent studies show that overfishing can exacerbate the effects of environmental change and that maintaining a healthy stock biomass, and a healthy ecosystem, is important to ensure the resilience of anchoveta populations (Anderson et al. 2008). There is also agreement that the economic and political costs of the collapse would have been much lower if the Peruvian government had limited capacity to sustainable levels rather than only limiting catches (Fréon et al. 2008). Given the magnitude of the profit disconnect and the power disconnect in Peru, it is not likely that knowledge of the potential for an oceanographic regime shift would have altered management substantially. This knowledge did not prevent a very similar cycle of governance response when environmental conditions changed and anchoveta populations rebounded in the 1990s (Aranda 2009; see chapter 8).

Few developing countries had responsibility for management of a fishery as large and important as the anchoveta stocks off Peru, but the interactions between the profit disconnect and the power disconnect caused smaller downward spirals of depletion in many developing countries (Royce 1987). The geographic scope of serial depletion expanded with industrial fishing fleets, and stagnation or collapse quickly followed in many areas. Again, the political aspect of this process is very important. Had governments been willing and able to curtail catch and effort earlier, these problems could have been minimized. As noted above, Gulland and others from organizations like the FAO recognized the problem of responsive governance and worked to quickly build management capacity in developing countries, but their work did not prevent the delaying effects of dampened problem signals. Indeed, as will be shown in the next subsection, even historically dominant fishing countries were not able to avoid the dynamic of serial depletion in spite of almost 100 years of management enforced by well-established maritime police forces.

7.4.2 Enclosure and Escalation

By the 1960s, in-shore fisheries in historically developing coastal states had already gone through several cycles of depletion and rebuilding (see figure 1.1 in chapter 1), so management in these areas tended to be fairly substantial. While responsive governance continued due to exogenous drivers of the action cycle, response usually occurred earlier in fisheries where management structures were already in place. However, the profit-power feedback described above had major impacts on fisheries outside of the 3-mile territorial sea. In fact, the growing disconnect drove multiple crises that ultimately helped push reluctant states like the US, Canada, and Britain to accept the 200-mile Exclusive Economic Zone (EEZ) in 1977. This subsection describes that process and ensuing management failures within the newly established EEZs.

International Conflict The dynamics of most major postwar fisheries differed from those of the Peruvian anchoveta fishery in that they fell outside the 3-mile territorial sea and were therefore open access internationally. During this time, the profit disconnect was increasing for the large-scale fleets targeting these fisheries, insulating some fishers from problem signals. As fishers and other industry actors expanded production they gained wealth and influence at the domestic level, much like Peruvian fleets. This translated into a power disconnect at the domestic level and considerable conflict over management measures at the international level. Both of these factors greatly delayed responsive governance.

The fisheries for herring in the northeast Atlantic are a good example. These stocks were a mainstay of European capture production for centuries. Scientists warned that there was some overexploitation of Atlantic herring prior to World War I and in the interwar years. In both periods, landings peaked at about 907,000 tonnes (ICES 2012). By 1950, landings of herring in the northeast Atlantic averaged around 2.5 million tonnes, and they spiked to about 3.76 million tonnes in 1964. This last spurt of growth was exceptionally rapid. It took only three years to increase landings by more than 1.5 million tonnes (FAO 2012a). While there were a few strong recruitment years in this period, the general consensus in the scientific literature is that this increase was directly attributable to greater levels of effort in the fishery associated with the government-subsidized rebuilding of coastal and distant-water fishing fleets. This massive increase in fishing effort and catch led to the serial depletion of herring stocks that has now been well documented by scientists, though it was not always apparent while overfishing was occurring.

Estimates show that by the early 1960s aggregate biomass of Norwegian spring-spawning herring, the primary stock in the northeast Atlantic, was 80% smaller than it had been in 1950. The stock increased slightly in the mid-1960s, but fishing effort spiked at about the same time and fishing mortality (an estimate of catch rather than landings) rose throughout the 1960s. The total population collapsed in 1972 (Hamilton et al. 2004). Near Britain, the Downs Banks stock collapsed in the late 1950s, but this was not clear to scientists due to the mixing of fish from different populations. Signals improved in the 1960s, when the bioeconomic collapses of the Dodger Bank fishery and then the Buchanan Bank fishery were clearly attributed to overfishing (Cushing 1980). Nevertheless, fishing effort continued to increase in the North Sea until the mid-1970s, when a complete collapse of all stocks devastated the British herring fishery (Burd 1985). The herring stock of the Celtic Sea and a stock off the west coast of Scotland collapsed at about the same time (Cushing 1980). We now know that these collapses can also be linked to multi-decadal shifts, specifically the North Atlantic Oscillation and the Arctic Oscillation, which create natural fluctuations called Bohuslän periods (Lehodey et al. 2006). Nevertheless, many recent studies show that fishing was the primary cause of the serial depletion of herring stocks but that the specific timing of the regional crisis was determined by climatatic conditions (Hamilton et al. 2004). Like the Peruvian anchoveta fishery, the economic scope of the crisis would have been much smaller if overcapitalization had been prevented through regulation.

In addition, although science was uncertain there were sufficient warnings to prevent the high costs of collapse if governance was proactive rather than responsive. Because herring were so important to Europe for so long, Bohuslän periods were recognized in the Middle Ages, though there was little understanding of their cause. It is also clear that scientists from the International Council on the Exploration of the Seas did not suspect climate effects in the 1960s, but they did alert decision makers to the serial collapse of herring stocks and increasingly recommended that catch limits and effort restrictions were needed for the fishery. However, members of the Northeast Atlantic Fisheries Commission, which had jurisdiction over the shared fisheries of the region, did not consider management measures until 1969. Even then, NEAFC took no substantive action to curb overfishing of any of these stocks until collapse actually occurred, at which point fishing was banned. The reasons for delayed management in these cases are linked to the lack of international enforcement mechanisms and concerns about free riding, but states also resisted catch reductions because their fishing fleets and related shore-side industries were still thriving. As in Peru, prices remained high due to high demand for fish meal, while innovation allowed the costs of production to decline in spite of smaller stock sizes. This profit disconnect was compounded by a power disconnect that mirrored the process described for anchoveta; the higher economic and social benefits they generated gave these industry actors greater political influence. This caused fishing states to choose the short-term benefits of continued overexploitation over the long-term benefits of catch and/or effort limitations (Nash, Dickey-Collas, and Kell 2009, 1881).

Similar processes played out in many other fisheries around the world, whether domestic or international. Whales were perhaps most profoundly impacted by the combined power and profit disconnects. The serial depletion of these species continued in the Antarctic after World War II because of high demand for edible oils and proteins generally. In the 1960s and the 1970s, the growth of demand for whale meat in Japan generated high prices. At the same time, Japan and the Soviet Union subsidized large-scale factory whaling fleets that were much more effective than their British and Norwegian counterparts. This gave both countries a vested interest in increasing effort as long as whaling remained profitable, so they worked to prevent the imposition of strong catch- and effort-reduction measures by the International Whaling Commission. Clark and Lamberson (1982) document these changes but they explain the heavy overexploitation of whales differently, as a rational choice to use up a resource stock with a lower rate of return over the next best alternative. The two perspectives are not mutually exclusive. The short time horizons described by Clark and Lamberson are indicative of responsive rather than proactive governance and they argue that a prolonged period of relatively high net revenues (normal profits) contributed to the heavy overexploitation of whales. In addition, whalers were the first distant-water fishers, and with factory trawlers their ability to target whales throughout the Southern Ocean further distanced whalers, related shore-based industries, and ultimately the governments that benefited from derived economic development from the repercussions of their actions. In short, the difference is primarily in language and perception, rather than the substance of the argument.

Distant-Water Fleets While domestic fleets could be insulated from the costs of overfishing by technological innovation and rising prices, distant-water fleets had the added competitive advantage of high flexibility in their choice of fishing areas. To cover high operating costs, these fleets also need to catch very large amounts of fish in relatively short amounts of time. This combination led them act as roving bandits—fishers who fish out one stock after another with no negative repercussions, at least not in the short to medium term. Olson (1990; 2000) introduced the term "roving bandits" in his seminal work on the application of game theory to international political economy. Berkes et al. (2006) extended the analysis to the operation of distant-water fleets and international fish processors, showing how high mobility led to the serial depletion of fisheries around the world. In this they echoed Gulland's (1971) observation that responsive governance was insufficient in view of the pace of economic growth and technological change in fisheries during the 1950s and the 1960s. However, in both studies, the authors focused on the managerial aspects of the problem, rather than on the political ramifications per se. That is, they identified management shortcomings but did not address the power structures that might prevent successful adoption of management solutions.

The political ramifications of the divergence between action and consequences for distant-water fleets played out in several ways. First, these fleets could greatly undermine domestic attempts to manage fisheries whenever coastal stocks could be targeted outside of national jurisdictions. Direct competition with distant-water fleets drove domestic fishers to abandon traditional management institutions and gave commercial fishing interests strong arguments when lobbying against limits on domestic fleets. This was a much larger problem prior to the adoption of EEZs in 1977, but it continues to be an issue for some fisheries. Second, distant-water fleets also undermined international management, both because their flag states were less likely to agree to reduced harvests—acting as "laggards" in Peterson's (1995) terminology—and because the fleets themselves were highly mobile and could move to avoid regulations. This latter issue is referred to as the "balloon problem" by DeSombre and Barkin (2013) and is very similar to the problem of roving bandits. It is also linked to the third problem, the ability of distant-water fleets to avoid management by engaging in illegal, unreported, and unregulated (IUU) fishing.

The western Atlantic cod fishery provides a good example of both the domestic and the international effects of the profit and power disconnects described above. Fished by European and North American fleets for hundreds of years, the Grand Banks off the coast of Newfoundland and Labrador are the largest fishing grounds in the region, with another smaller fishing area around the Georges Banks near the Gulf of Maine. In the 1950s, the Canadian fleet was the largest in the fishery. It was divided between a near-shore fleet of small vessels using nets and fixed gear and an offshore fleet of mid-size vessels using nylon driftnet gear called "draggers". A large distant-water fleet also fished the region, mainly comprised of large-scale trawlers from Europe (Mason 2002). Canadian fleets produced about 42% of landings in 1950, compared to a combined total of 51% of landings by European fleets, primarily from Spain (13%), Portugal (19%), and France (18%). The remainder of the harvest was landed by US fleets (4%) and Greenland (3%). In the 1950s, production increased from about 700,000 tonnes to about 1 million tonnes, and the share of landings for Canadaian and the European fleets declined somewhat as a result of increased production by fleets from countries like Iceland, Norway, and the Faroe Islands (FAO 2012a).

Both the scale and the distribution of production changed drastically with the introduction of factory trawlers in the 1960s. Production peaked at almost 1.9 million tonnes in 1968. By this time, Canada's share in the landings was down to 17%, even though their total cod production continued to fluctuate around 300,000 tonnes per year. Over the same time period, European fleets, now including a small group of factory trawlers from the United Kingdom, more than doubled their total production but managed to retain only a 41% share of landings. Again, production by small countries played a role, but the biggest change was the entry of fleets from the Soviet Union, Germany, and Poland. These fleets were not active in the area at all in the early 1950s, but in 1968 they produced more than 600,000 tonnes of Atlantic cod in the region, about 32% of total landings. Combined with Norway and a few others, distant-water fleets produced at least 75% of total landings in the western Atlantic that year (FAO 2012a).

Like other fisheries, scientists were most sensitive to biological problem signals. Starting in 1964, they warned that harvests in the region were approaching maximum sustainable yield for most fish stocks, including cod, and that some others might be overexploited. They recommended that some control of effort was necessary through the use of a total allowable catch system, with species-specific quotas allocated to different industry sectors (Anderson, Goeree, and Holt 1998, 82). However, during the 1960s the International Commission for Northwest Atlantic Fisheries (ICNAF) focused on reducing fishing mortality for juveniles and only passed gear restrictions and size limits to protect small fish. Enforcement was insufficient throughout the region, so these weak regulations were ineffective and problem signals increased. New regulations were passed in 1972, triggered by the negative economic effects of the near collapse of haddock on Georges

Banks (south of the Grand Banks), that was largely blamed on a huge influx of Soviet and German factory trawlers. The resulting total allowable catch scheme diverted the massive capacity working on the Grand Banks to other species, including cod. Fearing a repeat of the drastic depletion of haddock stocks, ICNAF instituted total allowable catches for most of its major species in 1974 (Cushing 1977, 234). These "preemptive quotas" were based on incomplete science, were usually set higher than current catch levels, and were very difficult to enforce because most fisheries were mixed rather than single species (E. D. Anderson 1996). Thus, the response to this crisis was both delayed and insufficient, largely because of the policies of distant-water fishing states and conflicts over access rights.

As coastal states, the United States and Canada were the primary instigators of these ICNAF policies. Responding to pressures from domestic fishers, who were sensitive to political and economic problem signals at the time, the US and Canadian governments were adamant about protecting their rights as coastal states. Canada asserted its right to establish and enforce management of Grand Banks cod in 1975, demanding that non-coastal states reduce their harvests by 40% from 1972 or 1973 levels (E. D. Anderson 1998, 82). With fleets that were insulated by their efficiency and flexibility as well as high subsidies, distant-water fishing states resisted management throughout the 1960s and only started to accept restrictions once their fleets began leaving the area for economic reasons in the 1970s. Here again, as economic problem signals were felt, the power structure was altered, though this time heightened problem signals helped to mitigate the power disconnect. Distant-water states were also concerned about the potential adoption of the 200-mile EEZ in the ongoing Law of the Sea negotiations. Given their reduced economic interest in the fishery and concerns regarding access to off-shore fisheries generally, distant-water states agreed to Canada's demands at the 1975 meeting. However, in 1976 both the United States and Canada announced intentions to establish 200-mile EEZs. This new policy was enacted in part so that they could protect coastal areas such as the Grand Banks (Cushing 1977). As of 1977, most of the major species harvested in the northwest Atlantic, including cod, fell under either US or Canadian management. However, Grand Banks cod could still be caught on the high seas just outside the Canadian EEZ. This undermined Canadian regulation of its domestic fleets, as well as international management of the fishery as a whole.

Multi-Level Feedback Effects Hypothetically, domestic management should have been more successful than international management, but it was not. Once it excluded foreign fleets from its new EEZ, the Canadian government began a large-scale development plan for its domestic fishing fleets, including those targeting Grand Banks cod. Private investors also sunk funds into the fleets, anticipating large payoffs now that distant-water fishers were excluded from most of the region. In 1981, Canada established a licensing system to limit new entrants into the fishery and instituted a TAC system for cod stocks (Mason 2002). There is considerable evidence that these TACs were based on overly optimistic scientific advice (G. D. Taylor 1995). Even so, in the 1980s the government frequently established TACs above the level recommended by scientists. This supported exogenous goals associated with the benefits of economic development and also occurred endogenously in response to lobbying from fishers (Mason 2002).

Here again, domestic fishers pushed against management because of the profit disconnect. Demand for cod was high during this period and subsidies ensured decreasing costs of production. Catch signals also appeared to be contrary to scientific advice at the time, which generated considerable distrust of scientists among fishers. So, while scientists were sensitive to biological problem signals, fishers were insulated by the profit disconnect and were angered by counterintuitive catch signals. However, fishers had greater influence over decision makers due to their numbers, long established personal connections, and the economic importance of their industry. They also still argued vociferously that any problems in the fishery were caused by foreign fleets fishing outside the EEZ. All of these factors led to prolonged delays in domestic management and increasing conflict at the international level.

As explained in chapter 5, once the United States and Canada established their EEZs they withdrew from ICNAF, and fishing states negotiated a new regional fisheries management organization, the North Atlantic Fisheries Organization (NAFO). Canada ejected most foreign fleets from its new EEZ, but Spanish and Portuguese fishers continued to target Grand Banks cod on the high seas. Thus, to fully manage Grand Banks cod, Canada still had to negotiate with distant-water states within the NAFO framework. Under continuing pressure from its rapidly growing domestic fishing interests, Canada consistently insisted on reductions by distant-water fleets. In their turn, Spain, Portugal, and the European Economic Community (EEC) protected their remaining vessels in the area by blocking Canadian proposals. Canada's demands escalated to a recommendation for the complete closure of fishing in areas on the high seas in 1986, but this was again opposed by the EEC. Non-members also started fishing in the region, further increasing fishing mortality in spite of strong efforts at exclusion by

NAFO members (E. D. Anderson 1996). In this case, fishers on all sides were insulated by some form of profit disconnect, but distant-water states had the power advantage because of the norm of decision by consensus in the structural context. Power disconnects at the national level were compounded by power disconnects internationally.

Many Canadian fishers and decision makers blamed distant-water fleets for the collapse of the Grand Banks cod fishery, but analysts paint a more nuanced picture (G. D. Taylor 1995). In fact, landings by distant-water fleets started to decline after the 1968 peak as vessels chose to exit the fishery and search for more profitable fishing grounds. By 1975, both EEC landings and Soviet landings were down to about 200,000 tonnes, or 30% of total landings. These numbers dropped precipitously after 1977. The Soviet Bloc's share went down to 5% of landings and then declined gradually until the fall of the Soviet Union in December of 1991. European harvests fell to 90,000 tonnes in 1978 and then oscillated around 60,000 tonnes through the 1980s. In contrast, Canadian harvests doubled from 1977 (235,000 tonnes) to 1988 (468,000 tonnes), amounting to 75% of total landings (FAO 2012a). A combination of subsidies and escalating prices generated a profit disconnect for the domestic fishery that then reinforced the power disconnect, but the Canadian government also chose to focus effort on reducing foreign catches instead of domestic landings. Canadians held a deep-seated belief that they had a right to the cod stocks of the Grand Banks and that domestic fishers should not suffer reduced catches to accommodate distant-water fleets. The resulting delay of strong response allowed continued high levels of effort and stocks had declined to very low levels by the early 1990s, when environmental and ecological conditions shifted, causing the sudden and long-term collapse of the heavily overfished stocks (Mason 2002).

As shown in chapter 5, the presence of distant-water fleets is not always detrimental to international efforts to manage fisheries. However, quite a few examples exist that are similar to the western Atlantic cod case. For instance, in the 1960s Pacific halibut stocks again declined as a result of increased by-catch of halibut in distant-water trawl fisheries targeting other demersal fishes. This created a power disconnect at the international level. The International Pacific Halibut Commission could only regulate longlines targeting halibut and found that reducing longline landings to compensate for trawl catches was politically difficult (Skud 1973). Since halibut were by-catch, trawl fishers had no incentive to reduce their catches without such regulation, so the problem continued to escalate. After prolonged debate over an appropriate response by the IPHC, the industry eventually agreed to reduce trawl harvests to compensate for increased mortality in the line fishery. This only occurred after substantial declines in biomass and loss of profits for the fishery. Similar dynamics hampered management of Atlantic salmon, herring, haddock, and plaice before 1977. After 1977, distant-water fishing was a problem for swordfish in all oceans, and also for several other highly migratory species. As described in chapter 5, IUU fishing also increased in the 1980s and the 1990s, creating additional management problems on the high seas and within the EEZs of countries with limited management capacity.

7.5 Management Capacity

Many of the management failures documented in previous sections occurred not for lack of regulation per se but because existing regulations were either insufficient or not well enforced. Weak measures are often favored because of the power disconnect, but there are also instances when governments do not have sufficient capacity to monitor the state of their stocks or to fund sufficient research to develop sound scientific advice. Similarly, government agencies may not have sufficient resources to effectively enforce regulations. The enforcement problem depends on the difficulty of monitoring regulations, the prescribed punishment for violations, and the financial and human resources available to management agencies, as well as informal norms and values held by fishers. Both scientific and enforcement capabilities are elements of management capacity and can be thought of as resources in the structural context. As such, management capacity can be shaped by endogenous and exogenous forces. This section first describes how Higgs' (1987) ratchet effect is linked to endogenous cycles of growth in management capacity, and then delves into the history of capacity building through responsive governance. The focus is on enforcement capacity, since growth in scientific capacity has already been covered.

The budgets of states and the relative importance of fisheries are key determinants of management capacity, including dedication of resources to ensure compliance. As discussed in chapter 6, states with more resources overall may spend more across the board whereas states where fisheries are economically important are likely to invest a larger percentage of their budgets on fisheries. However, enforcement of conservation management measures is often less popular than provision of expansionary or exclusionary management measures, so the history of a fishery also plays a role. Specifically, we can expect that spending on compliance and management capacity will be driven by governance response to problem signals in what

Higgs (1987) termed "the ratchet effect". Much like Ludwig's Ratchet, Higgs' Ratchet says that, with each crisis, government response is institutionalized, increasing the scope and resources assigned to a specific bureaucratic unit. This effect of responsive governance is widely prevalent throughout the fisheries literature. Because of multiple iterations of crisis and response in coastal and offshore fisheries, historically dominant fishing states now have long-standing and well-funded management institutions, and developing countries are building up management capacity of their own, often with support of the UN Food and Agriculture Organization and other international bodies. Nevertheless, given the rapid expansion of the industry described in Part I, the difficulties of ensuring compliance also increased rapidly after World War II, often outpacing management capacity.

Ensuring compliance with marine fisheries management requires strong monitoring and enforcement mechanisms. Traditional approaches to monitoring fisheries regulations include port inspections and at-sea inspections, which may occur through interception or by placing an observer on board a vessel. Airplanes may also be used to monitor fishing activities at sea. The nature of the inspection depends on the regulations in force. Inspectors may examine gear to make sure it meets existing requirements for mesh size, hook size, line length, etc. They may also inspect the catch to ensure that fishers are not taking more than their quota of a specific species or size class. For this purpose, port inspections are problematic, since fishers can discard overage at sea. Many countries also require that captains maintain logbooks that detail the location, time, and composition of their catch. These may be compared at intervals to observed catch or landings. Where time-area closures are in use, some patrolling of closed areas or areas where fishing activities are limited is necessary. This can be undertaken at sea, but many developed countries now use satellite-based vessel monitoring systems (VMS) to track vessel activities. VMS is one of the few monitoring methods that can be applied universally with a low cost. Market- or trade-based monitoring, in which violations are documented by comparing reported landings to sales in local or international markets, is also increasingly common.

Enforcement is a multilayered issue. A high probability of getting caught cheating means little if the penalty is small relative to the overall benefit of fishing illegally (L. G. Anderson 1989, 262–263). Penalties generally include fines, confiscation of property, reduction of catch quota or exclusion from the fishery, and even jail time for extreme offenses. For internationally traded fish products, penalties can also include sanctions on exports from countries with high levels of IUU fishing. Usually these penalties are set by political or bureaucratic organizations domestically and by RFMOs internationally. Hypothetically, penalties should be set high enough to deter free riding by making the expected costs of getting caught higher than the expected benefit of cheating (Kuperan and Sutinen 1998). However, this does not always happen. Lobbying by fishers and lack of information about the true benefits of cheating can cause decision makers to set lower penalties (Jentoft 2000). In areas where corruption is common, fishers can pay bribes to avoid prosecution. The penalty would then be the value of the bribe rather than the amount established by the government. Similarly, legal loopholes in legislation may allow fishers to escape penalties altogether (L. G. Anderson 1989).

Several factors can make monitoring and enforcement more difficult. First, all else equal, the larger the fleet and the fishing ground, the more resources are required to monitor its activities. Second, monitoring is more difficult when fishers can choose from a wide array of port facilities. In both cases, inspectors will have to spread out, either at sea or in port, to try to cover the entire fishery. Where monitoring resources are limited, coverage of a fishery will be less extensive and the probability that any individual fisher will be caught violating laws will be lower. In most cases, the costs of monitoring preclude 100% coverage, though some approaches such as use of logbooks and VMS may be applied universally where literacy is high and the necessary technologies are available. Third, the nature of regulations themselves can be problematic for monitoring. More complex regulations are often more difficult to monitor. For example, an overall ban on a certain gear type, such as driftnets, is simple to monitor at port, but port inspection alone is not sufficient when spatial zoning allows for use of multiple types of gear in specific areas (L. G. Anderson 1989; Sutinen 1993; Kuperan and Sutinen 1998). This was an important problem in the case of the Grand Banks cod fishery because fishers legally carried nets with different mesh sizes and would use the smaller-mesh nets even in nursery grounds where they were prohibited (Anthony 1990). This type of problem can be dealt with in several ways, including at-sea inspections and monitoring landings for small fish, but all require higher investment of resources.

The effectiveness of port inspection has changed over time because of the economics of expansion described in Part I. Around the turn of the 20th century, concentration of vessels at ports increased, particularly during the transition from sail to steam and then petroleum. New, powered vessels required modern port facilities, which were relatively scarce, making inspection easier. Gradually, new port facilities opened in much of the world, and vessel technology improved to the point that distant-water fleets now have extensive flexibility in their choice of port location. Similarly, use of transshipment at sea, which is actually an old method attributable to the Dutch herring fishers and the Hanseatic League before them, further increased difficulties of monitoring catches and landings in large-scale commercial fisheries. Because of these problems, transshipment is often banned, particularly in international fisheries, and managers may limit the number of ports where fishers can legally land their catch (Aanes, Nedreaas, and Ulvatn 2011; Clark 2006). To avoid these regulations, distant-water fleets increasingly targeted fisheries in international waters or within the EEZs of states with low enforcement capacity as they spread out over the oceans. In some cases this triggered institution building in developing countries, as when Mexico built capacity to confiscate US vessels in their claimed waters. However, in much of Africa, and in other regions with low political and economic resources, fleets from Japan, Russia, and European Union countries fished near shore with impunity, often drastically depleting local stocks of fish (Carneiro 2011).

At the other end of the spectrum, artisanal and small-scale commercial fisheries present a different set of compliance problems. These fishers tend to be numerous and, though living within one country, are often spread out in many coastal communities. Furthermore, they have little need for highly specialized port facilities. Indeed, the small skiffs used by artisanal fishers and often powered by electric or gas motors can be landed on almost any beach or cove. Thus, land-based monitoring costs can be quite high for these types of fisheries, even if the area fished is relatively small (Berkes 2003). Although often viewed as more benign than larger-scale fleets, artisanal fishers can substantially reduce coastal stocks when they integrate new technologies such as onboard and outboard motors, monofilament nets and lines, and other advancements described in Part I. By improving speed and mobility, these innovations also increase the difficulty of monitoring and enforcement. Furthermore, the proliferation of centralized command-and-control systems reduced management legitimacy generally in the 20th century. Governments in developed and developing countries frequently disregarded existing local management institutions, replacing them with top-down measures that fishing communities often perceived as unjust and unnecessary. This occurred in both developed and developing countries but was particularly problematic in the latter, where time and resources were insufficient to build up strong enforcement institutions (Royce 1987, 206–207; Gezelius and Hauck 2011).

Last but not least, the attitude of fishers themselves is a critical determinant of the difficulty of monitoring fisheries regulations. Where regulations are perceived to be legitimate, monitoring is largely aimed at reducing incentives to free ride, or to benefit from the abstention of others (Hatcher 2000). In such cases, fishers may monitor the activities of others and may even engage in their own forms of enforcement when regulations are accepted as norms in the governance portion of the social context. Where regulations are perceived to be unfair or unnecessary, many fishers may choose to circumvent the law in protest. They may even work together to avoid monitoring by the authorities and, much like pirates of old, they may establish connections to onshore interests, such as wholesale dealers and local politicians, to facilitate their illegal activities. As Jentoft (2000) points out, this response is related to Hirschman's (1970) concept of "exit" from illegitimate governance structures. In such situations, monitoring is exceptionally difficult because entire communities work together to prevent detection. Such problems are frequently observed in the history of fisheries management and may be best reflected in this quote from Senator William Cohen of Maine in a congressional hearing on responses to the New England groundfish crisis in the early 1990s:

> The Gulf of Maine is far too large for the kind of enforcement that would be necessary to enforce a plan that the industry did not support. A critical mass of support within the industry is necessary for any plan to be successful. (United States Senate 1992 , 8)

This aspect of compliance is complicated by the fact that legitimacy often depends on the perceived rather than the actual effectiveness of management. For instance, in the Georges Banks cod fishery discussed by Senator Cohen, the legitimacy of scientific advice was undermined in the late 1970s, when NMFS recommended severe reductions in landings due to low biomass. Although the populations of adults were dangerously small, this problem was masked by a strong influx of juveniles in 1977, which made it appear as if NMFS was demanding unnecessary limits on fisher livelihoods (Anthony 1990). In contrast, lobster fishers in the same region were also distrustful of government regulations earlier in the century, but in this case environmental factors amplified the perceived success of regulations, reinforcing their legitimacy and reducing the need for government enforcement of management measures (Acheson 1988). Other examples of both positive and negative reinforcement of legitimacy can be found throughout this book. This is not to say that legitimacy alone is sufficient to ensure compliance, but it is a necessary complement to deterrence methods, as per Kuperan and Sutinen (1998). I will discuss legitimacy-building management methods in Section 8.5, because these approaches were largely developed toward the end of the 20th century.

In short, although managerial capacity to ensure compliance increased in developed and developing countries after World War II, the difficulties associated with monitoring and enforcement also escalated in the wake of new technologies and the increasing scope and scale of fishing operations. These changes were challenging for developed countries with well-established fisheries bureaucracies and policing authorities. Difficulties were even greater for many developing countries, which lacked strong governance institutions and human and financial resources to improve compliance, particularly when faced with encroachment by foreign fleets. Those countries that did manage to curtail foreign fishing operations still had considerable difficulty ensuring compliance by domestic fleets, as shown by the Peruvian anchoveta case, among others. Ultimately, these issues of compliance exacerbated the problems described in this chapter and further undermined the potential for effective management.

This chapter has shown how cycles of serial depletion expanded with the growth of the profit disconnect and the power disconnect from the 1800s through to the early 1970s, using a few key, detailed case studies to illustrate prevailing patterns. The political and economic responses that allowed fishers to push past local limits insulated them from problem signals, and created resistance to strong conservation management measures that limit catch or effort. Furthermore, this process amplified existing power disconnects and created new asymmetries by undermining or reversing existing conservation management measures based on exclusionary strategies. This allowed industrialized fleets to grow at key stages in their development—the transition from wind to steam to petroleum power—and led to cycles of growth in political influence as well as economic scope and scale. The power disconnect that started with industrialization expanded geographically as fishers spread out through exploration, wiping out long-standing, sustainable management practices at the local level. Several major commercial fisheries were severely overexploited in this process, including salmon in all of its ranges, oysters, cod, herring, and most groundfish in the North Atlantic, and sardines and anchoveta in the Pacific, though some of these declines were exacerbated by environmental factors. Nevertheless, with each new fisheries crisis, governments responded by building new institutions and increasing management capacity. As will be discussed in the next chapter, other countervailing forces arose with the continued serial depletion of global fisheries in the 1970s, igniting conflicts but also providing new solutions.

8 Countervailing Forces

The cases documented in previous chapters show how responsive governance allowed and even fostered the expansion of fishing effort and the overexploitation of fisheries, but also built up management institutions that proliferated with cycles of serial depletion. Although these responses still dominate the Action Cycle/Structural Context (AC/SC) in fisheries, this chapter chronicles recent shifts in the pattern of responsive governance that can be traced to changing actors, governance structures, and management solutions. Section 8.1 describes the growing political power of non-commercial interests in the latter half of the 20th century and lays the groundwork for understanding the complicated effects of these groups on fisheries management. Additional discussion of these interest groups threads throughout the rest of the chapter, since their impact cannot truly be separated from the larger process of responsive governance and related institutional evolution. While most think of environmental groups as the primary non-commercial interests, this narrative also highlights the importance of recreational fishing interests and continues the analysis of the impacts of epistemic communities—that is, groups of experts whose work shapes the common understanding of fisheries problems and solutions (Haas 1992, 3).

The next three sections examine the evolution of management goals and solutions from the late 1960s on. Section 8.2 describes growing movements for the protection of charismatic megafauna, which may be targeted, as in commercial whaling operations, or killed as by-catch, as in the tuna-dolphin controversy. It also highlights some of the most radical preservationist movements, but then shows how technological innovation can be used to protect charismatic species without large reductions in fisheries production. Development of by-catch avoidance technology is an expansion of the management resources available in the structural context that facilitates response by making effective regulation more expedient. Turning back to commercial fisheries for non-charismatic species, section 8.3 introduces the concept of economic "rationalization," which was first implemented as a practical response to problems of overcapitalization but was disseminated by a growing epistemic community of fisheries economists. It entails the creation of use rights that may or may not be tradable. By assigning access rights to fishers or vessels, these systems help to minimize the core common pool resource (CPR) dynamic but do not completely alleviate the profit disconnect[1] or the power disconnect, so we still expect to see delayed response in "rationalized" systems, particularly when exogenous factors drive continued economic expansion.

Section 8.4 describes how fisheries scientists, ecologists, and managers started looking for new management methods within a shifting structural context characterized by the empowerment of new environmentalist organizations that were sensitive to biological problem signals. The epistemic context also changed substantially as the number of scientists and managers who valued conservation over protection of fishing interests increased markedly. Some fisheries scientists started cooperating with conservation organizations to lobby for approaches that would take environmental and ecological considerations into account. Though still quite nebulous and overlapping, these alternatives include ecosystem-based management, space-based management, and various precautionary approaches. Changes in non-commercial actors were complemented in historically dominant fishing countries by exogenous declines in the importance of fisheries relative to other economic sectors and by a decrease in the importance of food security and economic development as policy priorities. This facilitated improvements in maximum sustainable yield (MSY)–based management and the adoption of new alternative approaches. The structural context shifted in the opposite direction in developing countries, where the economic growth of industrialized fleets increased their political power and reinforced the fisheries-based economic-development goals of policy makers. Nevertheless, alternatives to MSY proliferated in formal legal texts, including the FAO Code of Conduct for Responsible Fisheries and related International Plans of Action.

As described in section 8.5, compliance remains a key issue in fisheries globally. From World War II on, the need for new and better monitoring and enforcement measures escalated with the growing scope and scale of fishing operations. By the late 1900s, there was also increasing recognition of the need for legitimacy in fisheries management and a shift away from hierarchical government-led management to more democratic processes such as stakeholder engagement and fisheries co-management. These participatory approaches are generally limited to domestic issues, and the linkages between the perceived legitimacy of management institutions and stakeholder engagement in management processes are not widely recognized at the international level. Instead, regional fisheries-management organizations (RFMOs) continue to rely on coercion through trade restrictive measures (see chapter 5). Some global non-governmental organizations (NGOs) are taking a different route, working within the market to build up demand for sustainably harvested fish products. The goal of many of these eco-labeling or listing programs is to create price premiums and other economic incentives for fishers to participate in well-monitored sustainable fisheries while also reducing demand for high-value, heavily overexploited species. This approach has considerable potential to build legitimacy and narrow the profit and power disconnects, but it is still limited in scale due both to the level of consumer demand for these products and to the limited number of certifiably sustainable fisheries around the world.

8.1 Non-Commercial Interests

One of the most powerful countervailing forces in fisheries management began with the rise of the environmental movement in the United States, members of the European Union, and other industrialized countries during the late 1960s and the 1970s. This altered the structural context and ignited action by non-commercial interest groups, including recreational fishers, environmentalists, and epistemic communities (Meadowcroft 2012). Well-publicized collapses of major fisheries also helped to catalyze interest group concern, and the near-extinction of several species of marine mammals inspired strong grassroots movements to protect charismatic megafauna (Freidheim 2001). Many of these groups improved fisheries governance by introducing new solutions to fisheries problems, working to change the goals of fisheries management, and pushing for stronger implementation and enforcement of existing regulations. At the same time, non-commercial interest groups could also weaken fisheries governance by increasing levels of conflict in the policy making process, thereby reducing the perceived legitimacy of management measures within the fishing industry. This section describes the various types of non-commercial interest groups that affected fisheries management toward the end of the 20th century. The impacts of these disparate movements on the long-term pattern of responsive fisheries governance will be covered throughout the rest of the chapter.

8.1.1 Epistemic Communities

The role of epistemic communities in fisheries management was introduced in chapter 7. This subsection provides additional theoretical background and summarizes recent trends in the growth and diversification of these groups. Haas (1989) brought the term "epistemic community" into the study of international environmental regimes. Derived from the literature on the sociology of knowledge, this term refers to communities of experts that help to shape policy making through conjoint knowledge creation. Experts are usually scientists, but may also be managers or, as seen in many fisheries, scientist-managers. All experts have their own closely held beliefs that shape their research and interpretation of results (Jasanoff 1987). Some are open about their political leanings, such as members of the Union of Concerned Scientists, a conservation-oriented NGO whose membership includes scientists from many disciplines (UCS 2014). For others, their underlying values are less explicit but still identifiable. Furthermore, it should be noted that feedbacks occur between the values and beliefs of members of epistemic communities and the formal or informal goals that drive their work (Jasanoff 1987). In particular, these experts are usually asked to provide advice on specific management goals, such as maintaining stock biomass at levels that would support maximum sustainable yield or ensuring the best size distribution within the stock for MSY. Such requests tend to channel their research in specific directions and may reinforce their belief in said methods—particularly when management appears to be successful, as in the case of Pacific halibut described in the previous chapter. At the same time, epistemic communities also engage in scientific innovation and can thereby increase the range of possible management methods and, ultimately, alter institutional resources in the structural context.

Examples of such feedbacks were already described in chapter 7. In the 19th century, cycles of serial depletion in fisheries shaped the growth of epistemic communities whose members improved scientific understanding of the effects of fishing on marine stocks and developed new management approaches. In particular, some early scientist-managers showed that restrictive measures could benefit fishers in the long run. Breaks within this community emerged fairly early on and then widened with the increasing overexploitation of commercially important fisheries. After World War II, three broad groups of experts emerged: those who favored a "go slow" approach using rich datasets with population structure, those who preferred management "up to MSY" using simple models with low data requirements, and a small group who believed that regulation should be based on detailed ecological analyses and that fishing should be limited to levels that would have no negative effects on ecosystem function.

Three general trends in epistemic communities concerned with fisheries are notable in the 1960s and 1970s. First, escalating crises undermined the argument for fishing "up to MSY" and strengthened the claims of experts who favored either "go slow" or ecosystem approaches to management. Present-day analysts categorize these remaining groups as either *fisheries-centric* experts who recognize the importance of healthy natural systems to the continued well-being of social systems but tend to focus on the latter, or *ecosystem-focused* experts, who place a higher value on ecosystem resilience and function than on the well-being of fishing communities (Christie et al. 2007). Second, increasing economic hardships associated with fisheries collapses led to the growth of epistemic communities focusing on the social science of fisheries management. These include economists, sociologists, anthropologists, and many other experts. As shown in section 8.3, neoliberal economists developed new management tools focused on the establishment of access rights to curb the CPR dynamic, although there were also critical economists who pointed to the negative social effects of these programs. Social scientists from all disciplines also engaged in interdisciplinary work to develop alternatives to maximum sustainable yield and to improve the legitimacy of management organizations more generally, as explained in sections 8.4 and 8.5.

The third major trend of the past 70 years was the increasing politicization of science and management advice. Much of the earliest scientific work on marine fisheries was funded by governments and universities, but by the 1940s groups of fishers had begun to fund their own think tanks, such as the National Fisheries Institute. This tendency grew as the profit disconnect expanded across a broad array of fisheries, because the economic signals that fishers were receiving increasingly contradicted scientific analyses, generating considerable skepticism toward government and academic experts. The general decline in empathy for fishers and the realignment of management goals described above also contributed to fishers' willingness to pay for their own experts. In later decades, increasing concern over the state of marine species led conservationists, recreational organizations, and other non-commercial interest groups to fund studies and to employ their own experts, which resulted in further politicization—and polarization—of epistemic communities.

These interest-based experts do not usually participate formally in management processes and so are frequently ignored in studies of epistemic communities. However, as Haas (1992) pointed out, even informal groups of scientists organized around shared perspectives on a given issue may have substantive effects on policy, either at the domestic level or at the international level. This is certainly the case in fisheries management. On one hand, the proliferation of interest-based communities frequently undermined scientific advice and management by generating politically motivated uncertainty about the state of marine systems and the effects of various management measures. On the other hand, these communities have also developed novel management options that reduced the political and economic costs of management, facilitating earlier effective response.

8.1.2 Recreational Fishing Interests

Recreational fishers have been active participants in fisheries management for hundreds of years, but they primarily focused on freshwater and coastal fisheries until late in the 20th century. Modern lobbying efforts by groups of recreational fishers date back at least as far as 1870, when the American Fisheries Society was established. Elsewhere, Norwegian anglers created their own organization in 1871, anglers in the United Kingdom organized in 1903, and South African anglers set up an association in 1912. Recreational fishers from multiple countries established the International Game Fish Association in 1939 (Harrison and Schratweiser 2008, 325). These early groups of recreational fishers worked to protect their access to freshwater and coastal stocks and were also strong proponents of hatchery programs designed to rebuild or redistribute populations of freshwater and anadromous sport fish. The first known club dedicated to saltwater sportfishing, the Tuna Club of Avalon, was founded in 1898. Marine or "blue water" fishing was popularized around this time by Zane Gray, Ernest Hemingway, and other writers, although few people could afford to participate in the sport until after World War II. As technology became more affordable and as incomes increased in developed countries, more individuals could take part (Borch, Aas, and Policansky 2008, 270). Today recreational fishing is a multibillion-dollar industry and blue water fishing is open to many people through charter boats, which can be hired for relatively small sums.

As sport fishing proliferated around the world, so too did NGOs representing recreational fishing interests. These organizations engage in multiple activities, including (1) promoting sport fishing or related industries, companies, and tournaments; (2) establishing best practices for recreational fishing, including catch and release of overfished species; (3) educating sport fishers about a range of topics including threats to recreational fishing interests and recreational impacts on the fisheries they target; (4) collecting or facilitating collection and analysis of data on recreational fishing and the status of valued trophy species; and (5) taking political action to protect recreational fishing interests, usually through traditional approaches such as lobbying, campaign contributions, or grassroots movements. The most effective recreational fishing NGOs utilize multiple approaches and also build networks within the recreational community and with other stakeholder groups (Harrison and Schratweiser 2008).

While the interests represented by recreational fishing NGOs are quite diverse, their primary objective is conservation for use, particularly for use by sport fishers. This position can create interesting conflicts and synergies with other groups. The earliest such conflicts occurred between recreational and commercial interests, as sport fishers prefer more large fish, while commercial fishing usually reduces both the stock size and the average size of fish. To minimize these effects, recreational fishing groups frequently lobbied for restrictions on commercial fleets. Commercial fishers would demand prohibitions on recreational fishing, claiming that fish should be used for food and profit rather than sport. In developing countries, recreational fishers sometimes come into conflict with subsistence and artisanal fishers who depend on fisheries for their livelihoods (Lyman 2008). Both commercial and subsistence fishers often accuse recreational fishers of "playing with their food," an epithet that reflects deep normative divides as well as conflicts of interest. Responding to the politics of exclusion, fisheries managers sometimes chose to exclude recreational fishers to allow larger catches for commercial or subsistence fishers, although as often as not the opposite effect occurred, particularly for freshwater resources. In industrialized countries, many rivers, lakes, and streams were closed to commercial fishing but remain open to sport fishing.

As conflicts between recreational fishing organizations and other groups increased with the expansion of sport fishing, recreational NGOs developed new political strategies to maintain their access to preferred stocks. Cooperation with commercial or subsistence fishers is a viable strategy when the profit disconnect is narrow (or catch signals are strong for subsistence fishers). In such cases there is an alignment of goals between interest groups that can be beneficial. Economic incentives can also align when local populations are able to benefit from recreational fishing actives. In several regions, recreational fishers worked with governments and with local residents to develop sport fishing as an important source of foreign exchange and economic development. This, in turn, can increase both government and community interest in maintaining stocks of sport fish and the habitats they require. It also empowers local communities that are most sensitive to changes in the core problems of the action cycle, reducing the power disconnect (Lyman 2008, 229).

Similarly, recreational NGOs work to persuade decision makers and the public by measuring the economic benefits of sport fishing in local economies as well as the economic value of the recreational fishing industry as a whole. For instance, in 2002 the American Sportfishing Association commissioned a study of the economic value of recreational fishing in the United States. They found that retail sales of recreational equipment and related services were worth US$41.5 billion, while overall output was worth US$116 billion. The study also estimated recreational fishing's state and federal tax contributions (US$7.3 billion) and employment in the industry (> 1 million jobs). The European Anglers Alliance commissioned a similar study, finding that the annual value of the industry in Europe was about US$10–13 billion. Other groups have used similar studies to highlight the importance of recreational fishing in developed countries as they lobbied for expanded access to recreational fishing (Harrison and Schratweiser 2008, 324–325).

Recreational NGOs also invest in scientific studies and innovative techniques to reduce their biological impacts. Environmentalists and commercial interests sometimes attribute high levels of mortality or ecosystem disruption to sport fishers, demanding reduction in recreational as well as commercial catch. This tends to be more common for freshwater and coastal fisheries (Cooke and Cowx 2006; Kearney 2001). Recreational fishing NGOs use several tactics to mitigate these concerns without curtailing recreational use. One common response to all conservationist concerns is catch-and-release fishing, in which a fish is caught (providing recreational value) but then released alive (minimizing the biological impact if survival rates are high). This practice is increasingly popular among recreational fishers as a result of education efforts by NGOs and innovative new technology that makes it possible to produce realistic fiberglass replicas of released fish from photographs (Policansky et al. 2008; Peel, Nelson, and Goodyear 2003).

Recreational fishing NGOs engage in scientific activities to determine stock status for important species and to differentiate between recreational and commercial impact where possible. The Billfish Foundation's tagging program for marlins and related species is a good example. By leveraging recreational volunteers, they can fill in gaps of knowledge for species that have few interactions with commercial gear and are so widespread that scientific sampling of an entire range is not feasible (Billfish Foundation 2014). They can also use the information generated by such programs in lobbying activities that are aimed at protecting recreational access while curtailing commercial exploitation and by-catch mortality. Similarly, such NGOs may engage in technological innovation to improve survivability of released fish and to document that improvement in the scientific literature (see subsection 8.2.2).

Recently, animal welfare activists have characterized recreational fishing as an unnecessarily cruel sport. Sport fishing NGOs view these efforts as a major threat to their memberships because animal welfare groups demand cessation of all recreational fishing, rather than the conservation for use ethic espoused by most conservation NGOs (Harrison and Schratweiser 2008). Faced with such implacability, recreational fishers do not engage with animal welfare groups directly but rather work to minimize the effects of anti-fishing messages on decision makers and the public. In doing so, they emphasize the economic benefits of recreational fishing and cite studies establishing that its effect on the environment is minimal. They also work to expand interest in sport fishing generally and among underrepresented groups, particularly women. Much as in commercial fisheries, the influence of recreational fishers is increasing with the profitability and economic contribution of their industry, and it is not likely that animal welfare activists will prevail on governments to ban sporfishing in the near future.

8.1.3 Conservationists and Preservationists

In the 1970s, people in many industrialized countries began viewing animals differently. Concern increased for animal welfare generally and for elephants, tigers, rhinoceroses, and other species now referred to as "charismatic megafauna." Within this broad movement, environmental NGOs worked to educate the public about several charismatic marine species. For instance, in the 1970s and the 1980s, US, Australian, and European NGOs successfully shifted public opinion on cetaceans like dolphins and whales. Campaigns highlighting the intelligence and sociability of these marine mammals—as well as the brutality of fishing operations—motivated millions of people to protest against whaling and to boycott tuna because of the large amounts of dolphin by-catch in the fishery (DeSombre 2000; D. D. Murphy 2006; Moore et al. 2009; B. G. Wright 2007). More recently, environmental NGOs started to advocate conservation or preservation of less charismatic marine species and ecosystems more broadly. This shift in concern was largely driven by the increasing scale of fisheries collapses and the greater recognition of the ecosystem impacts of fishing in the 1980s and the 1990s.

Environmental NGOs vary widely in size, scope, goals, and strategies for goal attainment, but all tend to be sensitive to biological indicators and scientific advice rather than profit.[2] The most important distinction is between conservationist NGOs, which favor conservation for use by commercial or recreational interests and preservationist NGOs, which work for complete protection of species or ecosystems from human use (Friedheim 2001). In general, preservationists are most interested in protections for charismatic megafauna and often lobby for regulations that ban all harm to these animals, including fishing, habitat destruction, and other forms of "interference." Animal rights NGOs also fit into this category, but they favor preservation for all species, whether or not heavily depleted, including those that are not particularly endearing to the public or even to larger interest groups like birders or scuba divers. Animal rights groups have tried several methods to increase public concern for non-charismatic species, such as the "Sea Kittens" campaign mounted by the People for the Ethical Treatment of Animals (PETA), which was designed to convince the public that fish should be protected from both commercial and recreational exploitation (PETA 2014). In contrast, conservation groups accept the idea of conservation for use and usually press for well-enforced, science-based management. They also tend to be more willing to compromise to allow continued, sustainable use of marine living resources. This can include protection of species that are heavily depleted due to commercial fishing, like bluefin tuna and Antarctic toothfish, and reduced mortality of charismatic by-catch species, including marine mammals, sea turtles, and sea birds. The World Wildlife Fund and other large organizations often lobby for the sustainable management of less well-known species, subsidizing such campaigns with funds raised in broad appeals featuring charismatic megafauna.

Most environmental NGOs use education and outreach to increase public awareness of problems associated with marine conservation or preservation. Where government is relatively democratic, public attention is an important component of political power as well as a way of recruiting members and soliciting donations (B. G. Wright 2007). A few environmental groups base their public appeals on recreational values, but their membership is usually limited to recreational users, like scuba divers or surfers (Cisneros-Montemayor and Sumaila 2010). Most environmental NGOs appeal to individuals primarily on the basis of non-use values, which are formally classified as existence values, option values, or bequest values. That is, people join because they value the existence of specific species or the health of the oceans more generally, or because they want to be sure that people will be able to benefit from these resources in the future (this generation for option, future generations for bequest). Because these non-use values can engender considerable political activity in support of restrictive management, they can provide substantial political leverage.[3]

Many environmental NGOs fund scientific research to document overfishing or ecosystem disruption. They may also commission contingent valuation surveys or other economic estimates of the use and non-use values of the species and ecosystems they wish to protect (Greenpeace 2014; TRAFFIC 2014). Such reports are used for lobbying activities, are cited in testimony before political bodies, and are widely publicized to educate and motivate the public to join the NGO, contribute to the NGO, and/or take political action on a specific issue. When by-catch is the primary problem affecting a given species, conservation groups may invest in the invention and dissemination of by-catch avoidance devices. Furthermore, scientific reports funded by NGOs can provide important sources of evidence for legal claims, particularly in countries where it is legal to sue the government, as in the United States (see subsection 8.2.2). Transnational NGOs often use such reports to sway decisions by RFMO member states.

The strategies noted above are similar to those used by recreational fishing NGOs and other interest groups. However, some environmental NGOs also engage in more extreme tactics. Greenpeace is well known for its attempts to block the activities of whaling vessels and others targeting charismatic or overexploited stocks. Sea Shepherd USA engages in similar tactics and has been known to use violence in addition to peaceful protest. They also use media in innovative ways, and are the focus of a reality television program called "Whale Wars," which dramatizes both their cause and the disruptive tactics that they espouse. Some observers believe that Sea Shepherd is an eco-terrorist organization, although their tactics are not as violent as those used by more radical groups, such as the Earth Liberation Front. Even so, they cause considerable disruption to commerce, and they can create risks of physical harm. Activities of this sort may have short-term conservation benefits, such as stopping a whale hunt for a period of time, but the long-term goal of these organizations is to raise awareness of issues that are important to their membership. Violence can backfire, however, by alienating the public or increasing resistance from recreational, subsistence, or commercial fishing groups (O'Neill 2012, 121).

Although it might seem that environmental organizations are inherently antagonistic toward fishing interests, this is not the case. Conservation organizations often align with recreational fishers to promote measures to curtail commercial fishing. Commercial fishers may also align with conservation interests when their goals are similar. DeSombre (1995, 2000) shows that environmental and industry representatives form highly effective coalitions of "Baptists and bootleggers," or groups with very different values that work together because they have a common goal. As described in chapter 1, coalition formation can reduce the power disconnect by increasing the influence of more sensitive actors. Coalitions can speed up management, but governance is usually still delayed because actors tend to only form coalitions when their interests align in crisis situations. As expected based on the AC/SC framework, signals matter in the process of coalition formation. For instance, if fishers respond to a strong political signal of entry by "outsider" fleets, they may align with conservation interests that also wish to limit effort and are willing to do so through exclusion. On the other hand, if the profit disconnect is increasing, fishers will fight against conservation management measures and will refuse to form coalitions with conservation interests until economic signals align with biological problem signals. Of course, resources are also important, and coalition formation is much more likely if any group can identify management measures that can solve the core fishery problems without substantial political or economic costs.

8.2 Charismatic Megafauna

As awareness of environmental issues increased during the 1960s and the 1970s, a number of interest groups started lobbying for the protection of charismatic megafauna. Whales, dolphins, and other marine mammals are the best examples in the marine environment, though sea turtles also receive considerable attention and recreational fishing NGOs work to popularize large fishes like marlins and sailfish. In general, charismatic megafauna differ from other species because of the scope and magnitude of public interest in their welfare. That is, large numbers of people in multiple countries are willing to take political and/or economic action to help preserve these species from harm by human beings. Their "charisma" has many sources, but impressive size, perceived intelligence and sociability, and aesthetic considerations are of primary importance. Perceptions about threats are also crucial, and publicity about the depletion of a species and the brutality of capture methods contributes to high levels of public concern. While their popularity insulates them somewhat, charismatic megafauna are still subject to the ups and downs of media attention cycles, and adroit NGOs utilize focusing events to mobilize public opinion and push through legislation (Jones and Baumgartner 2005). This section describes that process in various parts of the world for fisheries targeting charismatic megafauna (subsection 8.4.1) and those in which such species are by-catch (subsection 8.4.2).

8.2.1 Targeted Fisheries

Historically, marine mammals were in high demand as components of many important products. Whales were targeted for their oil, meat, bone, baleen, and ambergris, while pinnipeds like seals and sea lions were largely exploited for their fur and, in some cases, oil. However, the invention of synthetic substitutes for many of these products lessened demand and the exploitation of marine mammals became less lucrative after World War II, reducing the profit disconnect (Clark and Lamberson 1982). In fact, the United Kingdom and Norway halted their commercial whaling operations in 1963, and in the 1970s many countries started dropping out of the International Whaling Commission (IWC) simply because they no longer harbored commercial whaling fleets (Holt 1985). Increasing environmental concern also contributed, particularly in the reduced demand for furs. Because of these developments, the profit disconnect narrowed considerably and overcapitalized whaling and sealing industries declined in many countries through the 1960s and the 1970s. During the same period, activism for regulatory protection of these species escalated as the environmental movement grew and publicity about the exploitation of marine mammals increased. Together, the decline of the industry and the rise of environmental interests substantially reduced the power disconnect for these charismatic megafauna. This dynamic culminated in a number of domestic and international laws prohibiting exploitation of marine mammals.

Bailey (2008, 296–297) describes the rise of the anti-whaling movement in the United States, Europe, and Australia. Initially, conservation was the primary motivation and interest was driven largely by the well-publicized near extinction of blue whales, as well as by steep declines in other species in the early 1960s. This was not the first time that whaling resulted in severe reductions in populations. In fact, gray whales became extinct in the Atlantic in the 18th century, but at that time the industry was highly profitable and very powerful while the public was either unaware of or unconcerned about the issue (Clark and Lamberson 1982). By the 1970s, public concern regarding whale conservation was high, profit and power disconnects were minimized, and environmental NGOs like Greenpeace and the World Wildlife Fund began pushing members of the IWC to improve management, as mentioned briefly in chapter 5. With the decline of the whaling industry and upsurge in concern about the heavy overexploitation of whales, states pressured by NGOs managed to adopt a moratorium on whaling in 1982 that is still in effect, largely due to a stalemate between pro-whaling and anti-whaling forces (Andresen et al. 2001; Stoett 2011).

Both factions in the whaling debate view this stalemate as a failure. Pro-whaling interests, or those who favor conservation for use, are frustrated by what they view as the cooptation of an RFMO that could otherwise sustainably manage an important resource. On the other hand, anti-whaling or preservationist interests view continued legal fishing by Japan, Norway, and some other states under various loopholes in the moratorium as a failure to fully protect animals that are intrinsically valuable and should not be intentionally harmed (Andresen et al. 2001). Academics have advanced several explanations for the impasse. Bailey (2008) claims the rapid cascade of anti-whaling norms in the 1970s was halted due to divergences between the aspirational goal of whale protection and the practical issues of indigenous people's rights, the need for scientific whaling, and, ultimately, backlash against "green imperialism," or the imposition of the values of developed countries on developing country populations. Similarly, Sakaguchi (2013) applies a "reverse spiral" approach to understanding Japanese resistance to the anti-whaling norm. He shows that many in Japan were open to the anti-whaling campaign until threats of sanctions and other punitive posturing generated a backlash that entrenched the pro-whaling stance as a matter of national pride. In general, social values are the key to understanding the entrenched disagreements over commercial whaling and, with this norm solidly established in the structural context, it would be difficult for one side to persuade the other.

Although less global in nature, protections for pinnipeds and other targeted charismatic megafauna followed similar patterns to the whaling case. Development of synthetic alternatives and changes in consumer preferences caused a decline in demand for furs and related products, which, in turn, reduced the profit disconnect and weakened the fur industry politically. At the same time, steep declines in a large number of pinniped populations caused concern among conservationists, while media campaigns popularized now charismatic animals like baby seals and sea otters (Lavigne, Scheffer, and Kellert 1999). This resulted in widespread protections for these species as well as cetaceans at the domestic level. The US Marine Mammal Protection Act of 1972 is probably the most comprehensive piece of legislation protecting marine mammals, although US law does provide exceptions for indigenous hunting of these species. Globally, many populations of seals, sea lions, and other pinnipeds have recovered from overexploitation. Indeed, modern controversies surround the perceived need to cull pinniped populations to protect high-priced species like salmon from natural as well as human predation (Fraker and Mate 1999).

Attempts to protect sea turtles have been less successful, even though many countries banned harvesting of turtles and eggs and trade in products derived from all six species is prohibited under the Convention for the International Trade in Endangered Species of Flora and Fauna (CITES). Failure to rebuild populations is largely due to habitat loss and to continued by-catch mortality in other fisheries, although, like sharks, sea turtles are still subject to considerable illegal harvesting and trade because of high prices and a substantial profit disconnect (L. M. Campbell 2007; Mancini et al. 2011). Various NGOs work with local communities to create sea turtle based eco-tourism enterprises, which can generate incentives to conserve the animals. The idea is to reconnect profit signals to core problems in the action cycle. Similar approaches are used in the development of whale watching tours (Tisdell and Wilson 2002). All in all, the intense pressures for protections of charismatic megafauna from direct exploitation reflect both growing concern for these species and increasing awareness of the need to work with fishers or fishing communities, rather than imposing regulations on them from the outside. This need for accommodation is also clear in other issue areas, including protection of charismatic megafauna that are caught and killed in fisheries targeting less charismatic species.

8.2.2 By-Catch Issues

Direct targeting of marine mammals and other charismatic megafauna is now rare, except for a few species like whale sharks and other elasmobranchs (see section 8.5). However, these animals can also be killed or harmed when they are incidentally captured in commercial fisheries targeting other species. In such cases, the power disconnect is in full play. Environmental NGOs usually face fierce opposition from powerful industry interest groups whenever they propose by-catch reduction measures that would negatively affect harvests of targeted species. The result is usually a period of political stalemate between conflicting groups, which can result in the collapse of the by-catch species. Stalemates can be overcome through the development and dissemination of technological measures to reduce by-catch. This section shows how industry-environmentalist coalitions can be powerful agents of change. However, it also demonstrates the negative side effects of focusing on a single species or group rather than on an entire fishery or ecosystem.

The earliest and most iconic example of a by-catch problem involving charismatic megafauna was the so-called tuna-dolphin controversy, which started in the late 1960s. For reasons that are still unclear, adult yellowfin tuna tend to be found schooling under pods of dolphins in the eastern

Pacific Ocean (EPO). In the 1950s, commercial fishers sailing out of San Diego discovered this association and began using it to improve their search methods. Use of sonar and radar was limited at the time, so observing cues at the surface, such as areas of disturbance or the presence of seabirds, was the primary method for locating schools. Since dolphins travel at the surface and breach frequently, they were easily sighted and therefore helped fishers locate yellowfin in the EPO. In fact, fishers looked for three different species of dolphins, known colloquially as "spotters," "spinners," and "white bellies." Spotters are Pacific spotted dolphins (*Stenella attenuata*), spinners are eastern spinner dolphins (*Stenella longirostris* orientalis), and white bellies are common dolphins (*Delphinus delphis*). White bellies are known to rush out of encircling purse seine nets just before they close, taking the tuna with them, so most fishers set on schools associated with spotters and spinners rather than common dolphins. Fishers also tend to use the word "porpoise" rather than "dolphin" to avoid confusion with the dolphin fish (also known as mahi mahi; Felando and Medina 2011, 69).

In the early days of the fishery there was neither the incentive nor the technology to release dolphins without also reducing the catch of tuna, so most were hauled in with the fish and then discarded back into the sea. Although fishers tried to minimize mortality, large numbers of dolphins were killed and populations of spotters and spinners were drastically reduced. In the late 1960s, the research of a young graduate student, William Perrin, sparked public concern about dwindling dolphin numbers. Combined with movements to reduce harvests of targeted marine mammals generally, growing pressure for dolphin protection led to the passage of the Marine Mammal Protection Act (MMPA) in 1972 (Perrin 2009). The act required reduction of incidental mortality of marine mammals in commercial fisheries to "insignificant levels." The law gave the National Marine Fisheries Service (NMFS) responsibility for establishing a ceiling for dolphin mortality in the tuna fishing industry and implementing it through a permit system. These regulations caused incidental dolphin mortality to drop substantially in the 1970s, mainly because fishers began to use a "Medina panel." Developed by tuna boat captain Harold Medina in 1971, this innovation allowed many dolphins to escape before a tuna net was hauled in (B. G. Wright 2007). Fishers worked with NMFS to perfect this technology and other methods for reducing dolphin mortality in the early 1970s (Felando and Medina 2011, 33, 132–138). At the time, prices for canned tuna were rising due to increasing demand outside of the United States, and the carrying capacity of the US fleet increased in response to this price signal, mainly through replacement of smaller boats (151–152). So, although the profit disconnect was considerable, it was mitigated by the political leverage wielded by conservation groups, and fisher-led development of new technology facilitated governance response.

Even though new technologies allowed fishers to reduce incidental mortality of dolphins, many preservationist NGOs were not satisfied. Throughout the 1970s, these groups worked to ensure government compliance with a "zero mortality" interpretation of the MMPA. Tactics included multiple lawsuits against the US government and pressure for amendments to the Act. During this period, US captains started to change flags to avoid dolphin protection measures. Flag switching was also driven by the need to retain access to Latin American exclusive economic zones (EEZs) and was facilitated by the vertical integration and internationalization of the canning industry, which gave fishers a more dispersed set of options for selling and landing their catch (Constance and Bonanno 2000, 603). Viewing both the constant pressure from NGOs and the increasing competition with foreign fleets as threats to their livelihoods, remaining US fishers, represented by the American Tunaboat Association and other industry groups, lobbied the US government to increase domestic dolphin mortality limits and to push foreign governments to institute dolphin protection regulations to level the playing field (Felando and Medina 2011). These calls for exclusionary regulation by industry actors were answered during the Reagan administration in 1984, when the US government established a domestic dolphin mortality quota of 20,500 animals and banned tuna imports from countries that failed to adopt similar dolphin protections (Constance and Bonanno 1999, 604).

Extending dolphin protection measures to foreign fleets was important both to US-based environmental NGOs and to the remaining US fishers because by the 1980s most of the 100,000 dolphins killed per year in fishing operations were caught by non-US fleets. This alignment of interests was a direct result of the geographic expansion of effort described in chapter 2. It reduced the power disconnect in the United States and generated an influential coalition of industry and environmental groups that favored internationalization of US laws on dolphin protection. Even so, NMFS chose not to enact sanctions against Latin American countries with high levels of dolphin mortality until forced to do so via a lawsuit won by the Earth Island Institute in 1988 (DeSombre 2000, 68–69). NMFS delayed implementation due to worries that trade measures would cause a breakdown in negotiations at the Inter-American Tropical Tuna Commission (IATTC), the regional fisheries management organization in charge of regulating tuna fisheries in the EPO. Those fears were well grounded. The tuna-dolphin controversy brought IATTC negotiations to a halt in 1990, only a few years after disputes over access to Latin American EEZs were settled (192). Furthermore, Mexico and Venezuela brought suit against the United States under the auspices of the General Agreement on Tariffs and Trade (GATT), claiming that US sanctions were discriminatory. The GATT panel ruled in Mexico's favor, though no retaliatory measures were enacted, partly due to exogenous concerns, such as Mexico's interest in smoothing negotiations for the North American Free Trade Agreement. The European Community brought a second GATT suit against the United States in 1994, which it also won, but again no retaliatory measures were enacted, this time because of the finalization of an international agreement on dolphin by-catch (Joyner and Tyler 2000).

In addition to official US sanctions, pressure for international resolution of the dolphin-tuna controversy came from US consumers. In spite of growing markets elsewhere, the United States was still the world's largest consumer of canned tuna. Early in the 1980s, the US public was relatively apathetic about incidental dolphin mortality, in part because the threat of extinction had been reduced substantially by the avoidance technologies described above. This changed in 1988, when Sam LaBudde, an activist/scientist and founder of the Earth Island Institute, released a video of dolphin kills that he had recorded surreptitiously while working aboard a Panamanian-flagged tuna boat. This video reignited public concern and led to increasing demand for "dolphin-safe" tuna (B. G. Wright 2007). In 1990, StarKist Inc., one of the three corporations that dominated the canned tuna market, announced that it would buy only dolphin-safe tuna. The company also enthusiastically supported US embargos on Mexican tuna and orchestrated a nationwide public relations campaign to further convince consumers to purchase only dolphin-safe tuna (D. D. Murphy 2006). StarKist worked with the Earth Island Institute to develop the first dolphin-safe label and, with combined support from processors and environmentalists, legal protections for the label were enacted with the Dolphin Protection Consumer Information Act of 1990. Although other processors, particularly Bumble Bee, initially resisted use of the label, they eventually accepted it (Baird and Quastel 2011). This is a very interesting example of industry opening new markets in response to conservation concerns and using conservation as political leverage to exclude outsiders.

These dynamics in the action cycle had a profound effect on the structural context. Throughout this period of turmoil, the US State Department negotiated with other members of the IATTC to find a diplomatic resolution to the tuna-dolphin controversy. They were supported by a coalition of environmental NGOs that worked to convince Latin American governments to enact dolphin protections. Greenpeace-Mexico was a leader in this effort, and was joined by a number of other major conservation-oriented organizations, including the World Wildlife Fund, the Center for Marine Conservation, the Environmental Defense Fund, and the National Wildlife Federation. The first result of these efforts, was known as the La Jolla Agreement of 1992, which created the International Dolphin Conservation Program (IDCP). This agreement established a seven-year plan for reduction of dolphin mortality in the EPO from a limit of 19,500 dolphins in 1993 to less than 5,000 in 1999 and established complete elimination of dolphin mortality as an overall goal (IATTC 1992). Both the US and Mexico were signatories to this agreement, along with most other countries with fleets harvesting tuna in association with dolphins in the EPO and several conservation NGOs. The La Jolla Agreement was strengthened by the Panama Declaration in 1995. In 1998, members negotiated the creation of a permanent commission in the Agreement on the International Dolphin Conservation Program (AIDCP; DeSombre 1995, 205–206). Conservation NGOs continue to work with the IATTC and the AIDCP to ensure that dolphin mortality is minimal in the fishery, promoting improvements in technologies that reduce dolphin mortality, institution of 100% observer coverage on all vessels that catch tuna in association with dolphins, and the development of the AIDCP dolphin-safe label for canned tuna that was introduced in 2001 (IATTC 2001).

Although these activities are generally perceived to be successful, insofar as dolphin mortality is now near zero, some preservationist groups, including the Earth Island Institute (EII), continue to criticize all purse seining on dolphins (IATTC 2013a; Pala 2013). EII is particularly concerned about the psychological welfare of dolphins, and has sued the US government several times to require that the definition of "dolphin safe" should be limited to tuna harvested without encircling dolphins at all, rather than for fisheries where dolphins are encircled but released from nets (Gruszczynski 2012, 431). It is important to remember that EII initiated its own dolphin-safe label in 1990, which meets this more restrictive requirement, and that it has considerable resources invested in the maintenance and dissemination of this label in the United States, the United Kingdom, and Germany. Thus, the less restrictive AIDCP label could be seen as a competitor rather than a complementary label and EII's lawsuit therefore has exclusionary implications (Baird and Quastel 2011; Brown 2005).

After EII won its suit on dolphin-safe labels in the early 2000s, Mexico again brought a complaint against the US under the auspices of the World

Trade Organization. By this time, Mexico was the only country whose fleet consistently targeted tuna associating with dolphins. In the early 1990s, most other fleets had shifted to artificial fish aggregating devices (FADs), which are also associated with schools of tuna, though the size and species composition is different (D. D. Murphy 2006). Even so, Mexico was supported by the European Commission and by several other parties in its 2008 case. The initial WTO panel ruled against the United States, but an appellate body reversed that decision, so the Earth Island Institute label stands as a technical requirement that is allowable under WTO rules (Gruszczynski 2012, 433).

The dolphin-tuna controversy is representative of several similar cases. The problem of sea turtle by-catch in shrimp trawl fisheries follows almost the same pattern: a rise in public concern in the late 1960s and the 1970s, subsequent legal protection (listing under the Endangered Species Act), government-industry partnerships to develop avoidance technologies (turtle excluding devices, or TEDs), development of a "turtle-safe" label for shrimp products (by EII), additional NGO pressures to internationalize US measures through use of sanctions, and finally WTO rulings on the issue. Thus, an initial decrease in the power disconnect triggers government action, which instigates industry response to reduce the costs of by-catch avoidance and to challenge decisions through the legal system. Interestingly, on appeal the WTO ruled that US sanctions were acceptable as environmental protection measures under Article XX, but only if the US established programs to help all countries with shrimp trawl industries to implement TED technologies, thereby ensuring no discrimination against a particular country simply because of economic constraints (Lowe 1996; Joyner and Tyler 2000; L. D. Jenkins 2012). Though less controversial, later development of by-catch avoidance technologies for longline operations followed similar patterns. Marlin and other sport fish, as well as seabirds are by-catch in many long line fisheries (Hall, Alverson, and Metuzals 2000; Bagust 2005; Webster 2009). For sport fish, recreational fishing groups helped to develop technologies like monofilament lines and circle hooks that reduced by-catch mortality without reducing the catch of targeted species; a key requirement for uptake of avoidance technologies by fishers (Prince, Ortiz, and Venizelos 2002; Pascoe and Revill 2004). Seabird protection was somewhat different because it was a more international effort, rather than US-led. However, the pattern of response remained the same (Bergin 1997).

In the few cases in which avoidance technologies are not feasible, management outcomes tend to be very different. For the most part, without avoidance technologies coalition formation is limited, the power disconnect remains wide, and only minimal by-catch reduction regulations such as time-area closures and discard limits are implemented (Webster 2009). However, in the case of pelagic driftnets, by-catch was so severe that in 1989 the UN adopted a resolution recommending complete elimination of the technology by 1992 unless methods to make the technique more sustainable could be found (DeSombre 2000, 119). Pelagic driftnets hang vertically in the water column, drift unattended throughout the oceans, and can be as large as several football fields. Fish do not see the nets, which are made of clear monofilament plastic. Though small fish may swim through the nets, targeted species such as squid and tuna are caught in the mesh, usually by their gills. Non-targeted animals, including charismatic megafauna like whales, dolphins, turtles, seabirds, and billfish, also get caught up in these nets, as do high-priced commercial species such as salmon and swordfish (Richards 1994).

The wide spectrum of driftnet by-catch provided grounds for coalition building in favor of a ban. Because their targeted species were captured as by-catch in driftnet fisheries, fishers using different gear types had strong incentives to cooperate with environmental organizations to reduce driftnetting in the 1980s. In the United States, strong political concern resulted in several domestic laws prohibiting pelagic driftnets. Political concern was also high in other countries, such as members of the Forum Fisheries Agency, an intergovernmental organization for island states in the South Pacific (Islam 1991; DeSombre 2000). Although Japan, Taiwan, and other states with large driftnet fleets resisted the ban initially, a fairly narrow profit disconnect (prices for driftnet catch were very low), combined with continued political pressure, resulted in a reduction in the use of the technology over time (Wright and Doulman 1991).

While the successes described above may be heartening to environmentalists concerned about protecting charismatic megafauna, some are critical of such measures because they ignore ecosystem-wide impacts of fishing technologies. For instance, in addition to employing dolphin escapement technologies, tuna fishers avoided setting on dolphins by switching to fish aggregating devices. FADs exploit the tendency of small tropical tunas to associate with floating objects. This association was recognized historically but fishers did not use large numbers of artificially created FADs until international regulations were placed on the dolphin-based tuna fishery in the early 1990s. Within a few years, the fishery changed completely, with most fleets concentrating on production of skipjack tuna, which is an incredibly abundant and widespread species that can easily be harvested in association with FADs. However, FADs also attract juvenile yellowfin and bigeye tuna, along with many other species of fish, including large predators like sharks and billfish. Because schools associated with FADs are so diverse, the total biomass of by-catch is much larger than in dolphin-associated fisheries, and it is having negative impacts on several species that are not especially charismatic. The ecosystem effects are also large (Gilman 2011, 593). This brings us back to direct management of commercial fisheries and the effects of non-commercial interests, including fisheries economists and other epistemic communities.

8.3 "Rationalization"

Early discussions of fisheries management tended to focus on the problems of overexploitation and stock depletion. However, in the second half of the 20th century economists started to argue for the "rationalization" of fisheries management through the provision of access rights designed to eliminate the CPR problem that is at the core of the fisheries action cycle. As explained in chapter 5, establishing access rights is not a new concept in fisheries management; both formal and informal institutions for the distribution of endowments and entitlements to fisheries resources existed in many regions and eras. However, the Western concept of freedom of the seas proliferated in the 19th and 20th centuries, and many access-based regimes were destroyed. This allowed for significant overexploitation of fisheries and also resulted in substantial overcapitalization that widened both the power disconnect and the profit disconnect in cycles of serial depletion, as described in chapter 7. In the subsections that follow, I review the evidence on overcapitalization, explain the rise of various systems of government-regulated use rights, and review a range of critiques of these systems.

8.3.1 Overcapitalization and Overcapacity

As noted in chapter 7, economists began to study fisheries management during the 1950s. H. Scott Gordon (1954) put forward the concept of maximum net economic yield (MEY) and showed that, in most cases, MEY would occur at lower levels of effort than maximum sustainable yield and that the open access yield would be greater than either MSY or MEY. Thus, he was the first economist to describe the profit disconnect in fisheries. Gordon did not consider factors that widen the profit disconnect, but he did argue that the most common approach to MSY management at the time—the unallocated total allowable catch (TAC)—generated huge economic inefficiencies because it accelerated the race for fish. In this he identified two similar problems. First, limiting catch to MSY does not eliminate overcapitalization, which means that effort is higher than the level that would support MEY. Second, an unallocated TAC system incentivizes higher investments to capture a fixed amount of fish, so the level of overcapitalization is greater than it would be otherwise. Ultimately, overcapitalization also generates overcapacity, which occurs when the capacity of a fleet exceeds the level needed to produce maximum sustainable yield. Both overcapacity and overcapitalization generate political pressures to subsidize the industry (see chapter 6) and delay restrictive management that would reduce catch and/or effort (see chapter 7).

As evidence for his argument about unallocated TACs, Gordon (1954) pointed to overcapitalization in the Pacific halibut fleet, which was still seen as one of the best-managed fisheries in the world. Even though the International Pacific Halibut Commission set its TAC at "optimal yield," which they defined as MSY, the fishery remained open access. Anyone could harvest fish until the total allowable catch was reached, at which point the fishery would be closed for the year. This created incentives for individual fishers or boat owners to continually invest in larger, faster, more efficient vessels in order to capture a greater share of the TAC before the closure of the fishery. The long-run result was an increase in capacity and a decline in the duration of the open period, as the overcapitalized fleet could capture the TAC more quickly. For instance, the open season lasted about six months in 1933. Because of escalation in both the tonnage and the effectiveness of the fleet, the 1952 season lasted less than one month, even though the TAC was actually higher than it had been in earlier decades (133). Gordon describes similar problems in the Canadian Atlantic lobster fishery as additional evidence that management can exacerbate overcapitalization.

The Peruvian anchoveta fishery was an even more extreme case. The details of the first round of overcapitalization and collapse, including feedbacks between the power disconnect and the profit disconnect, are covered in chapter 7. The next round of responsive governance began with the rebound of Pacific anchoveta populations in the early 1990s. The increase in biomass was due to severely reduced fishing pressure and an oceanographic shift associated with the Pacific Decadal Oscillation. By this time, the fishing fleet had deteriorated due to prolonged economic recession associated with the extended period of low harvests in the 1970s and the 1980s. Indeed, by the end of the decade only 373 vessels remained from a fleet of 1,300 vessels during the 1969–1970 season, and 80% of those were old and outdated (Aranda 2009, 148; Paulik 1971, 172). This "natural" reduction in capacity provided the Peruvian government with an opportunity to proactively prevent overcapitalization and overcapacity, based on lessons learned from the collapse of the fishery in 1972.

In fact, several regulations were adopted to prevent overcapitalization in this period, including a vessel licensing system and requirements that those who purchased a new vessel must also decommission an old one. However, due to alignment between industry and government goals for economic development, these rules were not well enforced. In addition, as is common with these types of effort limits, fishers invested heavily in improvements to existing vessels or replaced smaller, outdated vessels with newer and larger ships. This phenomenon is known as "capital stuffing" (Townsend 1985). Nevertheless, the government continued to use unallocated TACs and a series of time-area closures to maintain catches at levels near MSY, but did not prevent fleet expansion. Fishing capacity increased sharply from 386 vessels in 1990 to 727 vessels in 1996 (Aranda 2009, 148). By 2009, the Peruvian anchoveta fleet comprised over 1,200 vessels. About half of these were larger (30–900 GRT hold capacity) steel-hulled industrial vessels with much greater fishing effectiveness than the old fleet, and about half were smaller (30–110 GRT hold capacity) semi-artisanal vessels that were not included in statistics until 1999. Estimates for 2005 show that the industrial fleet was at 69% overcapacity and the semi-artisanal fleet was at 75% overcapacity (Fréon et al. 2008, 402). This was similar to the level of overcapacity in the fleet in the late 1960s. Again, although originating in the common pool resource dynamic, this problem is exacerbated by the profit disconnect, which was increasing at the time as a result of a resurgence in demand for fish meal.

The Peruvian case is not exceptional. Fleets from historically dominant fishing countries were overcapitalized by the mid-1960s, and the majority of commercial fisheries were overcapitalized by the end of the 1970s (Crutchfield 1979). Christy and Scott (1965) document the level of overcapacity in the major fisheries of the 1960s. Aside from the Peruvian anchoveta, all documented fisheries were heavily targeted by distant-water fleets that grew rapidly after World War II. This growth was driven by developmental subsidies (chapter 6) and by the economic drivers of expansion described in Part I. Chapter 7 describes how these fleets undermined the management of most other large-scale fisheries in the postwar period. It also explains how, after the general acceptance of the 200-mile EEZ, overcapacity increased substantially as coastal states built up large fleets to replace excluded foreign fishers. The United States, Canada, Iceland, New Zealand, and many other countries sought economic development through encouragement of domestic fishing capacity. At the same time, the Soviet Union,

Japan, Spain, Poland, South Korea, and other countries worked to find ways to support continued operations by their distant-water fleets, including helping them (1) to negotiate access to "surplus" fisheries in the domestic waters of foreign states, (2) to develop the technological capacity to fish in deeper water much farther from shore, (3) and to increase production in their own coastal zones (Kaczynski 1979).

8.3.2 Proliferation of Use Rights

Considering the ample evidence of overcapitalization and cycles of over-exploitation associated with traditional effort restrictions, economists, managers, and fishers began looking for methods to reduce or prevent overcapacity in the 1960s. The first systems that allocated quota to specific vessels or fishers were instituted as a means to spread out greatly reduced catches among large numbers of fishers in the wake of stock collapse. However, low quotas were not always sufficient for economic success, so fishers began trading quota in a few fisheries. The results appeared to be successful, and a new epistemic community of neoliberal economists grew up around the concept of individual transferable quotas (ITQs). Those economists saw that ITQs were more efficient than prior management alternatives and that they allowed fishers to set the value of quotas themselves. With the support of this epistemic community, domestic-level ITQs proliferated in the 1970s and the 1980s. However, the availability of ITQs as a resource in the structural context did not necessarily reduce delays in management response. In fact, most ITQ systems were adopted during times of bio-economic crisis, when profit signals were strong, although the desire to exclude outsiders was also important. In the 1990s, social scientists also started to promote traditional use rights fisheries (TURFS) that would allow for community management on a smaller scale, and also advocated international tradable quotas as a way to improve management of international fisheries.

Domestic ITQs At the domestic level, after several cycles of crisis and response in various fisheires, governments started curtailing effort by imposing limits on the total number of vessels, taxes and fees to increase costs, and/or technological restrictions to prevent the capital stuffing described above. These measures often proved difficult to enforce, which made them politically appealing to many fishers. They also tended to increase the inefficiency of fleets. Although these types of measures are still common today, overcapacity problems persisted and managers of several different fisheries began to experiment with individual vessel quotas (IVQs) and individual fisher quotas (IFQs) starting in the 1960s. These programs assign a share of the total allowable catch to either a specific vessel or a specific fisher, respectively, and can be combined with licensing and other traditional management approaches. Such measures are now widely used to curtail both overcapacity and overcapitalization, but must be used in conjunction with vessel buyback and decommissioning programs to effectively reduce capacity in overcapitalized fleets. By providing fishers with a right to a specific share of the catch, well-implemented IFQs and IVQs can end the race for fish; however, they provide no incentive to sell off existing capital, and do not fully counter problems of the profit disconnect. In addition, economists still consider individual quota systems to be inefficient because the government rather than the market decides how much quota is assigned to each fisher or vessel.

In light of these concerns, Christy (1973) proposed that partially transferable, individual-level "fishermen catch quotas" would be a more efficient means of curtailing the race for fish. In such a system, more efficient fishers would be able to purchase quota from less efficient fishers, thereby reducing capacity and increasing the efficiency of the fishery as a whole. Moloney and Pearse (1979) used this idea and the literature on trading of pollution credits (e.g., Coase 1960) to develop a theory of "rationalized" fisheries management using individual transferable quotas (ITQs). They proposed that ITQs should be treated as property rights, such that ownership would extend in perpetuity, quotas would be perfectly transferable, and prices would be determined in open markets. In the 1980s, many economists started advocating ITQs as the best method of fisheries management, although most recognized that complementary measures might also be required. Currently, ITQs are used in more than 10% of fisheries worldwide and continue to proliferate (Chu 2009).

Iceland was the first country to institute an ITQ system, although this was a matter of necessity rather than economic theory. The development of this system started with the collapse of Iceland's herring fisheries in the late 1960s. Icelanders targeted two different stocks: the larger, spring-spawning Atlanto-Scandic stock and a smaller, localized summer-spawning stock.[4] Environmental conditions and overexploitation combined to cause a near simultaneous collapse of both stocks near Iceland between 1966 and 1969. Due to a climate regime shift and harvesting by international fleets throughout its range, the spring-spawning herring stock declined precipitously and the remaining fish stayed closer to Norway, so the stock could no longer be targeted by the Icelandic fleet (Cassou and Hamilton 2004, 328). The summer-spawning stock was not migratory, but similar climatic drivers of low recruitment combined with overfishing to cause a decline in biomass. Again, response was delayed by the power and profit disconnects, but Iceland tried to prevent a complete collapse of this stock by establishing an unallocated TAC system in 1966. This was largely ineffective; effort remained high even as landings and profits continued to decline. Finally, in 1972 the government placed a ban on all harvests of the summer-spawning stock (Arnason 1993, 205).

The collapse of the Icelandic herring fishery was a great economic blow to a number of coastal communities that depended on capturing and processing of herring. While the Atlanto-Scandic stock did not return until the 1990s, the summer-spawning stock began to recover with a strong year class in 1971 (Jakobsson and Stefansson 1999). When the government reopened the fishery in 1976, it recognized that capacity was much too high for the relatively small TAC available. Therefore, managers also established a limited entry system and imposed individual vessel quotas. The fleet was still large relative to the size of the stock, so the share per boat was quite small. To legally catch enough fish to survive, fishers created an informal quota trading system. In 1979, the Icelandic government formalized this process by establishing regulations for transfer of quota (Arnason 1993, 205–206). These measures allowed fishers who were determined to stay in the fishery to purchase quota from those who were willing to exit or to fish less, thereby incentivizing the reduction of fleet capacity and overcapitalization. From 1977 to 1990, catches of herring tripled while nominal fishing effort declined by about 20%, so the profits made by those fishers who remained in the industry increased (211).

Many scientists, economists, and decision makers viewed the resurgence of Iceland's herring fishery as a great management success, although the strong recruitment in 1971 also contributed significantly to rebuilding the stock and allowed for increasing TACs over the period. Much as with the Pacific halibut case in the United States and Canada, this perceived success translated into replication of the ITQ system in other Icelandic fisheries. The government established an IVQ system for capelin in 1980. This small pelagic fish is a good substitute for herring and many of the fishers displaced by the depletion of herring stocks started targeting capelin in the 1970s. Increased effort led to rapid overexploitation and growing overcapacity in the fishery. However, the government did not officially sanction quota trading until 1986 (Arnason 2006, 250). Management of the demersal fishery (i.e., for cod, haddock, and other groundfish), which was of much greater economic value than herring and capelin combined, was somewhat different. A growing power disconnect prolonged the period of ineffective management via unallocated TAC and number of allowable fishing days per vessel. However, after a period of gradual economic decline and a sudden sharp drop in the stock size, the government established an ITQ system for demersal fisheries in 1984. Pressured by enthusiastic economists and fishing corporations, the government created an inclusive ITQ system for all Icelandic commercial fisheries in 1991 (251). A few other early ITQ-like systems developed organically in the 1970s. For example, Norway established "fisheries concessions" for the remaining Atlanto-Scandic herring stock, which were tradable in practice, even though there was no formal legal mechanism for trades (Hannesson 2004).

In contrast to these early programs, most ITQ systems developed in the 1980s were established based on economic theory and were responses to less severe bioeconomic crises. Several countries started using ITQs in this period, but New Zealand and Canada are instructive cases. In 1986, New Zealand established a sweeping ITQ program covering most of its commercial fisheries. According to Dewees (1998), this program closely followed the ITQ design formulated by Moloney and Pearse (1979). It was created in consultation with the fishing industry, which was facing economic hardships. The profit signal was strong, and a vessel buyback program was included to reduce overcapacity quickly. In the Canadian Pacific halibut fishery, continued use of unallocated TAC management caused overcapacity to escalate throughout the 1980s (Dewees 1998). From 1982 to 1990, the length of the open season decreased from nearly two months to only six days. Concern about the safety of fishers as well as wastefulness in the fishery created political will to institute an IVQ system in 1991. In constant consultation with the industry, the Canadian government gradually phased in an ITQ system, allowing limited transferability in 1993 and then relaxing restrictions in subsiquent years (S136).

A number of other governments created ITQ management systems for their fisheries based on the perceived success of existing ITQ experiments along with continued theoretical arguments from economists. Alaska followed the Canadian example, establishing an ITQ system for Pacific halibut in 1995, and the US implemented ITQs for several other fisheries later that decade. The Netherlands was another early implementer of ITQs, while countries like the Falkland Islands and Peru instituted ITQs in the late 2000s. Chu (2009, 219) estimated that by 2008, 18 countries were using ITQs to manage 249 species, and several more have taken up the approach since then. In a global analysis, Chu identified a few rationales for the switch to ITQs. Most fit into one of two categories: perceived bioeconomic crisis and conflict among fishers over access to stocks. In a targeted analysis of the effects of ITQs on 14 major commercial fisheries, Chu found that ITQs generally result in reduced fishing capacity, but that biological and social consequences are more varied. This work is representative of several important critiques of the ITQ approach to management (see subsection 8.1.3).

Local and Global Variations There are two major variations on the practice of managing use rights, one for small-scale fisheries at the local level and the other for large-scale fisheries at the international or even global level. First, some social scientists recently started recommending the implementation of territorial use rights instead of ITQs for small-scale artisanal coastal fisheries, where monitoring is difficult due to the dispersed nature of fishing activities. Christy ([1982] 1992) coined the term and explained when and how TURFs might be preferable to ITQ-based systems. He points out that TURFs are particularly well suited to protect and improve livelihoods in small coastal communities where tradable quota systems could result in neo-feudalistic power structures. Multiple authors have built on this concept and provided examples of existing TURFs around the world. In addition to minimizing open access, TURFs could substantially reduce the power disconnect by giving local communities control of their resources.

Because TURFs do not completely eliminate the common pool resource dynamic, some economists are skeptical about their potential for cooperative management. However, Hannessson (2004) recognizes potential benefits in small-scale fisheries where administrative costs are often high due to the fact that fishers are numerous and are spread out along the coast rather than concentrated in specific ports, as with most larger-scale industrial fleets. Hilborn et al. (2005) are more optimistic about the benefits of collectively managed TURFs for small-scale coastal fisheries, although both Hannessson and Hilborn et al. point out that TURFs may be unsuccessful if the resource can be captured outside of a community's recognized territory. Berkes (1992) further points out that TURFs may be unsuccessful if governments fail to prevent incursion by outside fleets. Like any other fisheries management system, TURFs may also be subject to changing environmental conditions, but are particularly susceptible to coastal pollution, sea level rise, and ocean acidification. Nevertheless, given that over 85% of world fishers are small-scale producers, improving management of these fisheries is an economic as well as a humanitarian and ecological imperative (FAO 2012d). Furthermore, this approach can minimize both the profit disconnect and the power disconnect, improving management overall.

Second, there is increasing pressure to establish state-level use rights internationally. Theoretical grounds for this movement are rooted in the pre-EEZ period, when many more fisheries were exploited by international fleets. Christy and Scott (1965) proposed that states with fleets targeting shared stocks should negotiate state-specific quota-sharing arrangements to curb the problems of economic inefficiency and congestion on fishing grounds. This is common now in multiple RFMOs, but some, like the IATTC, continue to use unallocated TACs in combination with capacity limits and time-area closures for some fisheries (IATTC 2013b). So far, the IATTC approach has been effective for purse seine fisheries targeting tropical tunas, but this may be due to economic factors—particularly low profit margins in the fishery—as much as to management measures. Elsewhere, the effectiveness of management via state-specific quota arrangements is mixed at best. In some cases, demand for quotas causes states to agree to TACs that are well above scientifically recommended levels as a way to accommodate everyone. In other cases, states agree to scientifically recommended TACs but fail to enforce quotas at the national or the international level, so overexploitation and overcapitalization continue (Webster 2009, 2011). There have been some cases of successful international management with international enforcement mechanisms, as described in Chapter 5.

Barkin and DeSombre (2013) also recommend the use of international tradable quotas to reduce overcapacity in fisheries managed by regional fisheries management organizations. Although Christy and Scott (1965) recommended capacity reduction almost 50 years ago, RFMOs did not start tackling the problem until the 1990s, when large harvests by illegal, unreported, and unregulated (IUU) fleets started having a negative economic and biological impact on some of the most valuable international commercial fisheries. Here again, exclusion and reduced profits drove management measures. Regulation was piecemeal, however, and by the early 2000s it was clear that without a coordinated approach to capacity reduction RFMOs would simply be chasing IUU fleets from one fishery to another. In addition, overcapacity in member fleets continued to cause upward pressure on TACs in many RFMOs, such that TACs were often set higher than the scientifically recommended levels or were unenforced if set at MSY. In 1999, the UN Food and Agriculture Organization published an International Plan of Action for the Management of Fishing Capacity (IPOA-Capacity), and these problems are now widely recognized by most RFMO member states (Aranda, Murua, and De Bruyn 2012).

Indeed, this issue instigated formal cooperation among tuna RFMOs, particularly through regular meetings of all five tuna commissions. This is now known as the "Kobe Process," which started with a meeting in Kobe, Japan, in January 2007. Capacity was a primary issue of discussion at that meeting and, while all members could agree that capacity needed to be reduced, none were willing to pay the costs by committing to reduce their domestic fleets. Nevertheless, the tuna RFMOs did start negotiating capacity limitations for several overcapitalized fisheries, and some reduction has occurred. In addition, use rights were discussed at the Kobe III meeting in July 2011. Few use-rights mechanisms currently exist among tuna RFMOs, and only one allows quota trading. This is a program for transfer of quota among fishing fleets operating in the EEZs of small island states in the Western and Central Pacific Fisheries Commission. Quotas are not tied to vessels and are only granted on an annual basis, so the system provides flexibility in circumstances where the location of fish may be unpredictable, but it does not incentivize exit or capacity reduction (Aranda, Murua, and De Bruyn 2012).

8.3.3 Critiques of ITQs

Clearly, no approach to fisheries management is flawless. However, the debate over ITQs is particularly heated and deserves additional attention because it has broader political and economic implications. Much of the debate is normative, focusing on the appropriateness of efficiency as the primary goal of fisheries policy, but there is also disagreement about the economic gains from ITQ systems, most of which revolves around the potential for arbitrage and the macroeconomic implications of what is essentially a microeconomic management tool. The Iceland case provides an example of the range of positions on ITQs. Arnason (1993; 2000; 2006) is one of the most enthusiastic supporters of property rights systems in fisheries management generally. He makes sweeping claims about the "proven economic superiority" of ITQ systems, and calculates that total factor productivity for the Icelandic system increased by more than 91% from 1980 to 2002, with a growth rate of 3.5% over the period. He also claims that most of this growth occurred from 1994 on, after the establishment of the national ITQ system, rather than in the post-EEZ period, when the domestic fleet grew rapidly with government assistance after foreign fishers were excluded from the area (2006, 261). He pays little attention to distributional issues or to impacts on the wider economy.

On the other hand, authors like Eythórsson (1996; 2000) and Palsson and Helgason (1995) criticize the Icelandic ITQ system based on goal appropriateness. They argue that fleet efficiency should not be the only criterion for assessment of ITQs and that the social costs of the system are too high. For instance, Eythórsson (1996) describes deterioration in labor conditions, loss of jobs, several major strikes by fishers, and the "feudalization" of the fishing industry as negative outcomes of the ITQ system in Iceland. He also critiques the program on its own terms by analyzing broader economic and environmental outcomes, pointing to economic inefficiencies that are introduced with the consolidation of quota, including price ceilings established by processors who control large shares of important fisheries. This can be seen in the 50–70% difference between in-house processor prices paid to fishers and auction prices. Processors can pay the fishers who contract with them much less because their monopoly on quota shares gives them leverage when negotiating prices (277). Furthermore, Eythórsson criticizes the biological success of the program, since cod stocks continued to decline after the implementation of the ITQ system, as reflected by lower and lower TACs. Such criticisms are commonly directed at ITQ systems and are even recognized by proponents, although most still argue that the social benefits outweigh the social costs (Hannesson 2004).

Critiques of ITQs also focus on power relationships and exclusion. As described in chapter 7, most effort restrictions, and indeed, many other types of regulations, can generate exclusion of one group of fishers or another. Use of IVQs and IFQs involves implicit or explicit political decisions about participation in the fishery. Wilen (1988) describes how lobbying for exclusion shaped many of these programs throughout the 1980s. Hypothetically, ITQ systems are designed to reduce capacity by excluding less-efficient fishers, thereby enshrining efficiency as the only determinant of participation. However, implicitly, access to capital is a larger determinant of dominance in basic ITQ systems, which means that large corporations can push smaller fishers out of the market for ITQs even if there are not substantial efficiency differences between the two. In addition, ITQ systems can be explicitly modified to exclude new or foreign fishers, which can improve their political expedience. For example, excluding new entrants was a major driver of quota-based systems in Iceland, "New Zealandization" was an important component of ITQ management in New Zealand, and exclusion of non-Alaskan fishers and an "Alaska first" mentality were the major rationale for the 1995 introduction of ITQs into the Alaska halibut fishery as well as various salmon fisheries (Matthíasson 2003; Dewees 1998; Rogers 1979). Moreover, passage of the ITQ strategy depended on support from small-scale fishers, who used their political clout to ensure that the program would limit entry by large corporations. Thus, ITQs are not immune to political maneuvering and are still shaped by exclusionary practices as well as crisis response. This can widen the power disconnect and cause expansionary pressures on policy makers.

Politically motivated implementation of ITQs to exclude outsiders continued in the 21st century. For instance, Peru instituted an ITQ system for its anchoveta fishery in 2009. By this time the fleet was as heavily overcapitalized as it had been before the first major collapse of the fishery in 1972 (Fréon et al. 2008). However, there was an important difference between the mid-century and late-century contexts. During the first period of growth (1950–1972), the fishery's ownership structure was fairly horizontal, with many fishers owning their own vessels and a number of corporations controlling the processing plants. In the second period of growth (1992–present), capital was concentrated in the hands of the few companies that had survived the initial collapse; seven owned 50% of the fishing and processing capacity in 2007 (Schreiber and Halliday 2013, 4). These large companies favored capacity restrictions, particularly an ITQ scheme that would allow them to further consolidate access to the resource by buying out small fishers. Given pressures from these powerful corporations and general concern regarding overcapitalization, the Peruvian government expressed interest in establishing an ITQ scheme in 2003, but protests from smaller-scale fishers and processors—who recognized this attempt to exclude them—delayed implementation until 2009. Ultimately the government placated both sides; the established ITQ system did not apply to the artisanal fishery (Aranda 2009).

Recognition of the social and political implications of ITQs has generated considerable work on the design of ITQ systems to achieve goals other than efficiency. In countries where the structural context gives power to small-scale fishers, caps on the total percentage share of quota that can be owned by a single individual or company are now commonly used to prevent the quota agglomeration observed in Iceland and elsewhere. Canada is well known for these types of protections, which are important to indigenous peoples or "first nations" as well as to other small-scale fishing groups (Dewees 1998). At times, governments are forced to renegotiate ITQ plans to recognize the use rights of indigenous peoples, as when the Maori successfully lobbied for a larger share of the use rights under the New Zealand ITQ system. As part-time fishers, many Maori were excluded from important fisheries at the outset of the ITQ program because only full-time fishers received quota allocations. After a long legal and political battle, a portion of quota was set aside for the Maori, and they were given a 50% share in Sealord, the largest fishing company in New Zealand (Annala 1996, 55). Arrangements for the protection of small-scale and indigenous fishers' use rights are common in countries where such communities have considerable political power. However, in other countries these communities are marginalized and forced to find other sources of sustenance (Berkes 1986).

Although large-scale commercial fishers sometimes benefit from tradeable quota systems, commercial interest groups are not always in favor of ITQs or related IFQs and IVQs. Again, the structural context is critical. For instance, in Chile, attempts to establish an ITQ system under Pinochet in the 1980s were later undermined by industry lobbying once the military junta was replaced by a democratic regime. In this case, one of the biggest fishing companies in the country opposed the ITQ system because of past experience with ineffective implementation of IVQs, beliefs about the inexhaustibility of small pelagic species like sardine and anchoveta, and a desire to retain flexibility to freely enter new fisheries. With other industrial interests of like mind, this company successfully lobbied to weaken the ITQ system and related management measures proposed by the government. As a result, overexploitation and overcapitalization increased substantially in the 1990s, and a new IVQ system was implemented in the early 2000s utilizing a cooperative approach. Groups of fishing companies received shares of the TAC and then negotiated internally to establish specific quota arrangements (Hannesson 2004, 98). Similar cooperative systems were used in Alaska during the late 1990s and may be found elsewhere (Criddle and Macinko 2000). Where such systems protect the rights of economically sensitive fishers, the power disconnect is minimized, but ITQs can also increase the power disconnect by marginalizing these groups.

Treatment of recreational fishers in ITQs or other types of limitation systems may also be controversial. Traditional approaches to ITQ systems focus on maximizing scarcity rent for commercial fisheries. This would suggest that recreational fishers should be excluded from quota trading. However, recreational fishers have increasingly powerful lobbies in many countries and frequently protest such exclusion. They have also funded numerous studies to estimate the economic value of recreational fisheries, which do generate substantive revenue in several economic sectors, including tourism, vessel construction, and gear production. They therefore argue that the economic benefits of fisheries cannot be maximized without consideration for recreational fishers in individually based quota schemes. Where recreational fishers are powerful, as they are in New Zealand, these arguments are successful and some quota is usually set aside for them (Craig et al. 2000). This, in turn, often angers commercial fishers, who believe that the resource should be used for food or other products, rather than recreation. Here again, the effects on the power disconnect and responsive governance depend on the sensitivity of actors who are either excluded or included. Often recreational fishers receive biological problem signals earlier than commercial fishers, but profit disconnects can exist in both industries.

The second major critique of ITQ systems is that they do not always ensure the conservation of fish stocks. Criticism of this kind sometimes comes from environmentalists and recreational fishers as well as scientists and, at times, industry representatives. Even the proponents of ITQ systems recognize that additional measures are needed to deal with externalities such as habitat destruction or high grading (disposal at sea of lower-priced species found in a catch). Such externalities ensure that a profit disconnect remains in spite of well-established access rights. As noted above, there is also evidence that temporal myopia continues to skew fisher decision processes toward current rather than future revenues, which also leads to discrepancies between the private and social optima in ITQ systems (Asche 2001). Furthermore, because the TAC is still set by government decision makers, the biological success of an ITQ system depends heavily on managers' ability to establish and maintain effective TACs. As discussed in chapter 7, this requires both scientific knowledge and political will to curtail harvests, as well as strong monitoring and enforcement mechanisms. This can be difficult when profit signals push powerful fishers to lobby for expansion or when scientific advice fails to identify environmental changes in a system. Incorrect scientific advice should not be discounted, but it is much less common as a factor in biomass decline than political pressure. There are few examples of major scientific inaccuracies in modern times, such as when overoptimistic scientific evaluation of stock size caused overexploitation in spite of a well-enforced ITQ system in the New Zealand orange roughy fishery (Hannesson 2004). On the other hand, examples of TAC inflation, or the tendency to increase TACs above scientifically recommended levels, are widespread, and are usually due to pressures from the industry under conditions of high short-run profitability.

Indeed, even when access rights are established through ITQs, there is evidence that political pressures consistently drive TACs up when profits are high. Since most ITQ systems are established during periods of perceived crisis, early TACs are usually set in line with scientific advice. However, in a number of observed cases, after the economic and biological rebuilding of a fishery, managers decided to set the TAC above scientific recommendations to placate demands from industry interest groups. For instance, while the Icelandic government set TACs for summer-spawning herring that matched scientific advice throughout the 1970s and early 1980s, they set TACs at levels 10,000–25,000 tonnes higher than scientific recommendations in the 1990s and 2000s (Jakobsson and Stefansson 1999). In Chu (2009), nine of the 18 stocks studied in detail continued to decline five years after the establishment of the ITQ system, and six were still in decline ten years later. Some of the difference can be explained by biological and environmental variations, but in most cases continued overexploitation due to either TACs that were set too high or lack of monitoring and enforcement were the primary causes of continued decline. This type of pressure should not be observed if the CPR dynamic is the only driver of economic expansion, but it is entirely expected given insights regarding endogenous and exogenous drivers from the AC/SC framework.

Indeed, the importance of both the profit disconnect and the power disconnect when designing and implementing ITQs needs more study. Hypothetically, when property rights are fully enforced, fishers should be in favor of long-term management at MSY—or the lower MEY—but in practice the industry may still resist scientific recommendations that require fishers to curtail effort when profits are high. Thus, when the profit disconnect results in increased catches due to declining costs of production and/or increasing prices caused by growing demand, it may undermine management, just as in a system without property rights. This effect can be magnified by the power implications of ITQs. For basic ITQs, concentration of quota is a common outcome, often in the hands of a few large and sometimes multinational companies, which are well insulated from economic problem signals. This increases the power disconnect and can amplify pressures to drive up TACs.

There are reasons for concern at the international level as well. The annual resetting of state-specific quota in the Western and Central Pacific Fisheries Commission reflects the fact that political power in international fisheries negotiations is based on the concept of "national interest" in a given fishery. As described in chapter 5, in most RFMOs national interest is defined by a combination of coastal state status, developing state status, and historical catch in the fishery. Under a permanent ITQ system it is likely that quota shares—and political power—would shift from historically dominant fishing fleets to developing fleets with the geographic flow of fishing capital described in chapter 2. This gives historically dominant fishing states reason to reject international tradable quota systems, but it also has implications for the power disconnect. A system of international tradable quotas would marginalize historically dominant fishing states, which usually have higher levels of political concern for conservation, largely due to shifts in the structural context associated with long experience with overfishing and the growing influence of non-commercial interest groups. Such states also tend to have much better monitoring and enforcement capacity than developing states. As such, a shift of economic and political power from historically dominant states to developing states could have the unintended consequences of weakening management measures and undermining the potential benefits of international tradable quota systems. Thus, ITQs may have counterintuitive effects at the international level.

8.4 Alternatives to MSY

While economists sought to "rationalize" fisheries management with ITQs, epistemic communities of biologists and ecologists worked to advance fisheries science more broadly. A number of breakthroughs occurred in the 1950s and the 1960s that improved the realism of stock assessments by adding population structure, environmental control variables, and other key factors. By the 1970s, many fisheries scientists were disillusioned with the concept of MSY and began to develop alternative approaches. At first, these critics focused on the problem of complexity in fisheries systems by advancing theories of ecosystem resilience and by developing techniques such as adaptive management (subsection 8.4.1). In the 1980s, scientists began working on methods of multispecies management, to maximize total biomass production rather than single-stock biomass. At the same time, some environmental organizations started to expand their area of concern from charismatic megafauna to marine ecosystems as a whole. With combined theoretical advances from ecologists, lobbying by NGOs, and the continuing serial depletion of fish stocks, alternatives to MSY proliferated in the 1990s. Concepts like ecosystem-based management, space-based management, and the precautionary approach were incorporated into international law. Several countries also passed major legislation requiring application of these types of policies at the domestic level (subsection 8.4.2).

In most cases, both theory and policy advanced beyond practice during this period. Legislation often served as a major impetus for innovations that would facilitate practical implementation of MSY alternatives. These included scientific advances in the use of tools such as ecosystem benchmarks and environmental risk assessment. Given the difficulty of implementation, proponents of MSY alternatives advanced a range of hybrid approaches, with some suggesting that modest modifications to MSY are sufficient if well enforced and others advocating complete replacement of MSY with much more holistic—and complex—approaches. The practical evolution in alternatives to MSY included the development of the concept of stakeholder engagement, which should ideally allow the different groups with interests in an ecosystem to set their own priorities, thereby increasing the legitimacy of management measures. This, in turn, should help to ensure compliance with regulations and reduce costs of monitoring and enforcement. In combination with increasing political and economic problem signals, these innovations have led to several successful applications of alternatives to MSY. There have also been several notable failures, and implementation has not been as rapid as proponents would wish (subsection 8.4.3).

8.4.1 Accounting for Complexity

Epistemic communities helped build management capacity throughout the 20th century. As explained in chapter 7, by the 1950s there were three major types of experts. First, those who favored fishing "up to" a simple version of MSY derived from catch data that treated stocks as homogeneous populations. Second, those who favored the "go slow" approach to MSY and who usually preferred more detailed population models that incorporated population structure and habitat characteristics. And third, those who believed that MSY was not the correct approach at all, preferring highly detailed ecosystem assessments and low levels of fishing effort that accounted for predator-prey interactions. As shown below, by the 1970s, most experts had lost faith in the fishing "up to MSY" approach, having experienced multiple cycles of fishery overexploitation and collapse under this management norm. The search for alternatives to MSY started with the development of more complicated models of fish stock dynamics in the 1940s and 1950s but by the 1980s two new management approaches had been proposed: adaptive management and multispecies management. These approaches accounted for the complexities associated with marine systems but proved difficult to implement. Some problems with implementation were technical, others were political and economic. This subsection describes the evolution of these management options and related epistemic communities form the 1970s to the 1990s, when both were subsumed into broader constructs like ecosystem-based management.

Incorporating Ecosystems Many attempts to understand marine systems in the early 1900s were highly detailed studies of ecosystem and/or population dynamics. Scientists like Gilbert (1912), Russell (1931), and Graham (1935) pointed out that variations in fish abundance depended on many environmental and ecological factors, though all also agreed that overfishing was a major cause of stock decline. In contrast, scholar-managers, including Thompson (1919) and Schaeffer (1957), favored less data-intensive approaches focused on changes in catch per unit effort and, ultimately, maximum sustainable yield. Even though MSY dominated management from the 1950s on, application was not monolithic. Protections for juvenile and spawning populations were commonly used with catch or effort limits, and many scientists recognized the need to account for environmental fluctuations such as the El Niño Southern Oscillation and ecosystem effects such as predation or lack of food resources. Indeed, insights from biological oceanography and marine ecology were often included in MSY-based stock assessments and management.

Improvements in computational technologies facilitated the creation and parameterization of more detailed methods that incorporated more realistic assumptions about population dynamics and yield-recruit relationships (Beverton and Holt 1993). Beginning in the 1980s, integrated analysis and Bayesian analysis gave fisheries scientists tools with which to significantly refine statistical modeling techniques and to add greater detail to stock assessment models (Maunder 2003). Though these models were more advanced than the typical Gordon-Schaeffer model described in chapter 7, they still focused on the maximization of yield from a single stock of fish. In addition, these techniques were largely deterministic and produced static equilibrium points similar to MSY. As Walters (1986) pointed out, these approaches focused on a few primary variables, such as growth, recruitment, and mortality, even though they allowed those factors to change based on size class, population size, and environmental conditions. Thus, many of the postwar advances in fisheries science essentially provided modifications to MSY rather than true alternatives. One exception might be the use of a minimum escapement rule (allow a certain number of fish to "escape" the fishery) rather than a maximum harvest rule, but, the management goal remained the same: higher average yield for a single stock over time (Walters 1975).

However, in the late 1960s a few ecologists revived earlier critiques of single-stock management and introduced concepts from the burgeoning field of complexity theory into fisheries science. Their work was driven in part by observed sudden and unpredicted collapses in commercial fisheries, but also by studies of periodic upheavals in terrestrial ecosystems. Holling's (1973) theoretical paper was particularly important. He wrote about the persistence or resilience of ecosystems rather than fluctuations in the abundance of specific populations, which was the more common approach at the time. Specifically, he showed that while the composition of an ecosystem may be highly variable, it can still be resilient, retaining its overall function as forest, grassland, or reef. At the same time, Holling also showed that ecosystems could go through rapid and unpredictable transformations, shifting from one functional form to another. In technical terms, if pushed past an existing threshold, an ecosystem would shift from one domain of attraction to another. Later, Jones and Walters (1976) applied Thorn's catastrophe theory to fisheries collapses. Although catastrophe theory itself is not used much today, the concepts of "slow variables" that change gradually and "fast variables" that change quickly are still foundational in the ecological literature. Furthermore, disconnects between these variables are still associated with catastrophic change, though their mathematical representation is different from Jones and Walters' original approach (Holling, Gunderson, and Peterson 2002).

Two new alternatives to MSY-based management grew out of these theoretical advances in ecology: adaptive management and multi-species management. The latter developed relatively slowly, starting with multispecies models and questions of application to Antarctic krill fisheries in the late 1970s, but eventually evolved into a wide range of approaches in the 1980s and the 1990s. Adaptive management, developed much earlier, is more distinctive but less widespread. It should be noted, however, that the two are not mutually exclusive, and that some advocate their combined use, often treating one as a component of the other.

Adaptive Governance Walters and Hilborn (1978) introduced the concept of adaptive management as a means of dealing explicitly with the uncertainty associated with optimizing production from complex ecosystems. They argued that it is not possible to predict how such a system will respond to a given policy option because the system is constantly changing due to exogenous and endogenous factors. So, instead of management based on some static equilibrium condition (e.g., MSY), or even a dynamic approach to equilibrium, they proposed that optimal control theory be used to iteratively improve policy—in other words, policies should be designed as explicit experiments to provide scientists and decision makers with improved knowledge about the system. Furthermore, policies should be altered in response to that new information in an ongoing process of well-designed trial and error. While this may seem like a significant deviation from past management practices, Walters and Hilborn point out that fisheries management was already a process of passive trial and error in which policies centered on MSY would be revised with new information, particularly regarding the overfishing of the stock.[5] By choosing adaptive management, decision makers would simply be replacing blind, responsive trials with active experimentation based on sound research design.

Although adaptive management started in fisheries, it has had much broader application to terrestrial and freshwater systems (see, e.g., Milliman 1986; Imperial, Hennessey, and Robadue 1993). Walters and Hilborn's general discussion of adaptive management was informed by their earlier experiments with different types of management for Pacific salmon stocks. Several other fisheries management experiments were tried in the late 1970s, including the Inter-American Tropical Tuna Commission's explicit choice to allow overfishing of yellowfin tuna to ascertain the maximum sustainable yield for the stock (Walters and Hilborn 1976). However, there have been few applications to marine fisheries management since that time. No doubt the lack of interest is due in part to political factors. Walters and Hilborn themselves recognized that their approach could be blocked by political conflict—particularly in fisheries that were already overfished, where the best experiment might be severe reduction in catch (Walters and Hilborn 1978, 181). However, there are also important scientific considerations. In her seminal work on multispecies management, Pikitch (1992) points out that the biological and economic costs of such experimentation are not known prior to perturbation (e.g., severe restriction of effort or allowed overexploitation by fishers) and may in fact be greater than the benefits of improved understanding. Furthermore, she argues that it will be difficult to interpret the results of policy experiments in marine fisheries because the costs of data collection/monitoring are high, and because it is not possible to control for all environmental conditions, which are notoriously variable.

Nevertheless, work on adaptive governance has had a lasting impact on marine fisheries management by increasing the technical resources available to scientists and decision makers. Most notably, stochastic dynamic programming is used to develop computational techniques for generating an array of policy scenarios given underlying uncertainties in parameter values (Hilborn and Walters 1992). Even though Hilborn and Walters saw computational methods as a suboptimal solution, these approaches are an integral part of modern fisheries management and are particularly important for more complex approaches such as multispecies and ecosystem-based management. In later work, Walters (1984) provides a more refined conceptualization of adaptive management that fits well with these approaches and includes considerable use of computational models and related policy scenarios as a way to minimize the risks of experimentation. He also promotes the use of stakeholder engagement to determine goals and legitimize policy experiments, reducing political roadblocks.

Multispecies Management The second early alternative to MSY was multispecies management. Ecologists began to develop multispecies models of fisheries in the late 1960s, largely in response to observed oscillations in species abundance in long-run time series for specific regions. Riffenburgh (1969) is credited with the first multispecies model, followed by Agger and Nielsen (1972), Parrish (1975), and Lett and Kohler (1976). Anderson and Ursin (1977) further advanced the literature on multispecies modeling by adapting the age-structured population models of Beverton and Holt (1993) to include trophic interactions such as competition and predation. That same year, Larkin (1977) published his "Epitaph for MSY," which drew on multiple literatures, including Holling's work on ecosystems, Walters' work on adaptive management, and research on multispecies fisheries with by-catch problems. Larkin did not provide any alternatives to MSY, but his paper represents the dissatisfaction of a growing number of scientists and managers who saw a need for more holistic approaches. At the same time, increasing activities by conservation NGOs generated the first political demand for multispecies management.

Specifically, the prospect of a new fishery for krill in the Antarctic alarmed conservationist NGOs and scientists. Krill are a major food source for multiple species of whales. In fact, several species, including humpback whales, migrate to the Southern Ocean annually to feast on massive shoals of these tiny crustaceans. The conservationists feared that single-stock management of krill might result in less food and therefore lower survival rates for those whales, the populations of which were already severely depleted as a result of commercial whaling. May et al. (1979) provided some heuristic decision rules for multispecies management of the krill fishery and advocated a new management regime to implement their approach. NGOs also kept up the pressure on governments. They were already well organized for the battle to ban whaling that was raging at the International Whaling Commission and, since there was no krill industry to oppose them, they wielded considerable influence. Parties to the Antarctic Treaty System began negotiating a new management regime in the late 1970s. The Convention for the Conservation of Antarctic Marine Living Resources (CCAMLR) entered into force in 1982 and the resulting commission was tasked with multispecies management to balance commercial harvests of krill and other prey species with the needs of whales and other predators. Given these beginnings, it is not surprising that CCAMLR is one of the few RFMOs to embrace multispecies management and its descendant, ecosystem-based management (G. Parkes 2000, 83)

The concept of multispecies management evolved considerably in the 1980s. The greatest single advance occurred in 1984 with the Dahlem Workshop on the Exploitation of Marine Communities, which brought together some of the best minds in the areas of fisheries science, marine ecology, and resource economics. Attendees included biologist Robert May, fisheries scientists Raymond Beverton and Sydney Holt, fisheries management experts John Gulland and Serge Garcia from the UN Food and Agriculture Organization, marine ecologists Carl Walters and Daniel Pauly, resource economist Colin W. Clark, and many others who are well known in these disparate fields to this day. The report from the workshop contains group reports and single-author papers on such topics as known problems with ecosystem management, methods of dealing with uncertainty in fisheries to ecosystem modeling techniques, and the implementation of multispecies management. Throughout the report it is clear that participants agreed on the need for some form of multispecies management, but that they disagreed about the concept itself and about methods of implementation. As May (1984) pointed out in his introduction, the workshop and subsequent report provide information about what was lacking in the existing knowledge base and thereby identify needs for future research, rather than offering codification of a single approach to multispecies management.

Over the rest of the decade, many scholars and scientist-managers continued to work on the problems that had been identified at Dahlem. Pikitch (1992) provides an excellent overview of progress on multispecies management, or really the many similar approaches that fit under that umbrella. She starts by distinguishing between technological and biological interactions in fisheries. Technological interactions occur when different species are caught by the same gear. As has already been noted in this volume, management of gear selectivity has a long history, but for the most part fishers and managers were concerned with by-catch of juveniles of targeted species or the exclusion of "outside" fishing groups rather than by-catch of non-targeted species. Indeed, section 4.3 shows that high-grading, defined as the discarding of lower-priced species at sea, was common practice in large-scale fisheries. This could be seen as problematic even from a single-stock management perspective, since by-catch species were often targeted after higher-priced species became scarce. From an ecosystem perspective, however, loss of by-catch species could also have significant impacts on targeted species through trophic interactions. When by-catch species have low growth rates, their populations can be severely diminished by fishing, which, in turn, can cause an ecosystem to shift from one dynamically stable state to another, as described by Holling (1973) and others. As Pikitch (1992) points out, the technological interactions that cause by-catch problems are relatively easy to identify and control given sufficient monitoring to ensure low levels of discards at sea.

In contrast, biological interactions are difficult to comprehend because they involve ecosystem-level relationships between species, such as predator-prey dynamics and competition for space or food. These relationships are inherently complex and therefore are generally unpredictable in the moment. That is, although a general pattern such as a shift from an anchoveta-dominant regime to a sardine regime can be expected, the timing of the shift is not predictable. Furthermore, causal relationships are often impossible to trace, so policy effects cannot be determined. Although Pikitch (1992) cites numerous improvements in multispecies modeling techniques, including the use of sensitivity analysis to test for large discontinuities, she asserts that maximization of yield is not a feasible management goal for fisheries, even when multiple species are taken into account. Instead, she advocates "a more honest approach to management which recognizes that there are, and will continue to be, significant limits to our ability to predict the consequences of our actions and to control real systems" (Pikitch 1992, 129).

Indeed, Pikitch's work in the early 1990s reflects concerns that are similar to those posed by Holling (1973), by Walters and Hilborn (1976), and by participants in the Dahlem Workshop, along with many other authors who confronted the issues associated with ecosystem complexity in the 1970s and the 1980s. In many ways this struggle still continues, though a number of important changes have occurred in the intervening years. In particular, increased activities by environmental NGOs and epistemic communities allowed for the emergence of new management goals and new management solutions. These alternatives did not change the pattern of responsive governance, but they did alter the structural context substantially, and they also helped to generate more lasting and effective cycles of rebuilding in some cases.

8.4.2 Shifting Goals

In general, ecologists were ahead of managers in their decision to dismiss MSY as a useful management goal. A slight change to "optimal management" occurred in the 1970s because of concerns about the socioeconomic costs of regulation, but it was seldom feasible to distinguish between the "optimal sustainable yield" and MSY. Thus, MSY remained the *de facto* goal of fisheries management globally until the 1990s, when a new wave of severe collapses in commercial fisheries triggered goal reassessment among decision makers in many countries. The political will to move away from the MSY benchmark was reinforced by the growing concern and political power of non-commercial interest groups that were alarmed by fisheries crises and shifted their goals from protecting only charismatic megafauna to protecting ecosystem health and marine life generally. The growing epistemic community of ecologists and economists described above also played an important role, adding their political clout and providing the necessary theoretical work to advance new management alternatives. Although it is difficult to condense the exceptionally large and diverse work on these topics, this subsection sketches out the emergence of three important alternatives to MSY: the precautionary approach, geographic approaches including space based management, and ecosystem management.

The Precautionary Approach As described above, both multispecies management and adaptive management proved to be either politically or scientifically impractical. This caused several epistemic communities to search for new approaches that could compensate for high levels of uncertainty and the potential of unintended consequences related to ecosystem interactions. Endogenous and exogenous factors shaped this work in fisheries. The search for new alternatives to MSY was triggered by a second wave of major collapses in the late 1980s and the early 1990s, including the infamous failure of the groundfish fisheries in the northwest Atlantic (see chapter 7). The severe overexploitation of these stocks due to prolonged profit and power disconnects drew attention from environmental NGOs as well as from fisheries scientists and managers. The Georges Bank cod fishery provides a good example. In spite of drastic predictions by scientists, the New England Fisheries Management Council (NEFMC) refused to curb fishing effort for more than a decade. Finally, the Conservation Law Foundation (CLF) brought a lawsuit against the Council in 1992, alleging that the NEFMC was not carrying out its duties under US law (United States Senate 1992). The CLF won the suit, and the court mandated a full closure of the fishery. This intervention proved to be too little, too late, and groundfish stocks in the area have yet to rebound, probably as a result of a trophic shift in the ecosystem caused by overfishing amplified by environmental factors (Zhang, Li, and Chen 2012). However, this experience and others like it had a formative effect on the goals and strategies of conservation NGOs and related epistemic communities, altering their role in the action cycle.

Pursuit of alternative fisheries management goals was also driven by exogenous changes in norms about resource management more broadly. In the 1980s and the 1990s, environmental NGOs and some epistemic communities embraced ecosystem-focused approaches to management. This resulted from advances in ecosystem science, but also from exogenously changing values about nature in both developed and developing countries. In developed countries, post-materialist or post-industrialist values generated considerable political action favoring protection and rebuilding of terrestrial and marine ecosystems. This trend was reinforced by increased leisure time and nature-based recreation, including sport fishing, scuba diving, and related marine pursuits (Dietz, O'Neill, and Daly 2013). In developing countries, many local and indigenous communities had long traditions of conservation based on traditional ecological knowledge but were marginalized politically until the late 1980s, when a surge of activism helped to empower indigenous peoples in many parts of the world. These forces coalesced at the 1992 United Nations Conference on Environment and Development in Rio de Janeiro, where both the precautionary approach and ecosystem-based management were codified in international law.

The precautionary approach to resource management was developed as a response to similar problems of political delay in the area of marine pollution in Europe. The first legal implementation of a precautionary approach occurred in the 1980s. The concept then quickly proliferated to other issue areas (Hewison 1996). The idea of precaution resonated with conservation organizations and epistemic communities associated with fisheries. In their attempts to alter fisheries policy, both groups observed that scientific uncertainty was often used as an excuse to postpone regulatory action on fisheries. Even when there was a fair degree of consensus on management recommendations among most experts, fishers would reject information that contradicted their observed catches and related profit signals, saying that the science was too uncertain to warrant the high costs of reducing catch or effort. This was certainly the case in the Georges Banks fishery and in many similar fisheries. Recognizing this problem, conservation-oriented groups hoped that the precautionary approach would prevent management delays based on uncertainty.

A number of "precautionary approaches" exist, including what is now known as "the precautionary principle," which is codified in the Rio Declaration of 1992:

> Where there are threats of serious or irreversible damage, lack of scientific certainty shall not be used as a reason for postponing cost-effective measures to prevent environmental degradation. (United Nations Environment Programme 1992, principle 15)

Though formal precautionary approaches may vary, all are some version of this central idea that action should be taken in spite of scientific uncertainty. Precautionary approaches are difficult to implement because of the potential costs of "too much" precaution and potential tradeoffs between different issue areas. These will be addressed in subsection 8.4.3.

Space-Based Management Driven by similar concerns about the growing number of fisheries in crisis, epistemic communities also developed several approaches that focused on managing entire regions or areas rather than single stocks of fish. Some of these geographic approaches were developed exogenously and then appropriated by fisheries managers. For instance, coastal zone management was developed in the 1970s as decision makers struggled to come to terms with competing use values in ports and urban watersheds. In the 1980s and the 1990s, this approach was extended to include fisheries systems in what is now called Integrated Coastal Zone Management (ICZM; Douvere 2008). Like many of the management measures discussed in this section, ICZM is not well defined, in part to provide flexibility for implementation in multiple contexts. However, the three primary characteristics are: (1) improved coordination between regulatory organizations and sectors, (2) protecting coastal ecosystems from multiple anthropogenic factors such as coastal hardening, pollution, and overfishing, and (3) optimizing the utilization of coastal resources in a sustainable manner (World Bank 1996, 5). Applications of ICZM tend to work better in small-scale systems but are often stymied by organizational issues and by political wrangling.

Evolving from systems such as national parks and no-take marine reserve networks as per Murray et al. (1999), Marine Protected Areas (MPAs) are another geographic approach to management. An MPA is essentially an area of the ocean where destructive human uses are prohibited. Most MPAs are used to protect highly productive habitats (e.g., coral reefs or upwelling systems) or vulnerable marine ecosystems (e.g., sea mounts). In theory, MPAs should increase the abundance of all species within a protected area and also should improve fishing in adjacent waters as organisms spread outside of the zone (Alder, Zeller, and Pitcher 2002). Evidence on inshore MPAs generally supports this view, although there is considerable variation in effectiveness (see, e.g., White et al. 2008). Indeed, recent evidence shows that enforcement is lax in many MPAs around the world, among them the Phoenix Islands Protected Area (Pala 2013). Concerns have also arisen about the design of marine protected areas, which can be quite patchy, with different activities permitted in different zones. This opens MPAs to considerable political manipulation, as resource users may advocate zoning that is economically beneficial for them but ecologically nonsensical (Solomon et al. 2007; Torres-Irineo et al. 2011; Kearney et al. 2012).

Recently some NGOs and academics have begun to advocate the broader concept of space-based management (SBM), which combines lessons learned from the implementation of ICZM, MPAs, and more traditional management practices. Douvere (2008, 763) depicts SBM, or "marine spatial planning" in his terminology, as an iterative process consisting of planning and analysis, implementation, and monitoring and evaluation. The focus of this process is the development of zoning schemes that balance resource use and preservation. For instance, some areas might be set aside for specific users, such as recreational or commercial fishers, and others would be closed or protected from extractive practices. The main difference between SBM and ICZM appears to be in the geographic scope. Unlike ICZM, SBM can extend beyond the coastal zone (12 miles) to cover the entire exclusive economic zone (200 miles from shore) and has even been applied in international plans for management of large, open bodies of water such as the Wadden Sea. That said, SBM may also be applied to specific regions within an EEZ—for example, the Great Barrier Reef in Australia or the Scotian Shelf in Canada (767). Douvere describes SBM as an alternative to ecosystem-based management, but the line between the two is quite blurry. Each requires the protection of ecosystems and recommends use of spatial planning. The primary difference is in emphasis rather than in attributes.

Ecosystem-Based Management The third major innovation of the 1990s was the development of an umbrella concept called ecosystem-based management (EBM). Like the other approaches discussed here, EBM has many versions, but almost all are amalgams of elements from multispecies management, space-based management, and the precautionary principle (see e.g., Pikitch et al. 2004; Crowder et al. 2008; Dame and Christian 2006; Marasco et al. 2007). Some versions also include references to the use of adaptive management (Schwartz et al. 2002). Looking at ecosystem approaches specifically, Christie et al. (2007) identify two primary camps in this literature: fisheries-centric ecosystem approaches and ecosystem-focused approaches. In general, proponents of fisheries-centric approaches maintain the normative goal of maximizing either economic or biological yield from fisheries, but recognize that scientists must take multiple environmental and ecological factors into account to attain that goal. On the other hand, ecosystem-focused proponents are primarily concerned with preservation of ecosystem function, although they recognize the welfare of fishing communities as a secondary goal. Thus, the two camps share similar goals but place different emphasis on human systems vis-à-vis natural systems.

These various approaches to ecosystem-based management are shaped by endogenous and exogenous forces. On the one hand, EBM is the continuation of previous work grappling with the observed complexities of fisheries management, as described in subsection 8.3.1. In particular, the designation of marine protected areas is an approach to accommodating biological interactions that allows for the protection of habitats that are important for the maintenance of ecosystem functions. This approach can also be politically attractive if reduced catches in the protected area can be offset by increased harvests elsewhere, or if MPAs can be used as indirect exclusionary measures by powerful fishing interests. Of course, EBM goes beyond spatial management, and MPAs would still have to be combined with by-catch reduction strategies to lessen technological interactions (see subsection 8.4.3) and with modified catch limits to maintain ecosystem structure. Regarding uncertainty, use of the precautionary approach is particularly important for EBM because ecosystem science highlights uncertainties that are essentially ignored when calculating factors associated with MSY (Pikitch et al. 2004).

After the Rio Declaration, variations on the precautionary principle and on ecosystem-based management proliferated in international law. In fisheries, ecosystem approaches to management multiplied in the 1990s and 2000s. Variations on EBM and its constituent parts (precaution, geographic management, multispecies management, and adaptive management) are included in many international agreements and in the domestic laws of several countries. The precautionary principle was a cornerstone both of the UN Fish Stocks Agreement and the FAO Code of Conduct for Responsible Fisheries. The FAO produced several relevant documents to encourage the spread of EBM, including its Technical Guidelines on the Implementation of the Ecosystem Approach to Fisheries Management (FAO 2003). The FAO also maintains a website, called EAFnet, that disseminates best practices and archives case studies to provide practical advice to managers engaging in ecosystem approaches to fisheries management (FAO 2014).

As Bodansky (2004) notes, alternatives such as the precautionary approach and EBM tend to be popular aspirational goals because they appeal to a wide range of interests. Even the fishing industry can agree that fishery collapse should be avoided and that therefore precaution and ecosystem protection are laudable goals in principle (see, e.g., Hayden and Conkling 2007). However, in practice, implementation has been difficult for technical and political-economic reasons. There is still considerable debate about the efficacy of the more onerous aspects of alternative approaches, and a wide normative divide remains between fisheries-centric and ecosystem-focused epistemic communities. Furthermore, even when there is a clear mandate to implement EBM or related alternatives, response can still be delayed by the combined forces of the profit disconnect and the power disconnect. Indeed, these alternative approaches are still implemented as responses to crisis conditions rather than as proactive measures to prevent the core problems of the fisheries action cycle as described in chapter 1.

8.4.3 Implementation of Alternatives

There is a broad scientific and regulatory consensus that single-stock MSY alone is not sufficient for sustainable fisheries management and ecosystem-based management is now formally institutionalized as a regulatory goal in international law. Furthermore, important components of EBM are now in widespread use, including by-catch reduction mechanisms, use of MPAs to protect key ecosystems, and inclusion of environmental and ecological conditions in stock assessments. Nevertheless, the majority of fisheries around the world are still managed using modifications of single-stock maximum sustained yield rather than multispecies management, governance is still responsive rather than precautionary in most systems, and spaced-based management is still piecemeal and fragmented. Some of the barriers to implementation of EBM and its constituent concepts are technical, but, once again, the most important are political and economic. Implementation of EBM has costs as well as benefits, and decision makers must weigh both when formulating policy. Based on the responsive governance aspect of the AC/SC framework, it is not surprising that decision makers choose to apply those components of EBM that have low economic and political costs, or even net benefits as with MPAs, but eschew the more costly concepts that require direct limitations on fishers.

The Expedience of EBM The relative inclusiveness, complexity, and uncertainty associated with EBM present considerable technical difficulties for implementation. These are more limiting for ecosystem-focused approaches, which are generally more holistic, but are also important for fisheries-centric approaches (Christie et al. 2007). Frid, Paramor, and Scott (2006) review these barriers and point to two primary problems: lack of data and limits on modeling techniques. Even relatively simple versions of EBM require large amounts of information on specific species, on interactions between species, and on environmental effects on the system as a whole.

Often this information is not available in long time series, and collection of new data requires substantial resources that are frequently unavailable to budget-strapped management agencies. EBM also requires methods for delineating, modeling, and estimating the complex trophic interactions within ecosystems. Although several methodological tools and ecosystem models are now available, ecosystem science highlights uncertainties that are essentially ignored when calculating factors associated with MSY. Thus, ironically, as scientists work to make their models more realistic, their advice appears to be more uncertain. This is why a precautionary approach is crucial to EBM (Pikitch et al. 2004).

Increasingly, scientists and managers are working to develop ecological benchmarking schemas in order to reduce the data and modeling tasks associated with EBM. These benchmarks can be used to guide ecosystem-based management, much as MSY served as a guideline for single-stock management. Discussions of alternative reference points started in the early 1990s as part of the compilation of the FAO International Code of Conduct for Responsible Fishing. Caddy and Mahon (1995) expand on the use of multiple reference points in fisheries management, including target reference points like MSY or optimum size caught, and threshold reference points. Caddy (2002) proposed a traffic light approach to management that is now included in the FAO's (2003) Technical Guidelines on the Implementation of the Ecosystem Approach to Fisheries Management. Marasco et al. (2007) describe the development of such benchmarks for US fisheries, and Ye, Cochrane, and Qiu (2011) advocate use of ecological benchmarks in management of the South China Sea. Various NGOs, including the World Wildlife Fund, also push for use of benchmarking techniques (Schwartz et al. 2002). Sparholt and Cook (2010) further argue that MSY can still be a viable benchmark as long as it is calculated in a multispecies context.

In spite of these technical challenges, EBM would be more widespread and holistic if not for political and economic challenges. On the one hand, governance remains responsive for the most part, and decision makers are still largely concerned with issues such as employment, food production, and economic development. Furthermore, although fishers are running out of new fishing grounds to exploit and have not made many stepwise advances in technology in recent decades, gradual technological improvement and increasing demand continue to generate profit and power disconnects throughout the world. This is particularly apparent in developing countries, where the fishing industry is still seen as an important source of economic development and where management capacity is limited. Indeed, chapter 7 showed that fishers and secondary actors are still moving from developed countries, where the industry is small relative to other sectors and where management institutions are well developed, to developing countries, where the industry may be able to lower costs of production, wield greater political power, and avoid strong management measures.

Furthermore, certain aspects of EBM, including multispecies management, carry higher political and economic costs relative to other management schemes. Accounting for predator-prey relationships or even other human actors in a system usually results in much lower catch or effort limits than the MSY approach. As long as the profit signals are weak and the profit disconnect is increasing, such large cuts are usually protested by industry groups, and therefore decision makers may avoid implementation or choose to implement only expidient EBM-related regulations, such as by-catch avoidance technologies and MPAs. When environmental interests have substantial power—as with charismatic megafauna—this is not a major barrier, but when these interests are relatively weak, or when their concern is only triggered by near collapse, then related power disconnects ensure that governance remains responsive and that politically costly measures associated with EBM are not implemented until the fishery reaches a critical state. Because ecosystems often overlap multiple political systems, there can also be bureaucratic barriers to the implementation of EBM, which may require cooperation between terrestrial and marine management units. Changes in the structural context at this level usually require broader reform movements and are not generated by response in one area or sector.

The effects of the challenges mentioned above are even visible in legislation that mandates use of the EBM approach. Language is modified to provide decision makers with flexibility and to place socioeconomic considerations on par with ecological concerns. For example, the formal definition provided by the FAO's (2003) Technical Guidelines on the Implementation of the Ecosystem Approach to Fisheries Management explicitly includes socioeconomic considerations:

> [A]n ecosystem approach to fisheries (EAF) strives to balance diverse societal objectives, by taking account of the knowledge and uncertainties of biotic, abiotic and human components of ecosystems and their interactions and applying an integrated approach to fisheries within ecologically meaningful boundaries. (FAO 2003, 6)

Providing for socioeconomic flourishing is part of the sustainability ethos, but this definition also leaves considerable room for interpretation, and both fisheries-centric and ecosystem-focused approaches could fit under such a large umbrella. The same concern for balancing costs and benefits is also observed in other areas. For instance, the Precautionary Principle as articulated in the Rio Declaration states that scientific uncertainty should not prevent the application of *cost-effective* management measures. Thus, the Declaration does not support precaution at all costs, but again allows for political balancing of costs and benefits. Both critics and supporters of precaution find this application to be problematic. Critics claim that the concept is in itself too nebulous. In Bodansky's (2004) terms, there is no indication of what managers should be precautionary about. Furthermore, according to Hilborn et al. (2001), precaution should be extended to fishing communities as well as to fisheries ecosystems. This sentiment is echoed by Gonzalez-Laxe (2005) and others who point out that different implementations of the precautionary approach may have higher or lower socioeconomic costs that should also be considered in policy making. On the other hand, proponents of precaution warn against dilution of the approach through incorporation of political concerns. After all, the primary goal of the approach is to depoliticize management decisions (Myers 2002).

Mismatches in scale between ecosystems and regulatory systems can also cause significant political barriers to the implementation of EBM generally and of spaced-based management more specifically. This is particularly important for large marine ecosystems (LMEs), which encompass a greater array of species, interactions, and environmental forces but also overlap many jurisdictional boundaries (Cochrane 1999). Juda (1999) provides a legal perspective on LMEs and points out that coordination across levels of governance is required for successful management. That is, often regional or even national governments need to help local governments to coordinate their efforts—without simply dictating regulatory approaches, since political, economic, and ecological contexts may vary at the local level. Similarly, Browman and Stergiou (2005) cite mismatches of scale as a major difficulty in the implementation of EBM for LMEs. They also note that the piecemeal nature of the process—which relies heavily on project-specific pressures from NGOs—undermines effectiveness. Torell (2009) and Rudd (2004) reach similar conclusions about the effect of political fragmentation on management of LMEs. These continued signs of contestation highlight the ongoing importance of the power disconnect in the implementation of alternative management paradigms.

EBM in Practice Recognizing these political and technological barriers to implementation, it is useful to investigate how EBM and related concepts have been implemented to date. Webster (2013) analyzes the implementation of EBM at the international level. Though 12 of the 16 existing regional fisheries management organizations (RFMOs) with sufficient records for the study espouse the ecosystem approach at some level, only the International Whaling Commission (IWC) and the Commission for the Conservation of Antarctic Marine Living Resources (CCAMLR) actually account for ecosystem factors, particularly predator-prey interactions, in their management measures. Most other RFMOs still manage stocks based on single-stock maximum sustainable yield (ten), with a few implementing closures to protect vulnerable marine ecosystems (four), engaging in habitat restoration to improve ecosystem function for targeted species (one), and instituting by-catch reduction measures for specific species (eight). Otherwise, the Inter-American Tropical Tuna Commission utilizes habitat models to include ecosystem effects in its calculations of MSY, the General Fisheries Commission for the Mediterranean is looking into scientific methods to institute an ecosystem approach, and the relatively new Western and Central Pacific Fisheries Commission engages in some ecosystem modeling, largely for reduction of by-catch.

This brief survey suggests that the ecosystem approach laid out by the FAO's technical guidelines is slowly infiltrating the realm of international fisheries management. However, it should be noted that most by-catch avoidance mechanisms are not motivated by desires to implement EBM, but rather are put in place to protect commercially valuable species or to protect charismatic megafauna that are incidentally harvested in other fisheries. For instance, most of the tuna RFMOs have some by-catch reduction mechanisms for juvenile bigeye tuna, which are incidentally harvested in fisheries targeting skipjack in most oceans. Adult bigeye are highly valuable and are targeted by longline fleets, mainly from Asia. To ensure larger harvests of adults, states representing longliners negotiated by-catch reduction measures that prevent large incidental harvests of juveniles in other fisheries (Webster 2009). Similarly, charismatic species like dolphins, turtles, and various seabirds are protected as by-catch because of pressures from conservation interest groups. In addition, by-catch reduction measures were adopted for white marlin and some other sport fish because of pressures from recreational fishers, as described in subsection 8.2.2. In fact, pressures from non-commercial interests are important in the application of all other EBM-associated measures noted above, and are also key in the implementation of EBM at the IWC and CCAMLR, both of which have high levels of membership by non-fishing countries. Furthermore, CCAMLR itself was established to reduce potential ecosystem consequences from the creation of a new fishery for krill, as noted in subsection 8.4.1.

At the domestic level, Australia, Canada, and the United States are viewed as early adopters of EBM. The US and Canada implemented multi-species management in several Great Lakes fisheries in the 1970s, but their first marine applications of EBM started in the 1990s (Slocombe 1998). Since that time, ecosystem approaches have been applied in different contexts, ranging from the highly developed coastal area around Hong Kong to the Great Barrier Reef in Australia to German fisheries in the North Sea (Gadgil et al. 2003; Burkhard et al. 2011; Buchary et al. 2003). In all cases, these fisheries were heavily overexploited, with high levels of conflict among resource users, and a relatively strong profit signals for fishers specifically. The UN Food and Agriculture Organization actively fostered such projects in developing countries, including compilation of a "knowledge base" on the implementation of its own ecosystem approach to fisheries (EAF) management, as well as multiple expert consultations and provision of technical assistance to countries wishing to implement the EAF (FAO 2014). There is also a growing literature on lessons to be learned from the implementation of EBM around the world, as well as increasing quantitative evaluation of the effectiveness of the approach in different countries or regions (see, e.g., Seppelt et al. 2012; Baumgartner, Jones, and Wilkerson 2011). These factors increase management resources in the structural context and may help to increase implementation of EBM, though they do not remove political and economic barriers.

In the 1990s, NGOs also started working to protect fisheries ecosystems as a whole, rather than just charismatic megafauna, thereby increasing pressure to implement EBM approaches at the domestic level. In fact, the World Wildlife Fund and Oceana strongly support ecosystem approaches and have published their own studies on practical implementation (Schwartz et al. 2002; Grieve and Short 2007; Enticknap et al. 2011). Unfortunately, their influence is often greatest when stocks are severely depleted and ecosystems are already in steep decline. This is when the potential for cooperation with industry interests is highest, and when NGO members are most willing to provide monetary and political support for a campaign. Still, the increasing involvement of non-commercial interest groups in policy making for non-charismatic species may help to counteract the power disconnect and to instigate management intervention earlier than might occur otherwise. Escalating involvement by non-commercial interests can also lead to political deadlock over implementation of EBM and related policies if undertaken in a confrontational manner that causes fishers to take a defensive rather than cooperative stance. In this regard, NGOs learned much from their experience with protections of charismatic megafauna (see section 8.2), and now most of these groups seek to work together with fishers to build acceptance of EBM regimes (see section 8.5).

Although EBM is embraced by many NGOs and epistemic communities, some decision makers are profoundly frustrated by the "fuzziness" of EBM and other alternatives to MSY. From this point of view, having a plurality of aspirational goals is seen as a major drawback. However, the increasing array of theoretical and practical management options should make it easier to find solutions that work in a given context. Greater attention to engagement with fishers expanded the potential policy set further, adding less costly and therefore more politically acceptable solutions to the mix. Indeed, a number of authors now argue for incorporation of incentive-based measures such as ITQs with ecosystem-based management (Brady and Waldo 2009). One of the most common findings in this emerging literature is that ecosystem-based management cannot really be achieved without consideration for the social systems associated with fisheries. In the words of Berkes (2012, 467), "biological science is a necessary but insufficient component of implementing EBFM for the twenty-first century." Berkes is part of a broader movement that is grappling with the complexities of human-environment interactions using interdisciplinary approaches to study coupled human and natural systems, or in similar usage, social-ecological systems (Degnbol et al. 2006). This is a major step forward because it deals with the power disconnect as well as with the CPR dynamic in the fisheries action cycle.

In terms of management practice, scholars from this field encourage the use of stakeholder engagement in the implementation of EBM approaches. *Stakeholders* are usually defined as a set of individuals and groups with interests in the management of a specific resource. Some limit use of the term to only environmental NGOs, community organizations, and industry representatives, but others also refer to scientists and decision makers as stakeholders. Stakeholder engagement can take multiple forms, ranging from informal interactions with key players to formal scenario generation workshops designed to help diverse groups of stakeholders articulate their perceptions, knowledge, and goals, question underlying assumptions, and, ultimately, to build trust and cooperation. The growing literature on the use of stakeholder engagement and the related concept of fisheries co-management in the implementation of both EBM-focused and MSY-focused approaches to governance will be covered in the next section, which parallels section 7.5 on compliance.

8.5 Co-Management and Fisheries Justice

In addition to the interest in stakeholder engagement, support is growing in the literature for the concept of fisheries co-management. Co-management generally involves all stakeholders in a fishery, including commercial and non-commercial interests. Like many other new approaches, co-management is a nebulous concept. It may be defined as power sharing among groups, as institution building, as stakeholder engagement, as problem-solving method, or as an approach to governance that combines these practices. Most recently, the concept of adaptive co-management has been emerging. This approach brings adaptive management (see subsection 8.4.1) together with co-management (Berkes 2009).

Though these ideas may seem unrealistic to some, the need for co-management arises from a clear set of necessities. First, as Berkes (2012, 467) points out, fisheries management is a "wicked problem" in which causal relationships are obscure and outcomes are highly variable. In such systems, no one model or approach is sufficient. Through co-management, managers can assimilate multiple perspectives and capitalize on knowledge gained by many different individuals. Second, co-management can be used to endow management with political legitimacy among various groups, most particularly fishers themselves, which helps to counter the power disconnect. As those who must pay the short-run costs of management, fishers' perceptions of legitimacy are critical for the successful implementation of any type of fisheries management, whether ecosystem based or not. Politically, fishers will lobby to weaken any regulations they find to be unfair or inappropriate and as long as such regulations remain in place, fishers will choose to avoid them as much as possible (Jentoft 2000). This was clearly evident throughout the cases presented in previous chapters and is even more important when ecosystem considerations are included in regulations, since the livelihood impacts of EBM may not be as obvious or acceptable to fishers as they are to non-commercial interests.

Conservation-oriented NGOs began to recognize the importance of stakeholder engagement and co-management in the 1990s, and it is now also an integral part of their work to expand use of EBM in fisheries. In fact, it is a cornerstone of the World Wildlife Federation's practical guidelines for implementation of ecosystem approaches (Schwartz et al. 2002). Given the controversy over protections for charismatic megafauna harvested as by-catch in commercial fisheries, it should be clear that a conciliatory approach is even more crucial in pursuit of EBM because technological solutions are lacking. Certainly, where by-catch is a problem, avoidance technologies can facilitate implementation of EBM, but catch reductions, institution of no-take zones, and the banning of some gears may be necessary as well. These are hard pills for fishers to swallow, particularly when the profit disconnect is increasing, so working with fishers to minimize related costs and to communicate potential benefits is critical for acceptance of such measures. Furthermore, moderate conservation groups may gain support from fishers by presenting options that appear more "reasonable" than those preferred by more radical preservationist interests, which generally lobby for complete cessation of fishing. Indeed, if conservation NGOs can build coalitions with industry interests on EBM—or even on MSY management of non-charismatic species—they can remove some of the political barriers to implementation and help to ensure that both biological and socioeconomic benefits are realized. This would be a "win-win" in the parlance of the literature on green business, though it might be seen as "selling out" by radical groups.

Like all of the approaches described above, co-management has its weaknesses. It is a useful tool for improving both the technical and social application of management, but it is only one method and it can be misused (Jentoft, McCay, and Wilson 1998). If implemented callously, as an attempt to build legitimacy for preselected management measures rather than to fully incorporate stakeholder perspectives into management, these tools can generate great disappointment among fishers and other groups. That disappointment, in turn, may undermine the legitimacy of the process and cancel out any potential benefits (Jentoft 2000). Indeed, Davis and Ruddle (2012) are highly critical of the application of co-management in developing countries, suggesting that it reinforces existing power imbalances, thereby "betraying" small-scale fishers. Armitage, Marschke, and Plummer (2008) also point out that co-management remains a political process and therefore power relationships should not be ignored. They provide a schematic for breaking down learning processes associated with co-management to better understand when it provides empowerment instead of political manipulation. Their work reinforces the perceptions of many proponents of co-management who recognize that learning and cooperation may not always be feasible outcomes in every fishery and that actual implementation may differ from aspirational goals.

Although the application of co-management in local and regional fisheries governance is growing, including widespread legal requirements in the EU, the need for legitimacy in international management is not well recognized (see, e.g., Berghöfer, Wittmer, and Rauschmayer 2008; McClanahan and Castilla 2007). Instead, states and epistemic communities both focus more on the need for coercive compliance mechanisms. For instance, Barkin and DeSombre (2013) advocate a variety of solutions to the problem of international fisheries management, including instituting international tradable quota systems, improved use of international monitoring and compliance mechanisms, and coordination of management through a global umbrella organization for existing RFMOs. Similarly, the 2001

FAO International Plan of Action to Prevent, Deter, and Eliminate Illegal, Unreported, and Unregulated Fishing encourages states to engage stakeholder participation in domestic management but does not address social justice issues at the international level. This attitude is reflected in RFMO use of trade-restrictive measures, which disproportionately affects developing coastal states, and in other international treaties, including the 2009 Agreement on Port State Measures to Prevent, Deter and Eliminate Illegal, Unreported and Unregulated Fishing (Webster 2009, 2013). This agreement allows national governments to develop protocols for port-state inspection of foreign-flagged vessels so that importing countries can monitor landings irrespective of the vessel's origin (FAO 2009). Unfortunately, none of these proposed solutions address issues of justice or legitimacy in international fisheries regulation.

Of course, co-management and stakeholder engagement are not feasible on an international scale, but this does not mean that questions of legitimacy can be ignored. Chapter 5 showed clearly that security on the oceans is closely tied to security and the rule of law on land. If the example of piracy is any guide, "fish pirates" will continue to thrive as long as some states are willing to harbor them or are unable to control them. Thus, improvement of management capacity in developing countries as described in Section 7.5 is still necessary for the success of international fisheries regulation. Similarly, international monitoring and enforcement is needed to reduce free riders. Nevertheless, as was noted in section 8.2, successful international conservation requires acceptance of the legitimacy of rules and regulations, not just the use of coercive measures. For example, cooperation on dolphin-safe tuna was achieved only after international NGOs partnered with local NGOs to change perspectives on dolphin bycatch in Mexico and other countries. Alternatively, the failure of attempts to institutionalize a protectionist ban on whaling was largely attributable to a reverse spiral of reticence on the part of communities in Japan and other whaling states, which felt wronged by the threat of sanctions and implied criticism of their social values. These impacts on the structural context are even more important in international regulation of large-scale commercial fisheries, because the profit disconnect and the power disconnect are much greater than in the cases of charismatic megafauna.

As at the domestic level, improving international legitimacy of management should not merely be "icing on the cake," but rather the "yeast that makes the dough rise" (Beetham 1991; Jentoft 2000). In other words, while it is important to engage the fishing industry in multiple countries to build support for international regulations, it is also important to ensure that those measures represent fair compromises rather than coerced concessions. As such, power disparities would need to be addressed, particularly between historically dominant and developing coastal states. This is a difficult problem, since states rarely choose to give up power, and because reducing the power of historically dominant fishing countries would increase the power disconnect in many RFMOs, as discussed in subsection 8.3.3. However, the transfer of power is already occurring due to the migration of fleets from developed countries where costs of production are high to developing countries where costs of production are low. The history of the Inter-American Tropical Tuna Commission may provide some guidance here, both because power relationships are more equal than in many other RFMOS and because states appear to have a high level of trust in the secretariat and its staff of professional scientists (Walsh 2004). As described in chapter 5, the IATTC is seen to be one of the most effective of the RFMOs managing highly migratory species. This is due in part to the perceived legitimacy of the regime, where developing coastal states wield considerable power, but may also be related to economic considerations, particularly the low profit margins for most of the species targeted in the area.

Indeed, although legitimacy and rule of law are important determinants of compliance with fisheries regulation, gains in these areas can be significantly reduced by exogenous changes in economic conditions that increase the profit disconnect. Increasing demand for high-status products like bluefin tuna sashimi and shark fin soup can generate high profits that, in turn, can fuel involvement of criminal organizations in the purchase and trade of illegally harvested fish stocks. This undermines legislation and generates an informal power disconnect as poachers avoid regulations through corruption and non-compliance. The combined effect is a ratcheting up of the cost of monitoring and enforcement protections for trafficked species. In addition to RFMO efforts to reduce IUU harvests of species such as bluefin tuna and Antarctic toothfish (Chilean sea bass), states concerned with conservation of sharks and related species also work to prohibit trade through the Convention on International Trade in Endangered Species of Flora and Fauna (CITES). Until recently, major fish-producing and fish-consuming states have been reluctant to allow commercially targeted species to be listed, but this may change as the crisis—particularly for sharks and other elasmobranches—escalates further (Webster 2011). In 2013, INTERPOL started a program called Project Scale that tackles illegal fishing as a criminal activity like drug smuggling or human trafficking. Again, while these supply-side measures are necessary, they may not be sufficient as long as demand for luxury marine-based products continues to widen the profit disconnect in many fisheries (INTERPOL 2013).

Although demand-side issues do not have high priority on the international agenda, a number of powerful NGOs are working to reduce demand by increasing global public awareness of the severe overfishing of high-priced species. This includes lobbying efforts in multiple countries and formation of partnerships with fishers who legally harvest species like bluefin tuna. Another approach involves reducing demand for "bad" fish and increasing demand for "good" fish through eco-labeling and stoplight lists for consumers. The Marine Stewardship Council (2014) and FishWise (2014) are NGOs that work with industry interests to provide labeling for voluntary management programs. The Monterey Bay Aquarium, the Dutch Good Fish Guide group, and similar organizations also produce and publicize guides for consumers that place species into "buy" and "don't buy" categories based on various normative environmental and health criteria (De Vos and Bush 2011; Monterey Bay Aquarium 2014). Many such campaigns and labeling efforts are international in scope and aim to reduce demand for heavily overfished species globally. However, the efficacy of labeling alone is highly questionable—particularly when growth in demand outstrips supply, expanding the profit disconnect, which, in turn, creates incentives to skimp on conservation requirements, as when the MSC had to scramble to find new suppliers once its label was picked up by the high-volume chain Walmart. The proliferation of different systems can also be confusing to consumers, who generally have little time and energy to devote to their seafood choices (Jacquet et al. 2009).

As Ostrom et al. (2007) point out, there are no panaceas in environmental governance. All of the innovative approaches to fisheries management described in this chapter involve tradeoffs, and none will solve fisheries management problems indefinitely. However, the proliferation of new actors and new ideas that began in the 1970s and escalated in the early years of the 21st century is a sign of growing interest in improving fisheries management across the board. Increasingly, research by interdisciplinary epistemic communities, engagement by environmental interest groups that learn to value human as well as natural systems, and the diffusion of genuine concern throughout fishing communities and other sectors of the fishing industry are all signs of an emerging political will for change in fisheries governance. Furthermore, improvements in science and technology, an increasing array of politically expedient solutions to core problems in fisheries, and the building of management capacity worldwide also inspire some optimism regarding the future of fisheries governance. Two primary questions remain. First, given current trends in both the profit disconnect and the power disconnect, will these countervailing forces gather quickly enough to prevent further escalation of fisheries crises in the short run? Second, will these same countervailing forces be strong enough to cause a shift from unstable responsive management to sustainable fisheries governance in the long run? These questions are considered in chapter 9.

9 The Management Treadmill

Looking beyond the tragedy of the commons, the Action Cycle/Structural Context (AC/SC) framework clearly shows multiple, escalating cycles of expansion in fisheries that were exacerbated by the profit disconnect and facilitated by the power disconnect. While fishers and the fishing industry proved well able to push past local limits through exploration, innovation, and opening new markets, these economic responses dampened cost, price, and catch signals while amplifying the core problems of overfishing, overcapitalization, and ecosystem disruption. Furthermore, the economic expansion of the fishing industry widened the power disconnect, undermining management as the sectors least affected by declining stocks developed ever greater influence over fisheries policy. Combined, these disconnects consistently delayed management response until powerful industry interests ran out of economic options—usually when stocks were severely overexploited, fleets were heavily overcapitalized, and ecosystems were in jeopardy. Such crisis conditions frequently triggered transition from ineffective to relatively effective governance cycles, as illustrated in figure 1.1. Furthermore, each relatively effective response in the action cycle built up new institutions in the structural context, but none proved to be permanent solutions to the core fisheries problems. Instead, the history of fisheries governance resembles a *management treadmill* that oscillates between effective and ineffective cycles, as shown in figure 9.1.

This management treadmill is not caused by the tragedy of the commons alone but emerges from the complex interactions associated with responsive governance. Even in ancient Sumer, population growth and increasing affluence drove up demand, which caused intensification of effort in spite of the fact that the common pool resource (CPR) dynamic was prevented by well-established fishing guilds. Thus, exogenous economic forces undermined sustainable institutions, resulting in the collapse of several important commercial fisheries in the region. Chapter 5 also shows

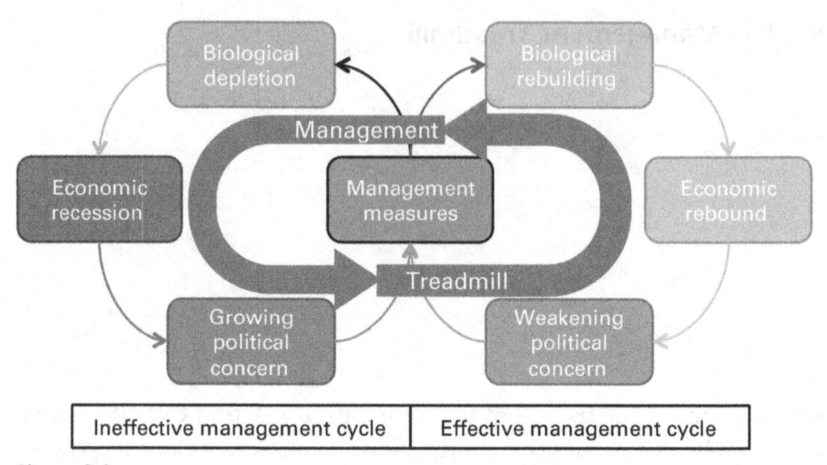

Figure 9.1
The management treadmill and responsive governance.

how colonialism and globalization weakened local management, replacing longstanding institutions with extractive open access regimes. As described in chapter 7, patterns of serial depletion persisted in the 20th century, in spite of the institutionalization of science-based management in the late 1800s. Furthermore, there was little evidence of learning from past experience of crisis and response. This is clear from the serial overexploitation of salmon fisheries and the cycles of overexploitation illustrated by the Peruvian anchoveta and North Atlantic cod cases. Thus, while the tragedy of the commons is important, responsive governance is also a deep problem in fisheries governance and so understanding the resulting management treadmill is crucial for sustainability.

This chapter describes the management treadmill and considers the potential for sustainable management of marine systems in the future. Section 9.1 explains how endogenous and exogenous drivers create and fuel the management treadmill. In isolated fishing communities, the institutionalization of rules and norms developed in response to problems can limit the operation of the management treadmill. The result here is a stable cycling of management or *dynamic equilibrium treadmill* that is fairly sustainable. However, open systems are subject to exogenous drivers like the growth of demand, globalization, or climate change, all of which make management more difficult, even when institutions are created to reinforce each governance response. In these cases, the management treadmill is constantly speeding up because the profit disconnect is widening and new problems are always arising. In this *growth cycle treadmill*, crises escalate with each turn of the action cycle and higher levels of political will are necessary to generate effective response, so collapse and ecosystem shifts are increasingly likely. Because so many drivers are exogenous, the growth cycle treadmill also suggests that, in the long run, sustainable fisheries management depends on sustainability in the global political economy as a whole, as well as on mitigation of endogenous drivers of expansion like the tragedy of the commons.

The next two sections delve more deeply into the management treadmill and develop expectations about the future evolution of fisheries governance. Section 9.2 reviews trends in the profit disconnect and other drivers of fisheries expansion and provides rough predictions regarding future iterations of the management treadmill. It shows that fishers are running out of options to avoid the costs of the tragedy of the commons as production reaches or exceeds global limits. However, many fishers will be able to remain active globally if they can exploit new markets and new technologies. At the same time, exogenous drivers such as population growth, economic growth, and globalization are likely to escalate in the foreseeable future, driving an escalating growth cycle treadmill. If demand increases faster than supply, prices will increase faster than the costs of production and the profit disconnect will widen, even for currently low-priced species such as sardines and anchovies. Unless these exogenous factors are mitigated through governance or other forces, the global management treadmill will speed up and more extensive and exhaustive crises are likely to occur.

In light of this conclusion, section 9.3 reviews the evidence on mechanisms that prompted switching from ineffective downward spirals of depletion to effective cycles of rebuilding. This analysis opens the "black box" of management decision making illustrated in figure 1.1, expanding our understanding of transitions in the management treadmill through synthesis of specific cases from Part II. It also describes policies that could be implemented to improve signaling in the action cycle so that response occurs earlier, before crisis sets in, which would reduce the negative consequences of overfishing (too few fish), overcapitalization (too many fishers), and ecosystem disruption. These policies can include changes to the structural context or the action cycle—particularly development of management options that are at once effective and expedient, reduction of expansionary management measures and other regulations that widen the profit disconnect, and institution of norms of transparency and legitimacy that narrow the power disconnect in fisheries. Unfortunately, these policy solutions involve many tradeoffs, and even when effective they provide only a temporary slowing of the management treadmill.

Ultimately, transition to sustainable fisheries management requires moving beyond the tragedy of the commons—and beyond a pure focus on fisheries generally—to consider the sustainability of the entire economy. Section 9.4 reviews key aspects of the literature on sustainable development. It draws parallels between the findings on fisheries and other environmental issues, arguing for additional work using the AC/SC framework. This brief analysis supports recent calls for reconsideration of structural elements in the study of the international political economy of the environment, as per Clapp and Helleiner (2012). It also highlights some important leverage points revealed in the very wide and diverse literature on sustainable development, including the relationship between the profit disconnect and environmental impact through four factors: population growth, economic development, globalization of trade, and technological innovation. It further discusses linkages between the power disconnect and issues of fairness and accountability, which are critical determinants of economic development and sustainability at multiple levels of analysis. Section 9.5 concludes by pointing out that, from an AC/SC perspective, a crisis can also be an opportunity, and that, because of the complex forces at work in fisheries management, understanding the relationships between action cycles and structural contexts is crucial for sustainability.

9.1 Understanding the Treadmill

Over the long term, fisheries governance clearly cycles between periods of effective and ineffective management, as illustrated in figure 9.1 above. In the simplest version of the management treadmill, biological depletion leads to economic recession, which creates incentives to implement politically costly measures that will help to solve underlying bioeconomic problems in the action cycle. Given enough political will and/or management options that are expedient and effective, this dynamic will eventually lead to a transition from ineffective (left side of figure) to relatively effective management (right side of figure). However, the bioeconomic improvement that results from effective management weakens political will to maintain effective management measures, which, in turn, undermines regulation, eventually sending the system back to its ineffective state. I call this the *crisis rebound effect* (CRE). However, when governance is ineffective, bioeconomic problem signals escalate again, eventually leading to a buildup of political will and switching back to an effective management cycle.

Although the CRE was observed in a number of cases, the reality of the management treadmill is usually much more complicated than is depicted in figure 9.1. Each of the factors in the AC/SC framework can have some impact on the speed of transition from ineffective to effective management and back again. Factors that widen the profit disconnect or the power disconnect delay switching to effective management and can even force a relatively effective management system back into an ineffective cycle. On the other hand, governance institutions and the development of more expedient management measures can facilitate transition to effective management and may even counter the crisis rebound effect for long periods of time. Exogenous forces can similarly speed up or slow down the management treadmill, locking in either effective or ineffective management cycles for years or decades. However, like ecosystems, the political-economic systems associated with fisheries are never completely static, and there will always be some pressure to switch from one side of the treadmill to the other. In this section I provide a general overview of the different types of management treadmills observed in various cases before moving on to explore possible future variants of current trends in greater detail.

Based on evidence from the application of the AC/SC framework throughout this book, the management treadmill manifests in four ideal types: endogenous dynamic equilibrium, exogenous dynamic equilibrium, endogenous growth cycle, and exogenous growth cycle. These can be further classified by the nature of the relevant problem drivers, whether political-economic or environmental. Table 9.1 summarizes these relationships. First, there are two types of dynamic equilibrium treadmill, in which governance moves back and forth between ineffective and effective cycles without significant escalation in either the severity of crisis or the magnitude of response. In other words, a dynamic equilibrium treadmill runs slow and steady. Dynamic equilibrium conditions can be driven by the endogenous crisis rebound effect described above and by exogenous biological and oceanographic cycles that cause fluctuations in targeted biomass. Exogenous cycles in dynamic equilibrium include seasonal migrations of fish stocks that cause temporary gluts of specific species and multi-decadal cycles that generate long-term transitions in the availability of specific stocks, such as the transitions between sardine-dominated regimes and anchoveta-dominated regimes described in chapter 7.

Both endogenous and exogenous dynamic equilibrium treadmills can be dampened through the establishment of institutions that alter governance in the structural context. The CRE can be counteracted through development of informal endowments and entitlements as explained in chapter 5 or by means of conservation-based rules such as taboos against harvesting juvenile fish or fishing during spawning times. Laws, governing bodies, and

Table 9.1
Dimensions of the management treadmill in global fisheries

	Endogenous	Exogenous
Dynamic equilibrium	Political/economic: Crisis rebound effect	Environmental: Biological and oceanographic cycles
Growth cycle	Political/economic: Subsidies, exploration, engineering, opening new markets	Political/economic: Globalization, economic growth, population growth, new technologies, subsidies
	Environmental: Trophic cascades, habitat destruction	Environmental: Climate change, pollution, seabed mining, etc.

related formal institutions can also prevent the CRE. These institutions are usually created in response to conflict among fishers or experience with the negative effects of overfishing, and are developed by fishing communities or governments as part of the responsive governance process. They create incentive structures and social norms that reinforce management measures even in periods when rebuilding generates economic rebound. Similarly, fishing communities can increase their resilience to natural cycles by diversifying their food and income sources and by utilizing traditional ecological knowledge to make the most of resources that are available at any given point of time. Most of the historiography covered in section 5.1 fits into one or both of these dynamic equilibrium categories. In these fisheries, communities were relatively stable, isolated, and low-tech, so they had the power to shape their own resource institutions and received clear problem signals to guide the development of governance.

In contrast, fisheries that exhibit growth cycles tend to be more open systems that are exposed to multiple exogenous drivers. In these cases, management becomes more difficult over time and governance efforts must increase in scope and scale each time the system switches back to the ineffective side. Continuing the metaphor, the speed of the growth cycle treadmill fluctuates around an increasing trend. Because management is more difficult in each round, again all else equal, the level of crisis required to instigate a shift from ineffective to effective cycles is usually higher. Thus, the risk of collapse increases with each cycle of the management treadmill, even as the magnitude of governance mechanisms is increasing. This stands in strong contrast to Higgs' (1987) ratchet effect, which is also based on the concept of responsive governance. However, Higgs assumes that crises dissipate, leaving behind institutions that are no longer needed but rather persist due to bureaucratic inertia. Here institutional persistence is absolutely necessary to dampen the growth cycle, and institution building may be the only way to prevent severe fisheries collapses.

There are two major forces that increase the difficulty of management in a growth cycle treadmill. First, economic factors can increase incentives to undermine management through political influence and non-compliance. Second, environmental factors can reduce the size of targeted populations, such that larger reductions in fishing mortality are necessary to hit a given benchmark. Most of the economic factors that drive the growth cycle treadmill are covered in Part I. These include endogenous drivers such as the CPR dynamic and fisher responses that widen the profit disconnect, as well as exogenous drivers such as globalization, population growth, economic development, and technological innovation. Politically, subsidies can be either endogenous or exogenous drivers of the growth cycle, as described in chapter 6. Environmental factors that undermine management can be linked to biophysical limits such as the endogenous effects of fishing on the environment, including trophic cascades and habitat destruction (see chapter 8), or exogenous environmental degradation caused by factors including climate change and seabed mining. Most of these exogenous environmental factors are caused by humans, and are increasing as a result of many of the same political-economic forces that drive fisheries expansion and the growth treadmill. Of course, humans can also slow down the management treadmill, as when consumers switched to substitutes for whale oil or furs, but in the vast majority of cases demand was increasing rather than decreasing and caused the management treadmill to speed up.

Section 9.3 describes factors that can increase the likelihood of switching to effective management when growth cycle effects are prevalent, making collapse less likely. Reducing the common pool resource dynamic can help, but for growth cycle treadmills it is also important to minimize the profit disconnect and the power disconnect. Reducing the profit disconnect requires reduction of subsidies and other political factors that obscure cost signals. Price signals are also important and consumers have considerable power to combat overfishing and other core problems if they are provided with information on the state of fisheries resources and the sources of the fish products that they purchase. Reducing the power disconnect is difficult, because the empowerment of non-commercial interests can simply lead to conflict and political stalemate rather than switching to relatively effective response. However, when moderate interest groups are aware that coalition building is necessary for their political success, empowerment can speed up response. Two factors improve overall response, even when both the profit disconnect and the power disconnect are wide: development of expedient yet effective management options and the use of legitimacy-building processes. Both types of measures increase the resources available in the structural context and make relatively effective management intervention more palatable to fishers and other industry interests. All of these factors reduce response times but they cannot eliminate the core problems in the action cycle or exogenous forces that drive the growth cycle treadmill.

At present, factors that speed up the management treadmill dominate global fisheries. If current trends in economic and environmental conditions continue, fisheries management will become much more difficult in the future. Economic drivers of fisheries expansion are likely to escalate over the next century, further ratcheting up the need for management. Moreover, environmental conditions continue to deteriorate in most fisheries, even with the implementation of habitat restoration and ecosystem-based management approaches. Climate change will pose unprecedented challenges for fisheries managers. There is considerable uncertainty regarding future production of greenhouse gases, direct effects on global climate, and indirect effects on global fisheries. Nevertheless, some impacts are already observed. Sea level rise, ocean acidification, and changing water temperatures are stressing coral reef communities, kelp forests, and other key marine ecosystems (MacNeil et al. 2010; Jackson 2001). Escalation of extreme events such as hurricanes and the El Niño Southern Oscillation is also currently affecting marine communities (Michener et al. 1997). Pole-ward migration of marine species is occurring as well, and is likely to increase, forcing fishers and managers to adapt to new species compositions and related problems of by-catch and ecosystem dynamics (Cheung et al. 2010). Effects in the polar regions are most severe biologically, as cold-water species cannot migrate to find preferred water temperatures (Jeffers 2010; Anisimov et al. 2007; Ragen, Huntington, and Hovelsrud 2008).

Governance in certain regions is also likely to become more difficult due to climate change and technological innovations. The Arctic is not yet governed by any regional fisheries management organization and is likely to be the center of considerable controversy in the near future. Russian companies have already started experimental fishing above the Arctic Circle, and several arctic states are unwilling to enact precautionary limits to prevent overexploitation in the region (FAO 2012a; Cole, Ismalkov, and Sjöberg 2014). In the Antarctic, states that anticipate expanding fisheries into the area in the future are making management more difficult in the present. Recently, Russia, Ukraine, and China blocked the creation of one of the largest marine protected areas in the world by the Commission for the Conservation of Antarctic Marine Living Resources (CCAMLR). They did not debate the specifics of the proposal. Rather, these states questioned CCAMLR's authority to institute any marine protected areas at all. This position dovetails with parallel calls to open up the Antarctic to greater resource exploitation as the ice recedes and new technologies facilitate operation in such a harsh environment (Mathiesen 2013). Similarly, commercial-scale exploitation of the seabed is increasingly likely as cheaper mining technologies become available and as prices for metals and minerals found in the seabed oscillate around an increasing trend due to economic development and related growth in global consumption (Goldenberg 2014).

Combined, all of the above factors suggest that fisheries management will become increasingly difficult. That is, as long as demand for fish products is increasing, costs of fishing are decreasing, and environmental conditions are getting worse, the management treadmill for global fisheries will exhibit a growth cycle of tremendous proportions. This does not mean that current attempts at fisheries management are useless or futile. Short-term solutions are absolutely critical for long-term sustainability. Without the management measures and governance institutions that currently exist, collapse would be more prevalent and many marine ecosystems would be more degraded than they are now. Furthermore, the growth of countervailing forces described in chapter 8 is essential for slowing down endogenous drivers of the management treadmill. However, without long-term solutions to a wide array of problems associated with economic growth and development, even the best fisheries governance will not be sufficient to prevent further deterioration in global fisheries. This is a daunting prospect, yet lessons can be learned from the fisheries experience that reinforce and expand existing understandings of sustainable development.

9.2 Economic Limits and Drivers

This section explores the economic limits and drivers associated with the management treadmill in greater detail. It summarizes the analysis of the economics of expansion presented in Part I and predicts future trends given a "business as usual" approach in the structural context. There is considerable uncertainty in any projections of the behavior of a system as complex as global fisheries, but it is still important to consider the array of possibilities that we may face over the next century. It is clear from the evidence presented in chapters 2–4 that the fishing industry as a whole is exceptionally resilient to local economic costs associated with the core drivers in the fisheries action cycle, including the tragedy of the commons. Even though specific fisheries stagnated or collapsed and numerous fishers were forced to exit, the industry continues to thrive, and the profit disconnect is still widening for many groups of fishers and secondary economic actors. This is not to say that the industry can expand indefinitely. In fact, some evidence shows that fishing effort is approaching global limits in several areas, reducing overall flexibility. However, it is likely that exogenous forces will continue to drive up demand, ensuring that prices remain high relative to costs. Here I will consider the options that remain open to fishers as they search for economic responses in the action cycle and describe how exogenous forces are likely to continue to drive increases in fishing effort well into the future.

9.2.1 Endogenous Factors

Part I shows that the fishing industry is adept at avoiding the costs of overexploitation and overcapitalization. Exploration and innovation allowed them to reduce costs of production and to diversify either the area or the species targeted, and thereby build up additional resources in the structural context. Many small to medium-size fishing fleets operate in the coastal zone or offshore but still within their country's 200-mile exclusive economic zone (EEZ). These fleets are generalists that are highly adapted to local cycles in the marine system. They diversified by targeting multiple species, ensuring that they can harvest fish in the event of stock collapse or ecosystem change. At the other end of the spectrum, fewer and larger vessels patrol the high seas, using technology, communication, and knowledge gained from long experience to find fish in the mostly empty ocean. They know what kind of water their targeted species prefer, and they use satellite data to locate such conditions. They know when periodic upwellings or spawning events bring many fish to a specific time and space. They target a narrow range of species but have highly specialized means for finding their fish (Orbach 1977; Squires and Vestergaard 2013). Thus, they diversified geographically, and they rely on their mobility to survive geospatial changes in species abundance.

These exploration and innovation responses would not have been feasible without additional work to create markets, either in new places or by developing novel products. As secondary economic actors, processers and marketers amplified this response option significantly. The processing industry faces few constraints on economies of scale and is thus dominated by large transnational corporations. Therefore, processors can provide considerable access to capital for advances in fishing and processing technologies to improve the quality and the palatability of fish products. Wholesalers and retailers cover a wider spectrum of size and quality combinations, including high-end sushi restaurants, supermarket fish counters, fast-food operations, and wholesale fish markets. Such diversity reflects and reinforces the growth in markets for fish products, providing consumers with variety and novelty. This current system of fish processing and marketing evolved with capture fish production and is well adapted to deal with changes in the composition of catches. These businesses also work to identify or create new markets for products to keep demand (and prices) high enough to cover costs of production. Again, not all of these businesses are able to succeed, but innovation and diversification ensure that new processors and retailers will replace those that are pushed out of the global marketplace.

On the surface, it appears that fishers and the fishing industry are running out of economic response options, but this is not the case. Exploration to find new fishing grounds is probably the most limited option currently. The only new areas remaining for geographic expansion of fishing effort are near the poles, where ice is receding due to climate change, and in the deep ocean, where life is scarce and species are not well known. Once these areas are open to fishers, exploration will be exhausted as an alternative for escaping the local tragedy of the commons. Innovation has slowed down recently. Economies of scale are now dissipated in the capture fishing industry, and most major technological changes in recent decades occurred in the area of fish-finding technologies. Marketing, too, may be limited, not least because so many large-scale stocks are already overexploited and so few new, commercially exploitable species remain.

Nevertheless, there are still a number of endogenous economic responses that can dampen problem signals in the future. In the area of technological innovation, fishers are developing substitutes for petroleum, including renewable energy sources such as solar power, which may have high initial costs but much lower operating costs. Indeed, this transition could substantially widen the profit disconnect, since fuel is a large proportion of operating costs.[1] Furthermore, gradual technological and skill improvements continue in most fisheries. The resultant incremental decline in costs dampens a key problem signal that would otherwise trigger reduction in fishing effort over the long haul (Squires et al. 2010). This potential for future technological innovation suggests that fishers who can adopt these new technologies will be economically resilient to further biological depletion of existing fishing grounds, even if exploration is no longer a response option. Marketing will also persist as a response option in the future, though it may look much different from today. On one hand, fishers may market smaller quantities of fish products at higher prices, as described in the paragraph below. On the other hand, marketing will allow the industry to take advantage of increasing populations and growing affluence dampening price signals in the long term (see section 9.2.2).

The above responses insure that those who can adapt in the fishing industry will prosper, but the fisheries of the future will probably look quite different from the fisheries of today. Like an ecosystem, the economic system associated with global fisheries is always changing with shifts in available resources, such as capital, technology, and market demand. However, unlike an ecosystem, this economic system evinces cycling around an upward trend rather than a dynamically stable state (Ostrom 2002). Given the biophysical limitations of marine systems, it is impossible to maintain such growth indefinitely. Indeed, most of the growth in production of fish products since the 1990s was in aquaculture rather than in capture fisheries. Christy and Scott (1965, 151) explained the economics of this problem and documented growth in demand for fish products in developed and developing countries. They believed that supply was already limited and that price pressures would be observed for some species by the 1970s. However, the endogenous responses of fishers and exogenous technological shifts reduced costs of production, shifting the supply curve out, which kept prices relatively low in spite of growing demand. If aquaculture continues to expand, then price effects may be minimized. If it does not, then demand will outstrip supply and prices for all fish products, not just popular species, are likely to rise.

In the most extreme possible future, fisheries products could become luxury goods once again, priced so high that only a few can afford to enjoy them. Already this is true for the highest-priced species like bluefin tuna, Antarctic toothfish, and wild-caught salmon. In such cases, large portions of the industry would be forced to exit, but those who remain could benefit greatly from the higher prices if they are able to continue to avoid the costs of core fisheries problems at the local level. Ethically this scenario is not desirable, because it would result in considerable environmental destruction as well as human suffering, both in the loss of jobs and in the reduction in food supply. Nevertheless, because of ecological limits, this outcome is entirely feasible if demand for fisheries products continues to increase and costs of production continue to decline with technological innovation. Both of these trends will be shaped, at least in part, by the exogenous forces discussed below.

9.2.2 Exogenous Factors

The four primary economic drivers of expansion identified in chapter 1 are population growth, economic growth, technological change, and globalization of trade. The importance of these drivers is confirmed by the historiography presented in Part I. In many different fisheries around the world, the first two growth drivers generated increasing demand for fisheries products, which, in turn, canceled out and even reversed the price signal that would typically curtail effort in an isolated CPR. In contrast, exogenous technological change acted as a catalyst for innovation in the fishing industry, thereby reducing costs of production and allowing effort to increase more than it would have otherwise. Escalation of global trade affected both costs and prices. Through comparative advantage, trade lowered prices for inputs generally and also allowed fishers and processors to relocate to minimize costs of production. At the same time, trade and related innovations in shipping technologies opened up new markets for fish products, further increasing demand and reducing the price signal. Combined, these forces drove fishing effort higher than would have been possible with endogenous fisher responses alone. However, dependence on these exogenous forces also creates economic vulnerabilities for the global fishing industry.

Population growth and economic development both dampen the price signal that would otherwise limit fishing effort even under open access. Economic growth is associated with higher per capita income, which, in turn, generates greater demand for all products, including seafood and other items that are produced using living marine resources. Figure 9.2 shows several possible population projections developed by the United Nations.[2] Of these, only the low-fertility variant predicts future population decline. The others predict that the world's population will increase by at least 3 billion between 2010 and 2100. However, the mortality rate is constant across all four scenarios, so negative impacts of climate change and resource constraints are not reflected in these projections. Cleland (2013, 553) asserts that growth to at least 9 billion people by 2050 is inevitable given current population structures. The World Bank predicts that economic growth will continue well into the future for high- and middle-income countries, and is hopeful that economic reforms will encourage growth in low-income countries as well (World Bank Group 2014). Many observers would disagree with this finding, but if economic growth does continue at this pace, the increase in demand triggered by population growth will be magnified many times over by the income effect. This will obliterate the price signal, driving effort higher and making management increasingly difficult as the growth cycle treadmill speeds up.

As long as both economic growth and population growth continue, demand for fish products will increase in the near future, although the effects will vary depending on the type of product. On the other hand, it is

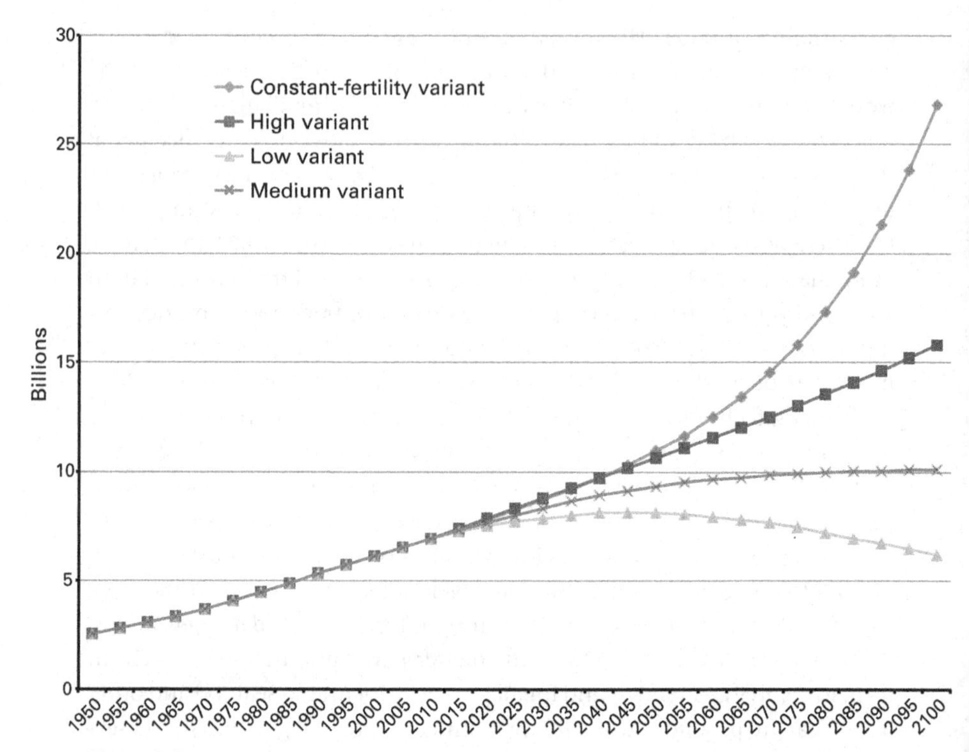

Figure 9.2
UN projections of four possible population growth scenarios. Source: United Nations 2013b.

possible that resource constraints under the higher population scenarios will choke off economic growth, such that fewer people will be able to afford fish even though there will be many more people on the planet. In addition to exogenous changes in the magnitude of overall demand, economic growth can also impact the relative demand for various marine products. Specifically, growth in middle and upper class income categories can lead to rapid increases in demand for luxury goods that are culturally designated as status symbols. Currently, this type of income effect is most pronounced in transitional economies like China and India, where demand for wildlife products, including goods like shark fin soup, bluefin tuna sashimi, and "Chilean sea bass" or Antarctic toothfish, has increased substantially over the last decade. This shift has contributed to higher prices for shark fins and other luxury marine products, which, in turn, has helped to spur exploitation and to drive down biomass for these species.

Paradoxically, economic growth also generates negative feedbacks that can reduce fishing pressure in the long run. The rise in operating costs associated with improved standards of living is one of the most important negative feedback loops in the system. Aside from the capital outlay for the vessel, gear, and associated technology, the biggest costs of production for fishers tend to be labor and fuel. Boat owners are reducing costs by hiring cheap labor from countries like Thailand and the Philippines. In fact, slavery is a known problem in the fisheries of these countries and a major method of keeping costs low (Skinner 2012). Similarly, processing facilities tend to be located in countries where wages are low to minimize costs of production (Campling, Havice, and Howard 2012). Many other companies use similar strategies, driving industry migration to low-wage countries. However, unless countered by government regulations, increasing demand for labor places upward pressure on wages, eventually erasing this source of comparative advantage. Thus, fishers and processers were forced to go through multiple cycles of expansion and migration in the past. Hypothetically, if economic growth brings prosperity to all regions of the world, wages will equalize at some point in the future and fishers will not be able to exploit cheap labor to keep costs down.

Fuel costs also increase with global economic growth, a problem that is particularly important for distant-water fleets. Since fuel can account for as much as 50% of operating costs, higher fuel prices can have a major impact on production. Some countries, like Venezuela, subsidize fuel prices for their fleets, selling marine diesel for as little as 19 cents per liter. For the most part, however, fishers must pay the market price for fuel. They can sometimes minimize this cost by fueling in ports where fuel is less expensive. In fact, for a time Venezuela allowed anyone to buy fuel at the subsidized price, but now sales there are limited to Venezuelan citizens. At such a low price, the trip might be worth it for many, but the tradeoff between fuel spent getting to a port and the savings on the cost of fuel does not usually work out so well. Depending on future prices for oil and the availability of substitute technology, fuel costs could substantially reduce fishing pressure by distant-water fleets and even by coastal fishers. Of course, open ocean fishing started out in boats fueled by renewable resources—the wind and human labor. However, to maintain current production in the petroleum-based economy, an equally fast and reliable fuel source would be required. As noted above, several fleet owners are currently investing in research and development for renewable power sources. Judging by the evolution of current vessel technologies, successful substitution will probably depend on exogenous innovations in the energy sector and in commercial shipping or naval technologies.

On the demand side, exogenous changes can also dampen the profit disconnect. In particular, the invention of substitutes can reduce the demand for marine products, which should reduce the price and, all else equal, the profit disconnect. One of the most important examples of substitution in the 20th century was the development of petroleum products that proved to be much better lubricants and fuels than whale, which was in high demand in the 19th and early 20th centuries (King 1975). Because of petroleum-based substitutes, the demand for whale products declined precipitously after World War II and many fleets stopped whaling altogether for purely economic reasons. This cleared a political path for conservation organizations to influence public perceptions of whales and national and international whaling policy (Young 2010). Petroleum is a fundamental part of the process of globalization. Plastics now substitute for tortoise shell, and we have synthetic approximations for marine chemicals used in beauty products. This helped to reduce pressures on some species, although the effect is subject to consumer preferences, and the trend toward natural ingredients in beauty products and away from plastics generally could lead to increased demand for non-food fish products.

Substitutes for food products derived from fish, which comprise the majority of demand, are harder to identify. Some substitution has already been observed in the growth of aquaculture production, as described above (FAO 2012d). Mariculture allows for cheaper production of high-value species such as salmon and bluefin tuna, though the latter is only in the earliest stages of culture from the egg and the product is primarily "ranched" through fattening wild-caught fish. As described in section 3.4, the effects of mariculture on demand can be surprising. For instance, growth in production of farm salmon did not reduce prices for wild-caught fish because producers and consumers differentiated farmed and wild markets. Wild-caught salmon now garners a price premium because of its luxury good status. However, prices for middle-of-the-road species like swordfish did decline with the introduction of farmed salmon because salmon was viewed as a substitute for swordfish by consumers (Webster 2009). Aquaculture takes its toll on prey species, which are harvested to feed these "tigers of the sea," and also creates many other negative externalities, even when practiced on inland waters (Naylor and Burke 2005).

As yet, aquaculture is not a sustainable substitute for capture fisheries generally and for high-priced species specifically, so some chefs and conservation advocates favor shifts toward vegetable-based proteins. However, this requires large changes in consumer tastes. Though vegetarianism is culturally important in many parts of the world and increasingly popular in the United States, Europe, and Oceania, meat consumption is increasing in East Asia and Latin America. Increased meat consumption affects fisheries directly, as fish is considered meat in these statistics, and indirectly, as fish meal is still an important input in other meat production industries. Optimistic estimates of a transition to 25% vegetarianism in countries where a positive trend is observed still result in high levels of future global meat consumption (Wirsenius, Azar, and Berndes 2010). Thus, if current trends continue, substitution of vegetables for meat generally and fish specifically will remain small relative to the effects of increasing population growth and affluence.

In fact, all of the limiting factors discussed above are overshadowed by the drivers of innovation and growth. Even if wages increase, fishers will probably find other ways to cut costs, or they will simply benefit from the concomitant increase in demand that is expected with growth in income. As population and affluence grow, demand will continue to increase in the foreseeable future, negating the price signal for fish products. The fishing industry as a whole will be able to survive, and some segments will thrive in spite increases in the core fisheries problems. Furthermore, capacity is likely to remain well above economically efficient levels without substantial regulatory intervention. A recent study by the UN Food and Agriculture Organization found that 12–15 million jobs in the fishing industry would have to be eliminated in order to reduce fishing pressure enough to rebuild depleted marine fisheries (Ye et al. 2013, 7). This is about half of the 24 million people employed as fishers in 2008, but only a small fraction of the 7 billion people now on the planet (3). Eliminating these jobs would be a political nightmare in coastal areas that are dependent on the industry, but continued business as usual could also cause the collapse of multiple fisheries around the world, generating food shortages as well as unemployment in dependent regions. These are the observed costs of the status quo of responsive governance.

9.3 Improving Switching Response Time

Given the economic trends described above, changes the structural context can help to mitigate the growth cycle treadmill by increasing the sensitivity of decision makers or providing them with expedient yet effective options so that response occurs earlier relative to the underlying core problems. This minimizes the negative social, economic, and environmental impacts of responsive governance and reduces the risk of catastrophic change in fisheries systems. In this section, I review key insights into the co-evolution of governance and the economics of expansion from Part II, and describe the array of political approaches to improving switching response time that this analysis revealed. Many of these measures, such as increasing the legitimacy of governance processes or building up management capacity, are already well known to fisheries scientists and policy makers. However, the AC/SC framework provides a better understanding of the linkages between regulatory action and switching response time, allowing for the identification of tradeoffs and unintended consequences that would not be obvious otherwise. It also shows how moving beyond the tragedy of the commons can help improve local management of fisheries problems.

The AC/SC framework predicts three primary delaying factors in fisheries governance: the CPR dynamic, the profit disconnect, and the power disconnect. From chapter 1, the CPR dynamic occurs when the combination of open access and rivalness creates incentives for fishers to ignore profit and catch signals because they fear that any fish left behind will be captured by others. It does not affect profit signals but rather alters fisher responses such that effort continues to increase even as net revenues decline. The profit disconnect occurs when the short-run economic optimum level of production is greater than the bioeconomic optimum or some other sustainable level of production. Wider profit disconnects minimize economic problem signals by allowing normal profits to increase in spite of the core problems of overexploitation, overcapitalization, and ecosystem disruption. The profit disconnect magnifies the CPR driver and is partly caused by temporal myopia (short sightedness) and negative externalities (costs not included in the price of a good). It can be widened by endogenous processes of exploration, engineering, and marketing, as well as by exogenous forces such as population growth, economic growth, and globalization of trade.

The power disconnect obscures political problem signals by marginalizing those who bear the costs of core problems in the decision making process and diminishing their ability to affect governance. Political marginalization of economically sensitive groups of fishers or of non-commercial interests groups, which tend to be motivated by biological and ecological problem signals, can create substantial power disconnects. A reinforcing feedback effect was also observed between profit and power in the historical evidence presented in Part II. Expansion was not a homogeneous process. Fishers with access to capital and technologies were able to thrive, and those who had few options were forced to struggle or find other work. This stratification exacerbated the power disconnect, as capital-rich industry sectors with little vulnerability to problem signals developed greater political influence than capital-poor coastal fishers, who were generally more sensitive to problem signals. Technological innovation by small-scale fishers helped to right this imbalance politically and economically but insulated coastal fleets from profit signals and therefore did not reduce the power disconnect per se.

Table 9.2 lists the mechanisms that can reduce these three sources of delay and improve responsive governance. The CPR dynamic is listed first because in the literature it is usually treated as the sole delaying factor for governance. However, my historiographical analysis suggests that the CPR dynamic does not greatly delay management unless amplified by the two disconnects. As expected from the AC/SC framework, the political problem signal of "threat to access rights" was often the first trigger observed in the historiography and proved to be the earliest endogenous driver of management response. In the right structural context, the drive to exclude and the related need to reduce conflict among fishers caused switching to occur well before overexploitation or overcapitalization, though it did not prevent a return to ineffective management in subsequent periods (see below). When the structural context favored sea tenure or a similar system, decision makers established formal endowments and entitlements that minimized conflict while reinforcing existing hierarchies. Where the broader governance context favored open access, informal systems for resource allocation were still commonly observed and could be maintained as long as local fishing operations remained relatively small scale and fishers retained sufficient political power to prevent incursion by outside fleets. However, these systems of access rights eroded with the expansion of fishing effort, as endogenous and exogenous forces widened both the profit disconnect and the power disconnect. Modern attempts to re-impose access rights on overcapitalized fisheries can cause political problems because of conflicts over distribution and perceptions of legitimacy.

Because of the profit disconnect, exclusion is only a temporary solution to the core problems in the fisheries action cycle. Even if access is restricted, fishers with endowments may increase effort substantially when costs are decreasing and prices are increasing. Therefore, measures to limit or reduce catch or effort are also important controls on the CPR dynamic. Here, preventing or minimizing overcapitalization is crucial, as shown in the case of Peruvian anchoveta. However, reduction of overcapacity contains many political pitfalls and can undermine the perceived legitimacy of governance institutions. Furthermore, in regional fisheries management organizations there appear to be diminishing marginal returns to exclusion. Measures

Table 9.2
Observed sources of delayed response and potential switching mechanisms

Sources of delay	Switching mechanisms
CPR dynamic	Establish access rights
	Reduce fishing capacity
Profit disconnect	Limit/reduce demand for fish products
	Limit technological advances (ineffective and inefficient)
	Reduce subsidies to fishers
Power disconnect	Empower actors who receive strongest problem signals
	Increase access to decision makers
	Align policy objectives to conservation
	Institutionalize leverage mechanisms
	Exclude fishers who receive weak signals/act as roving bandits
	Encourage coalitions of "Baptists and bootleggers"
All of the above	Develop more palatable management measures
	Build the legitimacy of policy process
	Build management capacity

adopted to curb illegal, unreported, and unregulated fleets also forced more states to join, increasing demand for access rights, which made agreement on reducing landings more difficult. Here again, multiple complementary solutions are necessary (see below). Domestically, establishment of individual or community property rights, through individual transferable quotas (ITQs) and traditional use rights fisheries (TURFs), can reduce open access problems and provide incentives for fishers to exit but can be undermined by other drivers of the profit and power disconnects.

In fact, strategies that dampen the CPR dynamic without considering other delaying factors only slow down the growth cycle treadmill. True sustainability still depends heavily on the ability to curb the profit disconnect and the power disconnect. This is a difficult task that requires action at both the fishery level and in the global political economy. The latter will be covered in the next section. Here, I look at fisheries-specific mechanisms for neutralizing the two disconnects. Section 9.3.1 shows that limiting or reducing demand for fish products can help minimize the profit disconnect and ensure early switching. Most factors that impact demand for fish products are exogenous and difficult to control, but non-governmental efforts to at least shift demand away from severely overfished species are increasing. Section 9.3.2 describes how empowering environmental NGOs and other actors who are sensitive to problem signals can help reduce the power disconnect, though this must be undertaken carefully to prevent the alienation of the fishing industry and resultant periods of stalemate. Lastly, section 9.3.3 covers three methods that can speed up response generally: development of expedient management measures that are also relatively effective, improving the legitimacy of fisheries governance systems, and building management capacity.

9.3.1 Reducing the Profit Disconnect

There is ample historical evidence that switching to effective governance occurs rapidly in response to economic crises associated with shifts in demand. For instance, in the Maine lobster fishery and the Pacific halibut fishery—both of which are seen as great management successes—switching was prompted by price drops caused by sudden changes in demand. In spite of years, or even decades of scientific agreement on the overexploitation of these stocks, only token management measures were adopted prior to economic crisis. Indeed, many attempts at restrictive management were spurred by the temporary decline in demand for fish products caused by the Great Depression. The resultant price signals generated considerable political will among fishers, who still preferred exclusionary or expansionary measures but were increasingly willing to accept restrictive measures as long as the economic crisis continued. In some fisheries, governments provided supportive subsidies, but time-area closures, size limits, and other restrictive measures also proliferated in the wake of the Great Depression. Governments stepped up negotiations on international management, as well, which was necessary for most important stocks at the time since states only had jurisdiction over the 3-mile territorial sea. In this light, the sudden proliferation of fisheries regulations in the 1930s was not a result of synchronous overexploitation of resources. Instead, global economic catastrophe created political will to build management institutions.

Given the cost of economic recession as a triggering event, it is important to find less harmful ways to reduce demand and thereby clarify the price signals received by the fishing industry. Chapter 8 describes how non-commercial interest groups worked to convince consumers to reduce consumption of specific fish products in order to clarify price signals. The dolphin-tuna controversy is the best-known example of the power of consumer behavior and eco-labeling to protect charismatic megafauna that are harmed in fishing operations. Numerous listing and labeling services now work to reduce demand for targeted species that are overfished as well. The success of these programs is debatable. Like many attempts to convince consumers to buy "green" products, lists and labels generally apply only to high-end seafood products and therefore affect a fairly small proportion of global fisheries. Furthermore, when labels are scaled up, sustainability is often sacrificed. The best example here is the Marine Stewardship Council (MSC) deal to provide sustainable seafood to Walmart and other large chains, which was roundly criticized because the MSC could only fill this demand by certifying numerous fisheries that only had plans to rebuild overfished stocks or that could easily be used to launder illegally harvested catches (Christian et al. 2013).

The primary problem with all labeling schemes is that current demand is above sustainable levels and so simply switching to well-managed options is insufficient. Since rationing and redistribution are not politically viable given the prevalence of free-market liberalism and the global reach of organized crime, governments are generally limited to policies of education and outreach to control demand. Exceptions occur in countries like China and Japan, where central planning is prevalent, but in these countries close ties with fishing industry groups and heavy cultural and geographic dependence on seafood generally prevent government injunctions to reduce consumption of fish products. Furthermore, evidence from fisheries in which production, trade, and sale are banned due to concerns about overexploitation suggests that even international sanctions are ineffective when prices for fish products are high. Shark fin soup is a major case in point. Finning is banned in all oceans, but the trade thrives in gray markets because demand for the soup—and the fins—is so high that potential profits far outweigh the costs of getting caught and punished. This is not just a fisheries problem, it is a global issue related to the wildlife trade more broadly.

Two other methods are used in traditional fisheries management to reduce the profit disconnect. These are already well studied, so I will only mention them. First, policies limiting technological advances may reduce the profit disconnect by causing cost signals to remain high. However, the historical record shows that these policies create inefficiencies in fish production and are also easily circumvented by fishers, and so are not useful except in the short run. Second, the removal of subsidies is a clear and direct method of clarifying profit signals. Here it is important to remember that fishers receive both endogenous and exogenous subsidies. Endogenous subsidies are related directly to fishing and may include direct payments, guaranteed loans, price caps on inputs, or rebates on costs of production. Exogenous policy factors that keep costs of production low include labor policies that prevent wage increases and subsidies to other industries that ensure low costs of production for many sectors of the economy (e.g. oil, transportation, etc.). Policies that remove these exogenous subsidies could have major effects on the profit disconnect, allowing for earlier response, but implementing them requires more extensive changes in environmental governance as explained in the next section.

It is also important to recognize that removing any type of subsidy impacts the power disconnect as well as the profit disconnect. Subsidies are often stoutly defended by recipients, so their removal is usually a difficult political battle in its own right. Using a combination of managerial tools, including buy-outs for fishers who are willing to exit the fishery and measures that help to increase profitability for remaining fishers, can make removing subsidies more politically feasible. There is also an endogenous process that reduces the power disconnect for subsidies; because subsidies do not solve the core problems in fisheries, they become more expensive over time, generating resistance from the public and decision makers. Demands are growing for subsidy reductions, as a result of public concern in many countries about budget constraints and due to international institutions favoring free trade, so this problem may recede in the future.

9.3.2 Reducing the Power Disconnect

The third source of delayed switching behavior is the power disconnect. Hypothetically, the power disconnect can be reduced by empowering the groups of actors that are most sensitive to problem signals so that they can have greater influence on governance. These sensitive actors usually include small-scale coastal fishers who have few options for economic response, and non-commercial interest groups who respond to biological problem signals derived from scientific analysis and media coverage of fisheries issues related to charismatic megafauna and (increasingly) fisheries sustainability. This simple idea is quite problematic. For one thing, when both sensitive and insulated actors are politically powerful, the result is most often political stalemate and inaction that ultimately benefit those who already profit from the status quo. In addition, attempts to redistribute power can threaten the legitimacy of governance, particularly among those who are already powerful. This generates intransigence among large-scale fleets that have considerable power to undermine management through non-compliance, and can have political ramifications that may further delay response.

Indeed, from the late 1800s on, observed attempts to alter power structures in fisheries resulted in political conflict and prolonged delay in management response. Most often, this delay favored powerful industry interest groups that were engaged in expansionary economic responses at the expense of smaller groups of fishers who bore most of the costs of the core fisheries problems. The National Sea Fisheries Protection Association's unsuccessful attempts to convince government officials to reign in steam-powered trawling near British coasts are a good example. Similarly, conflicts between large-scale and small-scale fishers resulted in considerable delay in the implementation of capital controls in the Peruvian anchoveta fishery and in other fisheries, even after extensive experience with stock collapse. More recently, the entry of non-commercial interest groups resulted in deadlock in many fisheries management arenas around the world, which again tended to benefit powerful industry interests. Nevertheless, shifts are taking place that favor groups who are more sensitive to problem signals, including expansion of legislative requirements for conservation and increasing public concern regarding marine fisheries.

Proactively, policy can be used to empower sensitive interest groups by granting them access to decision makers, enacting legislation that aligns policy goals with the goals of these groups, and providing them with recourse for leveraging public concern or other sources of power. Granting access can simply be a matter of increased interaction with decision makers, including visits with fisheries managers and lawmakers, but it may also be formally instituted through recurring meetings, stakeholder engagement processes, and, at the national level, lobbying reforms or related measures to ensure equal access by all constituents. Aligning policy goals also has formal and informal attributes. Formally, passing laws that make commitments to long-run sustainability and precaution is a first step toward goal alignment for conservation rather than expansion. As was noted in chapter 8, these formal institutions do not always generate the mandated action. Informally, hiring or electing decision makers who are conservation-minded may help to improve response time. The reverse is also true; hiring or electing decision makers who focus on food security or economic development can undermine formal attempts to implement conservation as a primary policy goal. Ideally, decision makers should in fact balance all of these goals, but this does not occur often and improved leadership can improve management generally as explained in the next subsection.

The third approach to empowering more sensitive actors, providing leverage, involves institutional shifts that increase the transparency of the decision making process, allow for legal recourse to challenge management decisions, and strengthen the democratic nature of the policy making process. Increasing transparency can be undertaken by fisheries bureaucracies without major shifts in larger governance structures. Exceptions may occur when laws or informal institutions that favor secrecy in government practices are in place. Other methods of increasing leverage require changes to the governance context at a higher level. For instance, in the United States stakeholders can sue the government when they believe that laws are not being implemented correctly. This option is not available to stakeholders in the United Kingdom (for example), and new laws would have to be enacted to provide this form of leverage there. Furthermore, in authoritarian countries like China, non-commercial interest groups can be denied the leverage of political protest, lobbying, or voting, and indeed have little power unless their members are affiliated with the ruling Communist Party. In these countries, a revolution of vast proportions would be required to empower non-commercial interests who are not members of the political elite. Nevertheless, as democratic institutions spread, the leverage of non-commercial interests is increasing globally and concomitant shifts in public perceptions have added to their power, so their access, alignment, and leverage are increasing.

Although this trend is important to improve reception of biological and ecological problem signals, history also shows that empowering sensitive groups can backfire and prolong the delay between bioeconomic problem signals and governance response. This dynamic is shaped by the total number of interest groups involved in the policy process and the conflicting values held by those disparate groups. When sensitive groups are simply granted access without redefinition of goals or reduction of access for less sensitive groups, the usual result is policy paralysis. Pulled in multiple directions, decision makers either postpone action or pass watered down regulations that attempt to satisfy all parties. In such cases, the exclusion of less sensitive groups could be helpful but may substantially reduce the legitimacy of the process. Furthermore, such exclusion can be highly detrimental when groups of commercial fishers are large enough to have an impact on the fishery through non-compliance. Excluding interest groups that are able to avoid problem signals may also raise legal issues. Open access or *mare liberum* is formally codified in international law and is foundational in many management regimes globally. In most parts of the world this makes it difficult for decision makers to exclude some groups of fishers but not others, particularly on the basis of vulnerability to the costs of overexploitation and overcapitalization.

The behavior of sensitive interest groups is also an important determinant of contested outcomes in the fisheries governance cycle. Different groups of fishers and industry actors tend to compete over access to fisheries resources and markets, but non-commercial interests frequently contest the underlying values associated with fishing. Preservationists value protection of specific species absolutely and challenge the "conservation for use" ethic that is accepted by many in the fishing industry today. Even moderate conservation groups that accept conservation for use in principle are sensitive to risks related to ecosystem disruption, and some of them have taken militant stances against the fishing industry in the past. These conflicts over values and confrontational approaches can cause severe, knee-jerk reactions among all commercial fishing interests, at times uniting them in active opposition to any proposed management as a matter of principle, even when they are sensitive to economic problem signals. Japan's stance on the anti-whaling movement is a good example of this response, but it is also observed at the domestic level in many other countries (Sakaguchi 2013). This antagonist dynamic has led to long standoffs between commercial and non-commercial interests that have delayed effective management response through policy paralysis, again benefiting powerful commercial interests that are able to continue their expansionary response in the absence of new management measures.

Hypothetically, in democracies widespread public support for preservation or conservation should break these stalemates, but I found no cases in which non-commercial interests convinced decision makers to impose costly management measures without also garnering the support of at least a few powerful industry interests. Even in the iconic tuna-dolphin case, non-commercial interests were successful only because they joined forces with major tuna processors to create the dolphin safe label. DeSombre (1995) famously compared coalitions of commercial and non-commercial interests in fisheries with the "Baptist and Bootleggers" concept from studies of prohibition. Encouraging such coalitions through stakeholder engagement or similar processes can reduce the power disconnect by empowering non-commercial interest groups while at the same time increasing the sensitivity of powerful fishers to problem signals. Furthermore, building these coalitions ensures that multiple goals are considered in the policy process, thereby advancing socioeconomic as well as biological or ecological sustainability. The AC/SC framework and the historiography of fisheries governance clearly demonstrate that such a balance is required. That is, it is not possible to have biological or ecological sustainability in the absence of socioeconomic sustainability, and vice versa.

9.3.3 Reducing All Delaying Factors

Three other interventions can have profound effects on the management treadmill given all three delaying factors. First, the development of management solutions that are less costly to implement and therefore more palatable to fishers can greatly improve response times. This is equivalent to expanding governance resources in the structural context. Inexpensive by-catch avoidance technologies that do not affect catches of targeted species provide one example of an expedient measure. Labeling strategies that provide price premiums for using those technologies can also make reduction of charismatic by-catch more palatable to fishers. Marine protected areas are becoming more popular as evidence develops that catches can increase outside of the protected area, making up for lost harvest from the closed region. Market-based approaches such as individual transferable quotas and traditional use rights fisheries also proved to be politically expedient yet effective in some cases, though this is not a uniform finding. The results depend on the power structure and the distribution of costs and benefits. ITQs were embraced in some countries because they were perceived as methods that would benefit both the fish and the fishers. Elsewhere, however, fishers resisted application of the same measures because of expected distributional effects that would privilege some groups over others. When they are expedient, measures like TURFs and ITQs may speed up switching, but it is important to note that they do not limit the profit or power disconnects.

Second, improved legitimacy in the policy process can clarify problem signals and improve responsive governance (Jentoft 2000). As noted in chapter 7, monitoring and enforcing fisheries regulation is difficult without the cooperation of fishers themselves, so maintaining their participation is critical. Furthermore, when regulation is seen to be illegitimate, fishers often develop their own parallel rules and norms of behavior that can undermine regulatory goals. Fishers do not resist management simply because of temporal myopia and financial greed. In fact, once fishers run out of options to circumvent economic costs, their concern for the sustainability of their industry increases substantially and they often seek conservation measures to keep the fishery going. However, adherence to these rules depends on their perceived legitimacy, which rests on perceived benefits and distributional fairness. Management measures that appear to cause arbitrary costs, that seem to privilege one group of fishers over another, or that challenge the basic values of conservation for use tend to undermine legitimacy.

In short, fishers learn to distrust management when the problem signals they receive do not match the management advice of scientists or resulting regulation. This mismatch is frequently due to the profit disconnect but may also be caused by conflicting interpretations of catch signals. For instance, in several instances fishers interpreted large influxes of juvenile fish as a sign that stocks were much healthier than reported by scientists, rather than recognizing that such surges can occur periodically due to exogenous improvements in environmental conditions. In cases where profit or catch signals do not reflect the true state of the stock, fishers usually perceive conservation management measures as unfair and unnecessary and therefore illegitimate. Disputes over the distribution of access rights or the costs of management can also be perceived as unfair by groups of fishers and thereby reduce the legitimacy of management measures. A good case in point is the fishery for Pacific anchoveta, where coastal artisanal fishers rejected attempts to exclude their vessels for the benefit of larger industrial fleets (as well as for the benefit of the anchoveta stocks). Their political response was two-fold. They continued to target anchoveta heavily, and they lobbied energetically to have their access rights reinstated. Again, caught between competing interests, Peruvian policy makers chose to allow the fishery, weakening all attempts to reduce overcapitalization of the fleet.

Keeping the complexities of legitimacy and policy making in mind, I'll highlight several methods that could be used to both improve acceptance of governance institutions and encourage earlier switching behaviors. First, any actions that reduce the profit disconnect can also reinforce legitimacy if not undermined by issues of distributional fairness. This is because reducing the profit disconnect helps to align the problem signals received by fishers with scientific and management understanding of the state of the stocks. Similarly, working to ensure that catch signals are well understood is useful for building legitimacy while reducing the power disconnect. Of course, cause and effect is difficult to establish in fisheries management. Many of the cases presented in Part II show that legitimacy was often judged on the basis of perceived success or failure that was actually due to exogenous environmental changes. This is why effective communication of scientific uncertainty is a counterintuitive yet crucial component of reducing the power disconnect. Some people fear that such admissions will reduce management effectiveness, since scientific uncertainty is often used as an excuse for delayed response. However, the evidence presented in this book shows that uncertainty is usually a red herring in such cases, and that other factors, such as the profit disconnect and mistrust of government, are much more important determinants of response.

Stakeholder engagement and fishery co-management are tools that can be used to build legitimacy while empowering actors who are more sensitive to problem signals. These methods may enhance the legitimacy of fisheries governance by ensuring that fisher concerns are addressed effectively and by communicating scientific uncertainties clearly. Furthermore, these approaches can help to identify measures that are palatable for fishers yet still relatively effective. It should be noted, however, that consensus-building approaches are not feasible in all contexts. Large numbers of claims on resources amplify conflicts over distribution, increasing the importance of relative gains and reducing the potential for convergence to an acceptable suite of management measures. Furthermore, fisheries management and conservation stir the passions of those involved, and discussions with outside groups may reinforce rather than reduce disagreements.

Thus, the stakeholder engagement process may facilitate the creation of commercial/non-commercial coalitions, but aggregate response rests on the interests and values of the groups involved, as well as on the prescience of group leaders and decision makers alike. Groups that favor conservation for use may find common ground with fishers, particularly when relatively palatable management measures such as by-catch avoidance devices are available. As noted above, groups that favor complete preservation are not likely to build such coalitions and therefore are not likely to achieve their goals as long as commercial interests are economically and politically powerful. Similarly, coalition formation is usually more successful if leaders are pragmatic and willing to negotiate. If leaders are not open to compromise—whether from commercial or non-commercial groups—then alliances cannot be formed and delayed response is expected. A variation on this dynamic is observed when extreme antagonists "scare" industry interests into alignment with moderate conservation actors. As supposedly objective actors, managers can help to bridge divides between groups of all types, whether the conflict is over distribution or over values. They can also escalate conflicts and reduce regulatory legitimacy by aligning with one side or another, which may be necessary in situations where compromise is not possible, and is indeed an important part of their role as decision makers.

Although leadership can also improve response, it should be noted that political entrepreneurs are a somewhat stochastic element in the action cycle. Leaders may sometimes advocate relatively effective measures, as they did when many scientist-managers called for a "go slow" approach to rebuilding fleets after each of the two world wars. They can also push for rapid proliferation of measures that are expedient but ineffective, even in relative terms. For instance, as shown in chapter 6, the widespread hatchery programs that were popular around the end of the 19th century were championed by several key leaders but were useless from a biological perspective. On the other hand, without these measures, funding for several national fisheries management bodies would have been much lower and development of these institutions would certainly have followed a different path. Similarly, fishing "up to" MSY, was championed by Wilbert McCleod Chapman but actually acted a rationale for overfishing and responsive governance, rather than as a break on the fisheries action cycle. These policies also exacerbated problems of overcapitalization, as shown in chapter 8.

Even so, cultivation of leaders who can build legitimacy and find ways to balance the concerns of competing interest groups may help to ensure earlier switching to relatively effective governance response. One novel observation based on the AC/SC application in Part I is that leaders are most successful from a conservation standpoint when they take advantage of windows of opportunity created by reductions in the profit disconnect. Indeed, access to the right solutions at the right time was as important as the presence of a strong leader in the observed cases of perceived management success. For instance, Horatio Crie was a fish commissioner in Maine for two decades prior to his great work establishing a new lobster management program. During this period he built networks to win the trust of lobster fishers and also kept up to date on scientific studies of the lobster fishery. Because of this careful groundwork, he was able to take advantage of the sudden growth in political will for management that occurred in response to the Great Depression and is given credit for the establishment of a conservation ethic in the fishery (Acheson 1997). Although waiting for crisis is not the best governance option, recognizing the effects of the power disconnect and the profit disconnect can allow leaders and other actors to strategically deploy resources at times when their efforts will be most effective.

Third, capacity building is another key policy response that can improve switching in a growth cycle treadmill. Kuperan and Sutinen (1998) point out the presence of small subgroups that persistently violate rules and norms even if they are generated through processes that are perceived to be legitimate. This suggests that monitoring and enforcement measures may always be necessary to ensure compliance by these "true" free riders, not least because their behavior can destroy legitimacy if rules are not enforced fairly. This is one of the oldest political responses to crises in fisheries management and played an important role in many of the switching responses recorded throughout this book. Capacity building was frequently linked to exclusionary tactics used by groups of fishers as well as to general concerns about fairness, either at the domestic or international levels. In fact, "not playing by the rules" is one of the few accepted rationales for involuntary exclusion of fishers under modern norms of open access, so strong monitoring and enforcement mechanisms can increase the likelihood of exclusion of some of the least sensitive fleets, while also contributing to reductions in fishing effort.

All of the interventions described above are necessary to ensure the sustainability of fisheries management in the future. None are sufficient. Even with the best possible leadership, highly legitimate governance institutions, extensive management capacity, and well-enforced property rights, effort limits, and related management measures, the growth cycle treadmill will continue as long as exogenous forces drive up incentives to undermine management. This conclusion is not just a logical extension of the AC/SC framework; it is also supported by evidence from the historiography. In fishery after fishery, effective governance institutions were eroded by exogenous factors. This dynamic is observed in small, isolated fishing communities as well as large, well-regulated commercial fisheries. Even the vaunted ITQ systems of Iceland and New Zealand were subject to the crisis rebound effect, particularly when it was amplified by exogenous price increases. Though fisheries-specific attempts to reduce demand can help to mitigate the effects of exogenous drivers, they cannot stop these pressures. The growth cycle treadmill will speed up as long as the global economy is based on the precept of infinite expansion.

9.4 Sustainable Development

Much like fisheries worldwide, the global economy is currently experiencing a growth spiral that may ultimately be limited by the availability of natural resources and the confines of the world system (Meadows, Meadows, and Randers 2004). Some economists believe that market forces will spur innovation and provide substitutes for the natural capital that we use now. If this happens, economic growth will not be limited by natural resources. Others believe that natural capital and created capital are complements rather than substitutes. If they are correct, then growth cannot continue and the future looks bleak indeed. In all likelihood, each view holds some truth, so a middle path is necessary to achieve sustainability (Clapp and Dauvergne 2011). Whichever group is right, these broad economic trends will have a deep impact on the bioeconomic resilience of fisheries in the future. The subsections below relate lessons from the AC/SC framework to the limits to growth literature and the much more difficult issue of sustainable development.

9.4.1 Implications for the Limits to Growth

Understanding the economics of response in global fisheries is important because it sheds light on several key debates regarding sustainable development. First, the AC/SC applications explain how economic responses that push past local limits affect sustainability more broadly, particularly when considering the Limits to Growth models of Meadows et al. (1974) and Meadows, Meadows, and Randers (2004). This work is widely criticized by economists who claim that price and cost signals will initiate a transition to substitutes long before physical limits kick in. However, the historiography presented in Parts I and II clearly demonstrates that endogenous and exogenous factors can neutralize these economic problem signals. On the one hand, when profitability is low due to overexploitation, the AC/SC analysis shows that technological innovation by the fishing industry can reduce marginal costs of production, keeping profits up in spite of declines in resource availability. Amplified by exogenous factors such as technological change and globalization of trade, innovation reduces incentives to curb production or seek out substitutes, amplifying the core problems in the action cycle. On the other hand, when production is high and effort is increasing, fishers can open new markets or take advantage of growing demand associated with population growth and economic development (or increasing affluence) in order to keep prices high. By dampening problem signals, both price and cost dynamics widen the profit disconnect, delay economic substitution, and prolong overexploitation.

Because the profit disconnect exists regardless of ownership structure, the above observations are important in many issue areas other than fisheries. Furthermore, exogenous factors that widen the profit disconnect affect the exploitation of *all* natural resources. For instance, chapter 3 described how exogenous technological and economic changes facilitated the shift from sail-powered to steam-powered to petroleum-powered fishing vessels. The latter transition in particular reflects the traditional economic perspective on resource use. Marginal costs of coal production were increasing toward the end of the 19th century, so entrepreneurs searched for cheaper substitutes. They found petroleum and ignited a new wave of global economic development. Yet coal is still utilized today and it can be a cheaper substitute for petroleum-based fuel sources. Just like fishers, energy producers explored for new sources, engineered technological innovations that allowed them to exploit reduced costs of production, and opened new markets for their products. At the same time, population growth and economic development continued on an upward trajectory, increasing demand for fuel sources of all types. Combined, these factors caused the profit disconnect between the short-run, privately optimal level of production and the socially optimal (i.e., sustainable) level to increase over time, causing extraction and related externalities to expand significantly.

In addition to these economic insights, the AC/SC literature formally incorporates the effects of economic problem signals on environmental governance. This element is missing from the field of resource economics, but it is extremely important to sustainability. To be fair, most authors in the sustainable development literature do view governance as a source for solutions to the problem of environmental impacts, but they generally treat it as an exogenous factor. The AC/SC framework internalizes political dynamics through the responsive governance lens and the power disconnect. AC/SC specifically points to the political delays created by the profit disconnect as reinforced by the power disconnect. Like fishers, mining corporations, energy companies, and other industries with direct effects on the environment have incentives to undermine regulatory attempts to rein in exploitation or production as long as profitability is high and they do not feel the economic effects of physical resource limits. This parallels the power disconnect in fisheries, as corporations with the least exposure to the environmental costs of exploitation often have the greatest influence over the policy process. Though the CPR problem is not important in these industries, management to reduce negative externalities can also be delayed as a result of the power disconnect. In fact, this is visible in the many instances of lobbying to reduce environmental regulation and of non-compliance with existing regulation documented by the news media and in the academic literature (Levy and Newell 2005).

These parallels between fisheries and other natural resources extend to the potential for industry cooperation on regulation when profit signals are strong and pressures from consumers and non-commercial interest groups are high. In fact, further comparison between different types of industries would help to refine the concepts of delaying factors, switching mechanisms, and the management treadmill. Given the importance of legitimacy in environmental governance generally and the scope and scale of modern industrial production, greater understanding of these implications from the AC/SC framework could be useful in solving a wide array of related problems. However, these findings would not necessarily diminish the expansionary forces associated with population growth, economic development, globalization of trade, and technological change. To reduce those pressures, it is necessary to dig deeper into the relationships between population and affluence and among trade, technology, and costs of production. Future work applying the AC/SC framework to these areas could be useful for identifying key switching mechanisms and institutional reforms outside of the fisheries context.

9.4.2 Switching to Sustainability

The growth cycle treadmill ensures that fisheries management cannot be truly sustainable until the entire global political economy is sustainable. Indeed, when systems are open to exogenous forces, one can argue that sustainability in any sector requires sustainability of the whole. This is explicitly accepted in the AC/SC framework, and resulting analysis points to the need for transition to a path of sustainable development so that individuals are able to thrive in the present without reducing the well-being of future generations. This is a wicked problem, but there is a wealth of theoretical and applied work on the topic (see Clapp and Fuchs 2009 for a review). The literature ranges from detailed case studies of green development projects, to proposals for revolutionary transformation to a green economy, to studies of institutional design to improve the operation of international environmental agreements. Rather than review this literature in its entirety, this section focuses on the role of political-economic empowerment, which can narrow both the profit disconnect and the power disconnect, increasing the sustainability of the global economy generally and fisheries specifically. Furthermore, the section does not provide solutions to the problem of sustainable development but rather supplies food for thought, particularly regarding the primary exogenous drivers in fisheries and so many other issue areas: population growth, economic development, and globalization of trade.

As described in section 9.3, the AC/SC framework highlights the importance of the political and economic empowerment of individuals or nation-states that experience the costs of core problems in the fisheries action cycle. In a broader context, Sen (1999) posits that economic development and political freedom are deeply intertwined precisely because democracy narrows both the profit disconnect and the power disconnect, allowing people to hold governments and businesses accountable for negative externalities and myopic decisions. In Sen's work, economic empowerment involves provision of education and the development of a highly diverse economy that provides a wide array of jobs rather than a resource-dependent economy that limits employment options. Political empowerment requires government accountability to all segments of society, not just the privileged elite. Hypothetically these two forms of empowerment are self-reinforcing but, as with fisheries management, switching from a downward spiral of corruption and dependency to an upward cycle of empowerment and sufficiency is not an easy task, or even a well-understood phenomenon.

That said, a closer look at the three primary drivers of the growth cycle treadmill is useful, since each can be mitigated somewhat by the political-economic empowerment of sensitive actors. First, empowerment of women is widely recognized as a major factor in declining birth rates and reduced population growth. In particular, birth rates decline precipitously when women are well educated, have access to good jobs and effective contraceptives, and are able to negotiate family size with their partners. Dissemination of social norms that expand female gender roles beyond household responsibilities and child bearing can also help to increase socially acceptable economic opportunities for women. The empowerment of women requires some level of economic development in order to provide resources for education and opportunities for employment. However, the tradeoff is lopsided—empowerment can have a large effect even at relatively low incomes, but if women are marginalized or repressed, as in Saudi Arabia and in several other Middle Eastern countries, high incomes do not reduce birth rates (see Dyson 2010 and W. T. S. Gould 2009 for further discussions regarding population and development).

Second, education and a strong civil society can increase economic development while also building social tools that are needed to ensure environmentally responsible decision making. For example, with larger and more diverse networks, well-educated individuals have greater opportunities to connect with employers and to organize for political action. Thus, a dense civil society can magnify opportunity and help people make use of their education. Increased interconnectedness also helps individuals to perceive and acknowledge the wider social costs and benefits of their decisions. Putnam (1993) documented this in his comparative study of northern and southern Italy. In northern regions, where civil society was strong, individuals tended to make decisions that would benefit the community as a whole, including curtailing corruption. This led to high levels of economic growth and political participation. In contrast, in southern Italy individuals had few connections outside the family and supported social norms of nepotism and corruption that might benefit strong clans but were detrimental to the economy and polity as a whole. Finally, civil society enables the institution and dissemination of a wider array of social norms, including appreciation of education and ecosystem services. From the AC/SC perspective, this generates more expedient yet effective responses to problem signals—for instance, providing fishers with alternative sources of employment or ensuring that sustainable technologies are available—and can therefore increase the likelihood of switching from ineffective to effective governance.

Third, to counter the increasing profit disconnect associated with the globalization of trade, substantial improvements in transparency and accountability are necessary. Globalization obscures problem signals by reducing costs of production and externalizing negative environmental and human rights effects. For instance, in fisheries, if a community both catches and consumes only local fish products, they pay all of the costs of the core fisheries problems and the profit disconnect is likely to be narrow. However, if international fleets catch the fish and local populations experience the costs, then producers and consumers in international markets benefit while the local community pays the price. Similarly, wages and other forms of compensation can be kept artificially low by government regulation to foster trade, or as in some fisheries, through the practice of slavery. This also keeps costs and prices low, ensures that the profit disconnect is wide, and increases production along with related environmental impacts. If fishers or consumers had to pay for all of these costs, then prices would be higher, quantity demanded would be lower, and fishing would be limited to more efficient and sustainable levels. Thus, globalization fuels growth but also drives environmental degradation like the overexploitation of fisheries resources. If external costs can be internalized, the profit disconnect would narrow, slowing down the growth cycle treadmill.

Interestingly, narrowing the profit disconnect associated with globalization of trade also requires reductions in the power disconnect. Coase (1960) postulated that externalities can be internalized through negotiations between those who generate the harm (usually businesses) and those who experience the harm (usually individuals, some of whom may also be consumers). However, Coase's theorem depends on a level playing field, where all parties negotiate on an equal footing. When power structures are asymmetrical, this assumption does not hold, and negative externalities are persistent, because the power disconnect is wide and those who are harmed by environmental degradation are marginalized while those who have the power to make decisions are insulated from that harm. Greater transparency and increased accountability of government can help to reduce the power disconnect and allow those who are negatively affected by externalities to push for regulatory protections that ultimately internalize costs and reduce the profit disconnect. Improved education and civil society can also empower marginalized populations, as described above.

Where government reforms are not possible, positive cycles of empowerment may be instigated by socially conscious consumers as long as there is transparency and accountability in the global economic system. If made aware of the negative externalities associated with their purchases, consumers can demand changes in production processes to reduce external costs. There are many examples of industry reform in response to this type of public outcry, including the dolphin-tuna controversy described in chapter 8, boycotts of Nike products after disclosures about sweatshop operations, and recent changes in labor practices in Apple's overseas factories. Eco-labeling and fair-trade labeling allow consumers to hold corporations accountable.

These types of initiatives have proliferated in recent years—so much so that the array of socially and environmentally friendly options available is somewhat perplexing. In addition, corporations are increasingly participating in voluntary environmental programs to protect brand value, improve efficiency, preempt regulation, and increase employee job satisfaction. Comparable price and quality also tend to increase consumption of "green" products (Gallagher and Weinthal 2012; Esty and Winston 2009; Winston 2014). For such socially responsible consumption to be effective, consumers must have clear information about the negative environmental costs associated with their purchases. Without this transparency, they cannot hold corporations accountable for their actions and may inadvertently pay premiums for products that falsely advertise environmental or human rights benefits.

Whether initiated by government reform or socially responsible consumption, sustainable development can only occur with a narrower power disconnect between individuals in developed and developing countries. This runs contrary to some theoretical work in academia and general public perceptions in much of the developed world. Inglehart (1995) proposed that people in developed countries care more about the environment because they are freed from the lower levels of Maslow's hierarchy of needs and have enough leisure time to appreciate nature. However, multiple empirical studies show that people in developing countries value the environment and, in some cases, express higher willingness to pay for environmental protection than individuals surveyed from developed countries (Dunlap and York 2012). To the extent that those people living in developing countries who value conservation and preservation are politically marginalized by elites with different, less environmentally friendly value systems, asymmetries in governance structures will magnify divergences between environmental problems, signals, and responses; in other words, sustainable development can be stymied by the power disconnect, much like fisheries management. In addition to fair trade labeling, many other intergovernmental, governmental, and nongovernmental programs are needed to provide people in developing countries with access to policy making and economic resources so they can express their concerns and influence the policy process. Unless these disconnects are reduced, sustainable development cannot be attained given the predominance of the responsive governance paradigm.

9.5 Conclusion

In modern times we often think of regulation as a means to conserve resources by combating the tragedy of the commons. This is a new perspective. As shown throughout this book, historical regulation of fisheries—or governance more broadly—was primarily used to allocate scarce resources, to reinforce existing power structures, and to reduce destructive conflicts among fishers. Starting in the Middle Ages, commercial fisheries were also protected and encouraged as a source of economic development and maritime power. During the Industrial Revolution, fisheries products became cheap staple sources of protein in inland as well as in coastal regions. Therefore, governments supported fisheries as an important component of food security. Although periodic local depletion of fish stocks engendered some early attempts to protect spawning and juvenile fish, conservation became a primary factor in fisheries governance only after hard lessons were learned from the collapse of several major fisheries in the 20th century. Finally, in the late 1960s and the early 1970s, countervailing forces began to emerge in the form of non-commercial interests and management institutions. Given that economic drivers of fisheries growth are likely to continue to increase in the future, the sustainability of global fisheries depends heavily on the continued evolution of these governance mechanisms.

Today most authors agree that fisheries are in crisis, but they disagree about the prospects for the future. Some believe that we have the tools to achieve sustainability while others assert that a revolution in governance and in consumer habits is required. The reality is likely somewhere in between. The oceans of the future will look different from the oceans of today. We will probably lose some species and even some types of ecosystems. Nevertheless, the pressures for better fisheries management are building and the range of options available to managers is more extensive now than ever before. With sufficient scientific knowledge, consumer concern, and stakeholder engagement, fisheries management can improve and expand in response to growing crises. Indeed, chapter 8 demonstrates that this is already happening and that the effects of countervailing forces are increasing. Unfortunately, endogenous and exogenous factors also continue to drive the expansion of the fishing industry, creating a growth spiral treadmill on which regulators are constantly catching up with crises rather than preventing them. Slowing this management treadmill requires solutions beyond the traditional realm of fisheries management.

Indeed, the Action Cycle/Structural Context framework highlights the connections between fisheries and other environmental issues. In this light, the sustainable management of global fisheries is deeply dependent on a successful transition to a system of political economy that favors sustainable development around the world. Some scholars are more optimistic about the prospects for sustainable development than others, but in general academics are increasingly seeking a middle path or a "third way" that relates context to action and identifies the conditions under which different approaches fail or succeed. Clapp (2011) identifies this movement in the study of the international political economy of the environment. A decade earlier, Young (2001) pointed out the importance of this approach in the area of global environmental governance. Even more broadly, Ostrom et al. (2007) make a persuasive argument for going beyond panaceas when searching for sustainability in social-ecological systems. All of these perspectives point to the importance of interdisciplinary research and the dangers of academic dogma. Epistemic communities, like commercial fishers, environmentalists, and so many others, are engaged in their own action cycles and limited by their peculiar structural contexts, but are striving to improve understanding and generate new solutions to worsening problems.

Finally, from the historical record, and from the writings of many wise scholars, it appears that "sustainability"—like "liberty" and "democracy"—is both a high ideal and a never-ending struggle. Some would say that it is a Sisyphean task. Yet history shows that progress can be made and that change, when it happens, can occur very quickly (Kuran and Sunstein 1999). There may not be a "happily ever after," but there can be improvement over time. Just as in fisheries, crises continue in many other issue areas but the solution set is expanding and the desire to implement sustainable governance systems increases with each new problem signal. Thus, liberty and democracy lie at the core of sustainability, because so many problem signals are obscured not by the profit disconnect alone but by its interaction with the power disconnect, which marginalizes those most harmed by environmental problems. Therefore, moving beyond the tragedy of the commons is a matter of empowerment and political accountability as well as of property rights and resource scarcity. In the face of such wicked problems, it is important to remember that human beings have been working toward these goals for centuries if not millennia. There have been many failures, but also great successes. The latter would not have occurred without the former. The worth of these successes and the weight of the balance of progress are matters for the individual to judge, but the AC/SC framework clearly shows that systemic change can occur in domestic and international environmental governance, and it provides useful insight into the conditions that make constructive change more likely.

Notes

Chapter 1

1. See for example: Simon (1995, 1997), Baumgartner and Jones (1991), and Gilovich, Griffen, and Kahneman (2013). All show that action is responsive when individuals and decision makers deal with complex problems such as marine fisheries governance.

2. According to Clark (2005), this applies for either normal or economic profits. The latter takes the opportunity costs of capital into account. This will be discussed in greater detail in later sections.

3. Usually we expect the maximum sustainable yield to be greater than the maximum economic yield, and therefore the latter is the bioeconomic optimum level of production because it is both efficient and sustainable. However, as will be shown throughout this book, determining that optimum is extremely difficult, not least because economic and biological factors are highly variable (Jentoft and Chuenpagdee 2009). Furthermore, most fisheries affect more than just the targeted stocks, so consideration for habitat destruction and ecosystem effects is an important though analytically difficult component of sustainability. See section 1.1 and chapter 8 for more.

4. This can also be thought of as the socially efficient level of production. The primary differences between the private and socially efficient or sustainable levels of production are (1) internalization of externalities, including environmental degradation, health effects, etc. that are caused by a production process and (2) a lower discount rate, which reflects society's greater concern for future generations.

5. Here again, there is ample support for the power disconnect in the literature, including Olson's (1971) seminal work on public goods and interest group formation, size, and political influence. The primary benefit of the new term is that it provides an umbrella for many similar concepts so that they can be incorporated into the analysis.

6. That figure rises to 16.6% if only animal proteins are considered (FAO 2012b, 3-5).

7. The data in figure 1.2 were collected from individual countries by the UN Food and Agriculture Organization. It is possible that a small proportion of the harvests reported were used for subsistence as well as for commercial sale.

8. "Fishes" means multiple species; "fish" is the plural form used for multiple individuals.

9. See Howarth 2009 for a discussion of various sources and implications of discount rates in ecological economics.

10. See section 1.4 for more on process tracing as a methodological tool and on the use of historiography in AC/SC applications. Webster (2015) demonstrates use of the AC/SC framework in hypothesis testing.

11. This model is based on the basic Gordon-Schaeffer model as described by Clark (2005). Similar descriptions can be found in any basic bioeconomics textbook. Various refinements of this model are discussed in greater detail below.

12. The original work by Gordon (1954) used normal profits, and this is still the assumption used for the base-line Gordon-Schaeffer model. See below for a discussion of this simple model when economic profits are considered.

13. There are exceptions to this stock effect. For instance, for schooling species the stock effect is not always observed, because as the size of a stock declines schools condense into smaller areas and that makes them easier to find (Bertrand et al. 2008; Clark and Mangel 1979). These types of dynamics often generate depensation, as described in detail by Clark (2005).

14. Here I depart from the typical Gordon-Schaeffer model, in which prices are assumed to be fixed. This is because there are many observed instances in which a sudden influx of fishers generates sudden price drops and related shifts in effort, so price dynamics are important determinants of response.

15. Olson (1971) extends Cournot's work on duopoly and oligopoly to show that when there are numerous firms in the market profits will always be zero at equilibrium because the benefit of restricting output to generate higher prices (a public good) is too widely distributed across firms. Of course, the concept of rents predates Olson's work by centuries to (see Smith [1776] 1976 and Riccardo [1817] 2004), but this definition fits with most characterizations of economic rents and captures the temporal aspects of the issue quite well. Rents can also be thought of as sustained positive net revenues (e.g., profits), over and above the producer's opportunity costs, that accrue as a result of social, environmental, or political differentiation from other producers.

16. There are also quasi-rents as per Schumpeter (1976), which are derived from differentiation based on innovation. As will be discussed below, fishers often innovate

in pursuit of such quasi-rents, and distinguishing between this process and the tragedy of the commons is difficult if not impossible without substantial restrictive assumptions. Thus, it makes sense to use the broader term *profit* throughout, although additional theoretical and empirical work to differentiate these forms of profit or rents in fisheries is useful, as work by Homans and Wilen (2005), Squires (1988), and others shows.

17. The discount rate is often equated to the rate of return on the next-best investment.

18. Though many believe that fishers are risk seeking, Smith and Wilen (2005) show that fishers are heterogeneous in their risk preferences and that there tends to be a positive correlation between the willingness to accept financial risk and the willingness to accept physical risk.

19. In technical terms, purchasing from multiple vessels or multiple fisheries smooths out the revenue curve for processors and marketers and gives them a higher expected return on investment than is possible for a single fishery or even the larger fishing companies that fish in multiple fisheries. Of course, this effect is limited by the structural context, since the number of fisheries from which a processor can purchase raw materials (or to which a boat builder can sell) is limited by the current state of processing and transportation technologies. Multiple authors have developed iterative decision processes to help processors mitigate high levels of uncertainty in fish catches, usually by increasing communication between fishers and processors so that the latter can coordinate options and effort responsively through linear programming or dynamic optimization (Jensson 1988; Randhawa 1995).

20. There is very little public information available regarding the internal decision making of corporations that process and market seafood; however, the majority are large corporations that produce many other products in addition to seafood (IBISWorld 2014). There is no reason to believe that these corporations would be exempt from the short-termism described above, though there are always exceptions—e.g., Unilever, whose CEO, Paul Polman, is notoriously opposed to short-termism (Winston 2014). In view of the scale of Unilever's operations, this longer-term perspective may conform to Clark and Munro's (1980) expectation that monopsony power in the processing sector could reduce the profit disconnect in a fishery.

21. In either case, innovation shifts the production possibility frontier out, allowing more to be produced with any given combination of inputs. This effectively reduces the marginal costs of production, shifts the open-access equilibrium out without increasing the sustainable level of production and therefore widens the profit disconnect.

22. See R. Hardin 1995 for an exhaustive discussion of conflict and group identity formation, power accumulation, and political or military action.

23. By identifying and explaining the broader patterns associated with the management treadmills, I engage in two types of theory building. The first is what Collier (2011) describes as the basic (Theory I) process of tightly connecting identified regularities; the second is the more advanced (Theory II) process, which includes explanations for the connected regularities defined under Theory I.

Chapter 2

1. See O'Connor, Ono, and Clarkson 2011 for recent evidence that there may have been pelagic fishing during the Holocene and for a review of the few sites that show any evidence of fishing with implements during that period.

2. Historically, there are many methods for measuring the size of vessels. In this book, the term *ton* refers to the approximate volume of water displaced, unless otherwise noted. *Tons (gross)* refers to gross tonnage, which is an index of the vessel's internal volume. In 1969, the international standard shifted from gross tonnage to the gross registered ton (grt), which is a slightly different measure of volume. Unfortunately, there is no direct method of conversion between any of these measures, so it is not possible to use a uniform metric throughout the book.

3. There is some conflict among historians regarding the beginning of the Industrial Revolution, but most now recognize that there were in fact several waves of industrialization that began in the mid-1700s (Allen 2006).

4. Unless otherwise noted, all calculation of real US dollar values was based on the most likely estimate provided by Officer and Williamson 2014a.

Chapter 3

1. It is not feasible to reliably convert these values into modern currency; however, this was a period of Dutch ascendency, and florins were a major currency of international exchange, much as the dollar or the pound is today.

2. This was a period of very low inflation in Britain, so most of the difference is due to increased capital investment (Twigger 1999).

3. Data are compiled from the Statistical Yearbooks of the UN Food and Agriculture Organization (FAO), which began collecting information on fisheries from as many countries as possible after World War II. Because methods differed from report to report and from country to country, these early data must be examined carefully, but the broad trend depicted here is not a result of changes in reporting.

Chapter 4

1. In this case, "cleaning" refers to the removal of the viscera (livers were kept for oil), the bones, and the head.

2. Unger (1978) argues that Beukelsz is a mythical character and that, though the Dutch did develop improved methods, this occurred somewhat haphazardly over about a century. He also points out that changes in Dutch supremacy in naval power and trade coincide with the growth of herring production, suggesting that control of seaways was an important factor. See section 5.2 for more on military power and fishery development.

3. In fact, cod and herring dominated global landings even in the 1950s and the 1960s (FAO 2012a).

4. Data were compiled from FAO 1955, and all FAO annual reports from 1960 to 1976.

5. Data on unprocessed fish sold fresh are not included in the figure. As described in section 4.2, the volume of fish sold fresh without processing is about 40% of total production, slightly less than the amount processed for human consumption (FAO 2012d).

6. *Whitefish* is a generic term for fish with flesh that turns white when cooked.

7. There is no equivalent dollar amount for this time period. However, using a base year of 1774, £500 would be approximately US$65,700 in 2010.

8. It should be noted that there is some overlap between the different types of processing with regard to food and non-food uses. On the one hand, frozen fish is still used as bait in some fisheries, though most is consumed as food. On the other hand, some fish meals and fish oils are used as dietary supplements and a few products included in the nei category may be consumed as food. In particular, there is little information on the scale of production of fry and fingerlings for use in aquaculture production (FAO 2012c).

9. Note also that divergences between the volume of imports and exports stem from different accounting methods as per FAO documentation (FAO 2012a).

10. Some of this change may be attributable to the fact that data collection and reporting were not as comprehensive early in the dataset (FAO 2012a).

Chapter 5

1. Dolphins associate with tunas only in the eastern Pacific Ocean. No clear scientific explanation for this behavior exists as yet, though there are many hypotheses.

Chapter 6

1. This calculation is based on the real price index for the British pound, which is the most conservative method available. Conversion based on labor value or other indices would result in costs that are much greater than reported here.

2. As indicative of the amount of salt used by processors at the time, Scottish processing facilities set up to implement the Dutch method used 7,500–15,000 tons of salt per year to produce 32,000–69,000 tons of pickled herring (J. T. Jenkins 1920).

3. Technically, the law stipulates that the bounty would be paid per "quintal" or 100 kilograms for dried fish and per barrel for salted and pickled fish.

4. All conversions of nominal US dollars to real 2010 US dollars were carried out using Williamson 2014. The real price index was selected as the most conservative conversion factor.

5. Here again, the real price index is used as a conservative conversion factor. If accounting for the labor value of this money in 1872, the value in 2010 would be $1.7 million while the income value conversion would be $3.7 million (Williamson 2014).

6. The abbreviation grt stands for gross registered tons. This standardized measure of ship size was adopted in 1969. It is an index of the internal volume of the ships capacity. See chapter 2, note 2 for more on its relation to gross tonnages and tons displacement.

7. The currency conversions here are based on Officer 2014 and Williamson 2014. As in previous conversions, the lowest estimated real value was selected as a conservative indicator of the change in value.

Chapter 7

1. Although the concept of limited government intervention in free markets was developed by the Physiocratic school in the 18th century, it was not embraced by governments until the 19th century (Sihley 2008).

Chapter 8

1. Most economists would expect that private property rights should completely eliminate overfishing, but this is based on long-run calculations regarding effort in investment. When fishers evince temporal myopia, as shown by Asche (2001), or if there are ecosystem-level externalities associated with fishing, property rights reduce but do not eliminate the profit disconnect.

2. NGOs tend to be non-profits but it should be noted that they are still motivated to maximize funding, which is necessary to maintain and build their political influence.

3. This fits well with Olson's (1971) work on interest groups more generally, particularly the propensity for privileged and moderate-sized groups to leverage the power

of numbers by working to motivate large latent groups to take action on specific topics.

4. There was also a spring-spawning Icelandic stock, though this fishery seems to have disappeared after a collapse in the 1960s (Matthíasson 2003).

5. This is similar to responsive governance but does not include political expediency as a policy selection factor, and it downplays the role of economic and political problem signals.

Chapter 9

1. Purchasing new technologies can be costly as well, but such investments are considered to be assets while purchase of operating inputs are tallied on the cost side of ledgers.

2. The constant-fertility variant shows what would happen if fertility remained constant over the projection period. In this case, population would increase to more than 25 billion. However, this type of growth is not likely given modern trends in birth rates and mortality (US Census Bureau 2013b). The medium-fertility variant is determined by using probabilistic modeling techniques to predict future changes in fertility levels, whereas the high and low variants are simple deterministic transformations of the medium value (+0.5 children for high; –0.5 children for low; United Nations 2010).

Works Cited

Aanes, Sondre, Kjell Nedreaas, and Sigbjøn Ulvatn. 2011. Estimation of Total Retained Catch Based on Frequency of Fishing Trips, Inspections at Sea, Transshipment, and VMS Data. *ICES Journal of Marine Science* 68 (8): 1598–1605.

Aas, Øystein, ed. 2007. *Global Challenges in Recreational Fisheries*. Oxford: Wiley-Blackwell.

Aberdeen City Council. 2013. "Aberdeen Ships: Fairtry." Accessed February 16, 2013. http://www.aberdeenships.com/single.asp?index=99213.

Acheson, James M. 1988. *The Lobster Gangs of Maine*. Lebanon, NH: University Press of New England.

Acheson, James M. 1997. The Politics of Managing the Maine Lobster Industry: 1860 to the Present. *Human Ecology* 25 (1): 3–27.

Ackman, R. G. 2003. A History of Fats and Oils in Canada. *Lipids* 38 (4): 299–303.

Adams, Walter Marsham. 1884a. A Popular History of Fisheries and Fishermen of All Countries, from the Earliest Times. In *International Fisheries Exhibition*, vol. 1, 461–573. London: William Clowes and Sons.

Adams, Walter Marsham. 1884b. The History and Literature of Fishing. In *International Fisheries Exhibition*, vol. 12, 119–129. London: William Clowes and Sons.

Admiralty and Maritime Law Guide. 2002. "Rules of Oleron." Accessed December 5, 2012. http://www.admiraltylawguide.com/documents/oleron.html.

Agarwal, S. C. 2007. *History of Indian Fishery*. Delhi: Daya.

Agger, Peder, and Else Nielson. 1972. *Preliminary Results Using a Simple Multispecies Model on Sprat (Sprattus Sprattus) and Herring (Clupea Harengus)*. Copenhagen: International Council for the Exploration of the Sea.

Akyeampong, Emmanuel. 2001. *Between the Sea and the Lagoon: An Eco-Social History of the Anlo of Southeastern Ghana, c. 1850 to Recent Times*. Woodbridge: James Currey.

Alder, Jacqueline, Dirk Zeller, and Tony Pitcher. 2002. A Method for Evaluating Marine Protected Area Management. *Coastal Management* 30: 121–131.

Allard, Dean Conrad. 1978. *Spencer Fullerton Baird and the US Fish Commission: A Study in the History of American Science*. New York: Arno.

Allen, Robert C. 2006. *The British Industrial Revolution in Global Perspective: How Commerce Created the Industrial Revolution and Modern Economic Growth*. Cambridge: Cambridge University Press.

Almond, Gabriel Abraham, and Sidney Verba, eds. 1989. *The Civic Culture: Political Attitudes and Democracy in Five Nations*. London: Sage. (1963).

Anderies, John M., Marco A. Janssen, and Elinor Ostrom. 2004. A Framework to Analyze the Robustness of Social-Ecological Systems from an Institutional Perspective. *Ecology and Society* 9 (1): art. 18.

Anderson, Christian N. K., Hsieh Chih-hao, Stuart A. Sandin, Roger Hewitt, Anne Hollowed, John Beddington, Robert M. May, and George Sugihara. 2008. Why Fishing Magnifies Fluctuations in Fish Abundance. *Nature* 452 (7189): 835–839.

Anderson, Emory D. 1998. The History of Fisheries Management and Scientific Advice: The ICNAF/NAFO History from the End of World War II to the Present. *Journal of Northwest Atlantic Fishery Science* 23 (1): 75–94.

Anderson, Knud P., and Erik Ursin. 1977. A Multispecies Extension to the Beverton and Holt Theory of Fishing with Accounts of Phosphorous Circulation and Primary Production. *Meddelelser fra Danmarks Fiskeri-og Havundersogelser* 7: 319–435.

Anderson, Lee G. 1989. Enforcement Issues in Selecting Fisheries Management Policy. *Marine Resource Economics* 6: 261–277.

Anderson, Simon P., Jacob K. Goeree, and Charles A. Holt. 1998. Rent Seeking with Bounded Rationality: An Analysis of the All-Pay Auction. *Journal of Political Economy* 106 (4): 828–853.

Andrade, Tonio. 2004. The Company's Chinese Pirates: How the Dutch East India Company Tried to Lead a Coalition of Pirates to War against China, 1621–1662. *Journal of World History* 15 (4): 415–444.

Anisimov, Oleg A., David G. Vaughan, Terry Callaghan, Christopher Furgal, Harvey Marchant, Travis D. Prowse, Hjalmar Vilhjálmsson, and John E. Walsh. 2007. Polar Regions (Arctic and Antarctic). In *Climate Change 2007: Impacts, Adaptation and Vulnerability*, edited by Roger Barry, Robert Jefferies and John Stone, 653–685. Cambridge: Cambridge University Press.

Anonymous. 1892. Tinned Fish. *British Medical Journal* 1 (1636): 981.

Anonymous, ed. 1955. "International North Pacific Fisheries Commission." Summarized from the papers of William C. Herrington and Wilbert McLeod Chapman. *Polar Record* 7 (49): 335–336.

Annala, John H. 1996. New Zealand's ITQ system: have the first eight years been a success or a failure? *Reviews in Fish Biology and Fisheries* 6 (1): 43–62.

Anspach, Lewis Amadeus. 1819. *A History of the Island of Newfoundland: Containing a Description of the Island, the Banks, the Fisheries, and Trade of Newfoundland, and the Coast of Labrador*. London: T. and J. Allman and J. M. Richardson.

Anthony, Vaughn C. 1990. US New England Groundfish Management Under the Magnuson-Stevens Fishery Conservation and Management Act. *North American Journal of Fisheries Management* 10 (2): 175–185.

Aqorau, Transform, and Anthony Bergin. 1998. The UN Fish Stocks Agreement—A New Era for International Cooperation to Conserve Tuna in the Central Western Pacific. *Ocean Development and International Law* 29 (1): 21–42.

Aranda, Martin. 2009. Evolution and State of the Art of Fishing Capacity Management in Peru: The Case of the Anchoveta Fishery. *Pan American Journal of Aquatic Sciences* 4 (2): 146–153.

Aranda, Martin, Hilario Murua, and Paul De Bruyn. 2012. Managing Fishing Capacity in Tuna Regional Fisheries Management Organizations (RFMOs): Development and State of the Art. *Marine Policy* 36 (5): 985–992.

Armitage, Derek, Melissa Marschke, and Ryan Plummer. 2008. Adaptive Co-Management and the Paradox of Learning. *Global Environmental Change* 18 (1): 86–98.

Armstrong, Terence. 2009. Soviet Sea Fisheries since the Second World War. *Polar Record* 13 (83): 155–186.

Arnason, Ragnar. 1993. The Icelandic Individual Transferable Quota System: A Descriptive Account. *Marine Resource Economics* 8 (3): 201–218.

Arnason, Ragnar. 2000. Economic Instruments for Achieving Ecosystem Objectives in Fisheries Management. *ICES Journal of Marine Science* 57 (3): 742–751.

Arnason, Ragnar. 2006. Property Rights in Fisheries: Iceland's Experience with ITQs. *Reviews in Fish Biology and Fisheries* 15 (3): 243–264.

Asche, Frank. 2001. Fishermen's Discount Rates in ITQ Systems. *Environmental and Resource Economics* 19: 403–410.

Asche, Frank, Lori S. Bennear, Atle Oglend, and Martin D. Smith. 2012. U. S. Shrimp Market Integration. *Marine Resource Economics* 27 (2): 181–192.

Asgeirsdottir, Aslaug. 2008. *Who Gets What? Domestic Influences on International Negotiations Allocating Shared Resources*. Albany: SUNY Press.

Atlantic Shipping. 2013. "Atlantic Shipping A/S—Freezer Trawlers." Accessed March 19, 2013. http://www.atlanticship.dk/sw/frontend/show.asp?layout=4&menu_parent=206920&parent=206920&valuta=&orderby=&subgroup=4.

Augerot, Xanthippe. 2000. *An Environmental History of the Salmon Management Philosophies of the North Pacific: Japan, Russia, Canada, Alaska and the Pacific Northwest United States*. Ann Arbor, MI: Bell & Howell Information and Learning.

Bagust, Phil. 2005. Interactivity, and the 'Big Bird' Races of 2004-2005. *Australian Journal of Communication* 36 (4): 1–17.

Bailey, Jennifer L. 2008. Arrested Development: The Fight to End Commercial Whaling as a Case of Failed Norm Change. *European Journal of International Relations* 14 (2): 289–318.

Baird, Ian G., and Noah Quastel. 2011. Dolphin-Safe Tuna from California to Thailand: Localisms in Environmental Certification of Global Commodity Networks. *Annals of the Association of American Geographers* 101 (2): 337–355.

Baird, Spencer Fullerton. 1889. *The Sea Fisheries of Eastern North America: Extracted from the Annual Report of the Commissioner of Fish and Fisheries for 1886*. Washington, DC: US Government Publications.

Bakun, Andrew, Elizabeth A. Babcock, and Christine Santora. 2009. Regulating a Complex Adaptive System via Its Wasp-Waist: Grappling with Ecosystem-Based Management of the New England Herring Fishery. *ICES Journal of Marine Science* 66 (8): 1768–1775.

Barkin, Samuel J., and Elizabeth R. De Sombre. 2013. *Saving Global Fisheries: Reducing Fishing Capacity to Promote Sustainability*. Cambridge, MA: MIT Press.

Barkin, J. Samuel, and George E. Shambaugh, eds. 1999. *Anarchy and the Environment: The International Relations of Common Pool Resources*. Albany: State University of New York Press.

Barnett, Patricia G. 1943. The Chinese in Southeastern Asia and the Philippines. *Annals of the American Academy of Political and Social Science* 226: 32–49.

Barrett, Scott. 2003. *Environment and Statecraft: The Strategy of Environmental Treaty-Making*. New York: Oxford University Press.

Baumgartner, Frank R., Christian Breunig, Christoffer Green-Pedersen, Bryan D. Jones, Peter B. Mortensen, Michiel Nuytemans, and Stefaan Walgrave. 2009. Punctuated Equilibrium in Comparative Perspective. *American Journal of Political Science* 53 (3): 603–620.

Baumgartner, Frank R., and Bryan D. Jones. 1991. Agenda Dynamics and Policy Subsystems. *Journal of Politics* 53 (4): 1044–1074.

Baumgartner, Frank R., and Bryan D. Jones. 2009. *Agendas and Instability in American Politics*. 2nd ed. Chicago: University of Chicago Press.

Baumgartner, Frank R., Bryan D. Jones, and John Wilkerson. 2011. Comparative Political Studies of Policy Dynamics. *Comparative Political Studies* 44 (8): 947–972.

Beaujon, Anthony. 1884. The History of Dutch Sea Fisheries: Their Progress, Decline and Revival, Especially in Connection with the Legislation in Earlier and Later Times. In *International Fisheries Exhibition*, vol. 9. London: William Clowes and Sons.

Beetham, David. 1991. *The Legitimation of Power*. New York: Palgrave Macmillan.

Bendor, Jonathan, Daniel Diermeier, David A. Siegel, and Michael M. Ting. 2011. *A Behavioral Theory of Elections*. Princeton, NJ: Princeton University Press.

Berghöfer, Augustin, Heidi Wittmer, and Felix Rauschmayer. 2008. Stakeholder Participation in Ecosystem-Based Approaches to Fisheries Management: A Synthesis from European Research Projects. *Marine Policy* 32 (2): 243–253.

Bergin, Anthony. 1997. Albatross Longlining—Managing Seabird Bycatch. *Marine Policy* 21 (1): 63–72.

Berk, Zeki. 2009. *Food Process Engineering and Technology*. New York: Elsevier.

Berkes, Fikret. 1986. Local-Level Management and the Commons Problem: A Comparative Study of Turkish Coastal Fisheries. *Marine Policy* 10 (3): 215–229.

Berkes, Fikret. 1992. Success and Failure in Marine Coastal Fisheries of Turkey. In *Making the Commons Work: Theory, Practice, and Policy*, edited by Daniel W. Bromley, 161–182. San Francisco: ICS Press.

Berkes, Fikret. 1999. *Sacred Ecology: Traditional Ecological Knowledge and Resource Management*. New York: Taylor and Francis.

Berkes, Fikret. 2003. Alternatives to Conventional Management: Lessons from Small Scale Fisheries. *Environments* 3 (1): 5–19.

Berkes, Fikret. 2009. Evolution of Co-Management: Role of Knowledge Generation, Bridging Organizations and Social Learning. *Journal of Environmental Management* 90 (5): 1692–1702.

Berkes, Fikret. 2012. Implementing Ecosystem-Based Management: Evolution or Revolution? *Fish and Fisheries* 13 (4): 465–476.

Berkes, Fikret, and Carl Folke. 2002. Back to the Future: Ecosystem Dynamics and Local Knowledge. In *Panarchy: Understanding Transformations in Human and Natural Systems*, edited by Lance H. Gunderson and C. S. Holling, 121–146. Washington, DC: Island.

Berkes, F., T. P. Hughes, R. S. Steneck, J. A. Wilson, D. R. Bellwood, B. Crona, C. Folke, 2006. Globalization, Roving Bandits, and Marine Resources. *Science* 311 (5767): 1557–1558.

Bertrand, Arnaud, François Gerlotto, Sophie Bertrand, Mariano Gutiérrez, Luis Alza, Andres Chipollini, Erich Díaz, Pepe Espinoza, Jesu's Ledesma, Roberto Quesque'n, Salvador Peraltilla, and Fracisco Cavez. 2008. Schooling Behaviour and Environmental Forcing in Relation to Anchoveta Distribution: An Analysis across Multiple Spatial Scales. *Progress in Oceanography* 79: 264–277.

Bestor, Theodore C. 2004. *Tsukiji: The Fish Market at the Center of the World*. Berkeley: University of California Press.

Billfish Foundation, The. 2014. "The Billfish Foundation." Accessed February 8, 2014. http://www.billfish.org.

Bjørndal, Trond. 2009. Overview, Roles, and Performance of the North East Atlantic Fisheries Commission (NEAFC). *Marine Policy* 33 (4): 685–697.

Bjørndal, Trond, and Jon M. Conrad. 1987. The Dynamics of an Open Access Fishery. *Canadian Journal of Economics* 20 (1): 74–85.

Blake, H. P. 1884. Improved Facilities for the Capture, Economic Transmission, and Distribution of Sea Fisheries. In *International Fisheries Exhibition*, vol. 10, 417–468. London: William Clowes and Sons.

Bodansky, Daniel. 2004. Deconstructing the Precautionary Principle. In *Bringing New Law to Ocean Waters*, edited by David D. Caron and Harry H. Schreiber, 381–391. Boston: Martinus Nijhoff.

Boerma, L. K., and J. A. Gulland. 1973. Stock Assessment of the Peruvian Anchovy (*Engraulis Ringens*) and Management of the Fishery. *Journal of the Fisheries Research Board of Canada* 30 (12): 2226–2235.

Borch, Trude, Øystein Aas, and David Policansky. 2008. International Fishing Tourism: Past, Present, and Future. In *Global Challenges in Recreational Fisheries*, edited by Øystein Aas, 268–291. Oxford: Blackwell.

Borisov, P. G. 1964. *Fisheries Research in Russia: A Historical Survey*. Jerusalem: Israel Program for Scientific Translations.

Brady, Mark, and Staffan Waldo. 2009. Fixing Problems in Fisheries–Integrating ITQs, CBM and MPAs in Management. *Marine Policy* 33 (2): 258–263.

Brander, K. M. 2007. Global Fish Production and Climate Change. *Proceedings of the National Academy of Sciences of the United States of America* 104 (50): 44–46.

Breisach, Ernst. 2007. *Historiography: Ancient, Medieval, and Modern*, 3rd ed. Chicago: University of Chicago Press.

Brody, Samuel D., Sammy Zahran, Arnold Vedlitz, and Himanshu Grover. 2008. Examining the Relationship between Physical Vulnerability and Public Perceptions of Global Climate Change. *Environment and Behavior* 40 (1): 72–95.

Bromhead, Don, Jennifer Foster, Rachel Attard, James J. Findlay, and John J. Kalish. 2003. *A Review of the Impact of Fish Aggregating Devices (FADs) on Tuna Fisheries: Final Report to Fisheries Resource Research Fund*. Canberra: Bureau of Rural Sciences.

Bromley, Daniel W., ed. 1992. *Making the Commons Work: Theory, Practice, and Policy*. San Francisco: ICS Press.

Browman, Howard I., and Konstantinos I. Stergiou. 2005. Politics and Socio-Economics of Ecosystem-Based Management of Marine Resources. *Marine Ecology Progress Series* 300 (1): 241–296.

Brown, J., L. Fernand, K. J. Horsburgh, A. E. Hill, and J. W. Read. 2001. Paralytic Shellfish Poisoning on the East Coast of the UK in Relation to Seasonal Density-Driven Circulation. *Journal of Plankton Research* 23 (1): 105–116.

Brown, James. 2005. An Account of the Dolphin-Safe Tuna Issue in the UK. *Marine Policy* 29 (1): 39–46.

Buchary, Eny Anggraini, Wai Lung Cheung, Ussif Rashid Sumaila, and Tony J. Pitcher. 2003. Back to the Future: A Paradigm Shift for Restoring Hong Kong's Marine Ecosystem. *American Fisheries Society Symposium* 38: 727–746.

Buck, Eugene H. 1995. *Atlantic Bluefin Tuna: International Management of a Shared Resource*. Washington, DC: Congressional Record Service.

Buck, Susan J. (1985) 2010. No Tragedy on the Commons. Reprinted in *Green Planet Blues: Four Decades of Global Environmental Politics*, 4th ed., edited by Ken Conca and Geoffrey D. Dabelko, 46–54. Boulder: Westview.

Burd, A. C. 1985. Recent Changes in the Central and Southern North Sea Herring Stocks. *Canadian Journal of Fisheries and Aquatic Sciences* 42 (Suppl. 1): 192–206.

Burger, Joanna, Alan H. Stern, Carline Dixon, Christopher Jeitner, Sheila Shukla, Sean Burke, and Michael Gochfeld. 2004. Fish Availability in Supermarkets and Fish Markets in New Jersey. *Science of the Total Environment* 333: 89–97.

Burke, Lenore, and John Phyne. 2008. Made in Japan: The Japanese Market and Herring Roe Production and Management in Canada's Southern Gulf of St. Lawrence. *Marine Policy* 32 (1): 79–88.

Burkhard, Benjamin, Silvia Opitz, Hermann Lenhart, Kai Ahrendt, Stefan Garthe, Bettina Mendel, and Wilhelm Windhorst. 2011. Ecosystem Based Modeling and Indication of Ecological Integrity in the German North Sea—Case Study Offshore Wind Parks. *Ecological Indicators* 11 (1): 168–174.

Burnett, D. Graham. 2012. *The Sounding of the Whale*. Chicago: The University of Chicago Press.

Burns, G. F., and H. Williams. 1975. *Clostridium Botulinum* in Scottish Fish Farms and Farmed Trout. *Journal of Hygiene* 74 (1): 1–6.

Caddy, J. F. 2002. Limit Reference Points, Traffic Lights, and Holistic Approaches to Fisheries Management with Minimal Stock Assessment Input. *Fisheries Research* 56 (2): 133–137.

Caddy, J. F., and K. L. Cochrane. 2001. A Review of Fisheries Management Past and Present and Some Future Perspectives for the Third Millennium. *Ocean and Coastal Management* 44 (9–10): 653–682.

Caddy, J. F., and R. Mahon. 1995. *Reference Points for Fisheries Management*. FAO Fisheries Technical Paper 347. Rome: Food and Agriculture Organization of the United Nations.

Caddy, John F., and Luca Garibaldi. 2000. Apparent Changes in the Trophic Composition of World Marine Harvests: The Perspective from the FAO Capture Database. *Ocean and Coastal Management* 43 (8–9): 615–655.

Campbell, James Duncan. 1884. The Fisheries of China. *International Fisheries Exhibition*, vol. 5. London: William Clowes and Sons.

Campbell, Lisa M. 2007. Local Conservation Practice and Global Discourse: A Political Ecology of Sea Turtle Conservation. *Annals of the Association of American Geographers* 97 (2): 313–334.

Campling, Liam. 2012. The Tuna 'Commodity Frontier': Business Strategies and Environment in the Industrial Tuna Fisheries of the Western Indian Ocean. *Journal of Agrarian Change* 12 (2–3): 252–278.

Campling, Liam, Elizabeth Havice, and Penny McCall Howard. 2012. The Political Economy and Ecology of Capture Fisheries: Market Dynamics, Resource Access and Relations of Exploitation and Resistance. *Journal of Agrarian Change* 12 (2–3): 177–203.

Carneiro, Gonçalo. 2011. "They Come, They Fish, and They Go" EC Fisheries Agreements with Cape Verde and Sao Tomé e Principe. *Marine Fisheries Review* 73 (4): 1–25.

Cash, David W., William C. Clark, Frank Alcock, Nancy M. Dickson, Noelle Eckley, David H. Guston, Jill Jäger, and Ronald B. Mitchell. 2003. Knowledge Systems for Sustainable Development. *Proceedings of the National Academy of Sciences of the United States of America* 100 (14): 8086–8091.

Cassou, Steven P., and Stephen F. Hamilton. 2004. The Transition from Dirty to Clean Industries: Optimal Fiscal Policy and the Environmental Kuznets Curve. *Journal of Environmental Economics and Management* 48 (3): 1050–1077.

Catarci, Camillo. 2004. The World Tuna Industry—an Analysis of Imports and Prices, and of Their Combined Impact on Catches and Tuna Fishing Capacity. In *Second Meeting of the Technical Advisory Committee of the FAO Project: Management of Tuna Fishing Capacity: Conservation and Socio-Economics, March 15–18, 2004*, FAO Fisheries Proceedings 2, edited by William H. Bayliff, Juan Ignaciode Leiva Moreno, and Jacek Majkowski, 233–278. Rome: Food and Agriculture Organization of the United Nations.

Catsat. 2013. "Catsat—Worldwide Satellite Service to Help Fishing." Accessed February 25, 2013. http://www.catsat.com.

Chapman, Wilbert McLeod. 1966. Politics and the Marine Fisheries. In *Fisheries of North America*. Washington, DC, April 30-May 5, 1965. Washington, DC: US Department of the Interior.

Chavez, Francisco P., John Ryan, Salvador E. Lluch-Cota, and Ñiquen C. Miguel. 2003. From Anchovies to Sardines and Back: Multidecadal Change in the Pacific Ocean. *Science* 299 (5604): 217–221.

Cheung, William W. L., Vicky W. Y. Lam, Jorge L. Sarmiento, Kelly Kearney, Reg Watson, Dirk Zeller, and Daniel Pauly. 2010. Large-Scale Redistribution of Maximum Fisheries Catch Potential in the Global Ocean under Climate Change. *Global Change Biology* 16 (1): 24–35.

Chicken of the Sea. 2013. "About Chicken of the Sea." Accessed April 28, 2013. http://chickenofthesea.com/company.aspx.

Christian, Claire, David Ainley, Megan Bailey, Paul Dayton, John Hocevar, Michael Le Vine, Jordan Nikoloyuk, Claire Nouvian, Enriqueta Velarde, Rodolfo Werner, and Jennifer Jacquet. 2013. A Review of Formal Objections to Marine Stewardship Council Fisheries Certifications. *Biological Conservation* 161: 10–17.

Christensen, Anne-Sofie, and Jesper Raakjaer. 2006. Fishermen's Tactical and Strategic Decisions. *Fisheries Research* 81 (2–3): 258–267.

Christie, Patrick, David L. Fluharty, Alan T. White, Liza Eisma-Osorio, and William Jatulan. 2007. Assessing the Feasibility of Ecosystem-Based Fisheries Management in Tropical Contexts. *Marine Policy* 31 (3): 239–250.

Christy, Francis T. 1973. *Fishermen Catch Quotas: A Tentative Suggestion for Domestic Management*. Law of the Sea Institute Occasional Paper 19. Providence: University of Rhode Island.

Christy, Francis T. (1982) 1992. *Territorial Use Rights in Marine Fisheries: Definitions and Conditions*. FAO Technical Paper 227. Rome: Food and Agriculture Organization of the United Nations.

Christy, Francis T., and Anthony Scott. (1965) 2011. *The Common Wealth in Ocean Fisheries: Some Problems of Growth and Economic Allocation*, 3rd ed. Washington, DC: RFF Press.

Chu, Cindy. 2009. Thirty Years Later: The Global Growth of ITQs and Their Influence on Stock Status in Marine Fisheries. *Fish and Fisheries* 10 (2): 217–230.

CIA. 1959. *Construction and Imports of Fish Factory Trawlers for the Soviet Fishing Fleet*. Washington, DC: The Central Intelligence Agency.

Cisneros-Montemayor, Andrés M., and U. Rashid Sumaila. 2010. A Global Estimate of Benefits from Ecosystem-Based Marine Recreation: Potential Impacts and Implications for Management. *Journal of Bioeconomics* 12 (3): 245–268.

Clapp, Jennifer. 2011. Environment and Global Political Economy. In *Global Environmental Politics: Concepts Theories and Case Studies*, edited by Gabriela Kutting, 42–55. New York: Routledge.

Clapp, Jennifer, and Peter Dauvergne. 2011. *Paths to a Green World: The Political Economy of the Global Environment*, 2nd ed. Cambridge: MIT Press.

Clapp, Jennifer, and Doris Fuchs. 2009. Agrifood Corporations, Global Governance, and Sustainability: A Framework for Analysis. In *Corporate Power in Global Agrifood Governance*, edited by Jennifer Clapp and Doris Fuchs, 1–26. Cambridge: MIT Press.

Clapp, Jennifer, and Eric Helleiner. 2012. International Political Economy and the Environment: Back to the Basics? *International Affairs* 88 (3): 485–501.

Clark, Colin W. 1973. The Economics of Overexploitation. *Science* 181 (4100): 630–634.

Clark, Colin W. 2005. *Methematical Bioeconomics: Optimal Management of Renewable Resources*, 2nd ed. Hoboken, NJ: Wiley.

Clark, Colin W. 2006. *The Worldwide Crisis in Fisheries: Economic Models and Human Behavior*. Cambridge: Cambridge University Press.

Clark, Colin W., and R. Lamberson. 1982. An Economic History and Analysis of Pelagic Whaling. *Marine Policy* 6 (2): 103–120.

Clark, Colin W., and Marc Mangel. 1979. Aggregation and Fishery Dynamics: A Theoretic Study of Schooling and the Purse Seine Tuna Fisheries. *Fish Bulletin* 77 (2): 317–337.

Clark, Colin W., and Gordon R. Munro. 1975. The Economics of Fishing and Modern Capital Theory: A Simplified Approach. *Journal of Environmental Economics and Management* 2: 92–106.

Clark, Colin W., and Gordon R. Munro. 1980. Fisheries and the Processing Sector: Some Implications for Management Policy. *Bell Journal of Economics* 11 (2): 603–616.

Cleland, John. 2013. World Population Growth: Past, Present and Future. *Environmental and Resource Economics* 55 (4): 543–554.

Coase, Ronald H. 1960. The Problem of Social Cost. *Journal of Law & Economics* 3: 1–44.

Cobb, John Nathan. 1906. *The Commercial Fisheries of Alaska in 1905*. Bureau of Fisheries Document No. 603. Washington, DC: Government Printing.

Cobb, John Nathan. 1919. *The Canning of Fishery Products*. Seatte: Miller Freeman.

Cobb, John Nathan. 1921. *Pacific Salmon Fisheries*. Bureau of Fisheries Document No. 902. Washington, DC: Government Printing Office.

Cochrane, K. L. 1999. Complexity in Fisheries and Limitations in the Increasing Complexity of Fisheries Management. *ICES Journal of Marine Science* 56: 917–926.

Cohen, Michael D., James G. March, and Johan P. Olsen. 1972. A Garbage Can Model of Organizational Choice. *Science* 17 (1): 1–25.

Coker, Robert Ervin. 1910. The Fisheries and the Guano Industry of Peru. *Bulletin of the Bureau of Fisheries* 663: 335–365.

Cole, Scott, Sergei Ismalkov, and Eric Sjöberg. 2014. Games in the Arctic: Applying Game Theory Insights to Arctic Challenges. *Polar Research* 1: 1–13.

Collier, David. 2011. Understanding Process Tracing. *PS: Political Science & Politics* 44 (4): 823–830.

Collins, E. J. T. 1993. Food Adulteration and Food Safety in Britain in the 19th and Early 20th Centuries. *Food Policy* 18 (2): 95–109.

Conrad, Jon M. 2010. *Resource Economics*, 2nd ed. Cambridge: Cambridge University Press.

Constance, Douglas H., and Alessandro Bonanno. 1999. Contested Terrain of the Global Fisheries: "Dolphin-Safe" Tuna, the Panama Declaration, and the Marine Stewardship Council. *Rural Sociology* 64 (4): 597–623.

Constance, Douglas H., and Alessandro Bonanno. 2000. Regulating the Global Fisheries: The World Wildlife Fund, Unilever, and the Marine Stewardship Council. *Agriculture and Human Values* 17: 125–139.

Cooke, Steven J., and Ian G. Cowx. 2006. Contrasting Recreational and Commercial Fishing: Searching for Common Issues to Promote Unified Conservation of Fisheries Resources and Aquatic Environments. *Biological Conservation* 128 (1): 93–108.

Cooper, William D., Robert G. Morgan, Alonzo Redman, and Margart Smith. 2002. Capital Budgeting Models: Theory vs. Practice. *Business Forum* 26 (1/2): 15–19.

Cornish, Thomas. 1884. Mackerel and Pilchard Fisheries. In *International Fisheries Exhibition*, vol. 6, 109–146. London: William Clowes and Sons.

Cox, Anthony, and Carl-Christian Schmidt. 2002. Subsidies in the OECD Fisheries Sector: A Review of Recent Analysis and Future Directions. Background paper, *FAO Expert Consultation on Identifying, Assessing and Reporting on Subsidies in the Fishing Industry, December 3–6, 2002*. Rome: Food and Agriculture Organization of the United Nations.

Cox, Anthony, and U. Rashi Sumalia. 2010. A Review of Fishery Subsidies: Quantification, Impacts and Reform. In *Handbook of Marine Fisheries Conservation and Management*, edited by R. Quentin Grafton, Ray Hilborn, Dale Squires, Maree Tait, and Meryl Williams, 99–112. New York: Oxford University Press.

Craig, John, Sandra Anderson, Mick Clout, Bob Creese, Neil Mitchell, John Ogden, Mere Roberts, and Graham Ussher. 2000. Conservation Issues in New Zealand. *Annual Review of Ecology and Systematics* 31: 61–78.

Criddle, Keith R., and Seth Macinko. 2000. A Requiem for the IFQ in US Fisheries? *Marine Policy* 24 (6): 461–469.

Crowder, Larry B., Elliott L. Hazen, Naomi Avissar, Rhema Bjorkland, Catherine Latanich, and Matthew B. Ogburn. 2008. The Impacts of Fisheries on Marine Ecosystems and the Transition to Ecosystem-Based Management. *Annual Review of Ecology Evolution and Systematics* 39: 259–278.

Crutchfield, J. A. 1979. Economic and Social Implications of the Main Policy Alternatives for Controlling Fishing Effort. *Journal of the Fisheries Research Board of Canada* 36: 742–752.

CSIRO. 2013. "Remote Sensing Home Page." Accessed February 20, 2013. http://www.marine.csiro.au/~lband.

Cushing, D. H. 1977. The Atlantic Fisheries Commissions. *Marine Policy* 1 (3): 230–238.

Cushing, D. H. 1980. European Fisheries. *Marine Pollution Bulletin* 11 (11): 311–315.

Dallas, Lynne L. 2012. Short-Termism, the Financial Crisis, and Corporate Governance. *Journal of Corporation Law* 37 (2): 267–362.

Daly, Herman E. 1987. The Economic Growth Debate: What Some Economists Have Learned But Many Have Not. *Journal of Environmental Economics and Management* 14 (4): 323–336.

Daly, Herman E. 1996. *Beyond Growth: The Economics of Sustainable Development.* Boston: Beacon Press.

Dame, James K., and Robert R. Christian. 2006. Uncertainty and the Use of Network Analysis for Ecosystem-Based Fishery Management. *Fisheries* 31 (7): 331–341.

Dangers in Tinned Tuna. 1970. *Nature* 228 (5278): 1247.

Davies, A. R. 1997. Modified-Atmosphere Packing of Fish and Fish Products. In *Fish Processing Technology*, 2nd ed., edited by G. M. Hall, 200–223. London: Blackie Academic and Professional.

Davies, R. W. D., S. J. Cripps, A. Nickson, and G. Porter. 2009. Defining and Estimating Global Marine Fisheries Bycatch. *Marine Policy* 33 (4): 661–672.

Davis, Anthony, and Kenneth Ruddle. 2012. Massaging the Misery: Recent Approaches to Fisheries Governance and the Betrayal of Small-Scale Fisheries. *Human Organization* 71 (3): 244–254.

Day, Francis. 1873. *Report on the Sea Fish and Fisheries of India and Burma.* Calcutta, India: Office of the Superintendent of Government Printing.

Day, Francis. 1884. Indian Fish and Fishing. In *International Fisheries Exhibition*, vol. 2, 441–534. London: William Clowes and Sons.

Degnbol, Poul, Henrik Gislason, Susan Hanna, Svein Jentoft, Jesper Raakjær Nielsen, Sten Sverdrup-Jensen, and Douglas Clyde Wilson. 2006. Painting the Floor with a Hammer: Technical Fixes in Fisheries Management. *Marine Policy* 30 (5): 534–543.

Delvos, Oliver. 2006. WTO Disciplines and Fisheries Subsidies—Should the "SCM Agreement" Be Modified? *Victoria University of Wellington Law Review* 37: 331–364.

Dercole, Fabio, Charlotte Prieu, and Sergio Rinaldi. 2010. Technological Change and Fisheries Sustainability: The Point of View of Adaptive Dynamics. *Ecological Modelling* 221 (3): 379–387.

De Sombre, Elizabeth R. 1995. Baptists and Bootleggers for the Environment: The Origins of United States Unilateral Sanctions. *Journal of Environment & Development* 4 (1): 53–75.

De Sombre, Elizabeth R. 2000. *Domestic Sources of International Environmental Policy: Industry, Environmentalists, and U. S. Power.* Cambridge, MA: MIT Press.

De Sombre, Elizabeth R. 2002. *The Global Environment and World Politics.* London: Continuum.

De Sombre, Elizabeth R. 2005. Fishing under Flags of Convenience: Using Market Power to Increase Participation in International Regulation. *Global Environmental Politics* 5 (4): 73–94.

De Vos, Birgit I., and Simon R. Bush. 2011. Far More than Market-Based: Rethinking the Impact of the Dutch Viswijzer (Good Fish Guide) on Fisheries' Governance. *Sociologia Ruralis* 51 (3): 284–303.

Dewees, Christopher M. 1998. Effects of Individual Quota Systems on New Zealand and British Columbia Fisheries. *Ecological Applications* 8 (1): 133–138.

Dietz, Rob, Dan O'Neill, and Herman Daly. 2013. *Enough Is Enough: Building a Sustainable Economy in a World of Finite Resources*. San Francisco: Berrett-Koehler.

Digital Globe. 2013. "Digital Globe › Solutions › By Industry › Marine Services › Sea Star Fisheries Information Service." Accessed February 11, 2013. http://www.geoeye.com/Corp_Site/solutions/by-industry/marine-services/seastar-fisheries-information-service.aspx.

Douvere, Fanny. 2008. The Importance of Marine Spatial Planning in Advancing Ecosystem-Based Sea Use Management. *Marine Policy* 32 (5): 762–771.

Downs, Anthony. 1972. Up and Down with Ecology—The "Issue-Attention Cycle." *Public Interest* 28: 38–50.

Drouard, Alain. 2009. The History of the Sardine-Canning Industry in France in the Nineteenth and Twentieth Centuries. In *Exploring the Food Chain: Food Production and Food Processing in Western Europe 1850–1990*, edited by Jan Bieleman, Yves Segers and Erik Buyst, 177–190. Turnhout, Belgium: Brepols.

Duff, R. W. 1884. The Herring Fisheries of Scotland. In *International Fisheries Exhibition*, vol. 6, 69–108. London: William Clowes and Sons.

Dunlap, Riley E., and Richard York. 2012. The Globalization of Environmental Concern. In *Comparative Environmental Politics: Theory Practice and Prospects*, edited by Paul F. Steinberg and Stacy D. Van Deveer, 89–112. Cambridge, MA: MIT Press.

Dyson, Tom. 2010. *Population and Development: The Demographic Transition*. London: Zed Books.

Edwards, Martin, Gregory Beaugrand, Pierre Helaouët, Jürgen Alheit, and Stephen Coombs. 2013. Marine Ecosystem Response to the Atlantic Multidecadal Oscillation. *PLoS One* 8 (2): e57212.

EIA. 2012. "Annual Energy Outlook 2013 Early Release." *US Energy Information Administration*. Accessed March 13, 2013. http://www.eia.gov/forecasts/aeo/er/tables_ref.cfm.

EIA. 2013. "Short-Term Energy Outlook." *US Energy Information Administration*. Accessed March 12, 2013. http://www.eia.gov/forecasts/steo/realprices.

Ellickson, Robert C. 1991. *Order Without Law: How Neighbors Settle Disputes*. Cambridge, MA: Harvard University Press.

Enticknap, Ben, Ashley Blacow, Geoff Shester, Whit Sheard, Jon Warrenchuk, Mike LeVine, and Susan Murray. 2011. *Forage Fish: Feeding the California Current Large Marine Ecosystem*. Monterey, CA: Oceana.

Esty, Daniel C., and Andrew S. Winston. 2009. *Green to Gold: How Smart Companies Use Environmental Strategy to Innovate, Create Value, and Build Competitive Advantage*, 2nd ed. Hoboken: Wiley.

Eythórsson, Einar. 1996. Theory and Practice of ITQs in Iceland: Privatization of Common Fishing Rights. *Marine Policy* 20 (3): 269–281.

Eythórsson, Einar. 2000. A Decade of ITQ-Management in Icelandic Fisheries: Consolidation without Consensus. *Marine Policy* 24 (6): 483–492.

Fallon, Liza D., and Lorne K. Kriwoken. 2004. International Influence of an Australian Nongovernment Organization in the Protection of Patagonian Toothfish. *Ocean Development and International Law* 35 (3): 221–266.

FAO. 1947. *Yearbook of Fishery Statistics*, vol. 1. Rome: Food and Agriculture Organization of the United Nations.

FAO. 1952. *Yearbook of Fishery Statistics*, vol. 4. Rome: Food and Agriculture Organization of the United Nations.

FAO. 1954. *Yearbook of Fishery Statistics*, vol. 5. Rome: Food and Agriculture Organization of the United Nations.

FAO. 1960a. *Yearbook of Fishery Statistics*, vol. 12. Rome: Food and Agriculture Organization of the United Nations.

FAO. 1960b. *Yearbook of Fishery Statistics*, vol. 13. Rome: Food and Agriculture Organization of the United Nations.

FAO. 1962. *Yearbook of Fishery Statistics*, vol. 15. Rome: Food and Agriculture Organization of the United Nations.

FAO. 1977. *Yearbook of Fishery Statistics*, vol. 44. Rome: Food and Agriculture Organization of the United Nations.

FAO. 1986. *The Production of Fish Meal and Oil*. FAO Fisheries Technical Paper 142. Rome: Food and Agriculture Organization of the United Nations.

FAO. 1995. *Code of Conduct for Responsible Fisheries*. Rome: Food and Agriculture Organization of the United Nations.

FAO. 1999. "International Plan of Action for the Management of Fishing Capacity." Accessed August 24, 2014. http://www.fao.org/docrep/006/x 3170e/x 3170e 04.htm.

FAO. 2001. "International Plan of Action to Prevent, Deter and Eliminate Illegal, Unreported and Unregulated Fishing." Accessed August 24, 2014. http://www.fao.org/docrep/003/y 1224e/y 1224e 00.htm.

FAO. 2002a. *The State of World Fisheries and Aquaculture*. Rome: Food and Agriculture Organization of the United Nations.

FAO. 2002b. Expert Consultation on Identifying, Assessing, and Reporting on Subsidies in the Fishing Industry, FAO Fisheries and Aquaculture Department, vol. 698. Rome: Food and Agriculture Organization of the United Nations

FAO. 2003. The Ecosystem Approach to Fisheries. In *FAO Technical Guidelines for Responsible Fisheries*, vol. 4, suppl. 2, chpt. 2.. Rome: Food and Agriculture Organization of the United Nations.

FAO. 2006. *The State of World Fisheries and Aquaculture*. Rome: Food and Agriculture Organization of the United Nations.

FAO. 2009. *Yearbook of Fishery and Aquaculture Statistics*. Rome: Food and Agriculture Organization of the United Nations.

FAO. 2010a. *The State of World Fisheries and Aquaculture*. Rome: Food and Agriculture Organization of the United Nations.

FAO. 2010b. *Yearbook of Fishery and Aquaculture Statistics. FAO*. Rome: Food and Agriculture Organization of the United Nations.

FAO. 2011. "FAO Fisheries and Aquaculture Atlas of Tuna and Billfish Catches, Overview." *FAO Statistics Division*. Accessed February 3, 2013. http://www.fao.org/fishery/statistics/tuna-atlas/en.

FAO. 2012a. "Fishstat J: Universal Software for Fishery Statistical Time Series." *FAO Statistics Division*. Accessed February 11, 2012, Dataset updated February 11, 2013. http://www.fao.org/fishery/statistics/software/fishstatj/en.

FAO. 2012b. "FAOSTAT." *FAO Statistics Division*. Accessed November 27, 2012, http://faostat 3.fao.org/home/index.html#METADATA_CLASSIFICATION.

FAO. 2012c. "Small-Scale and Artisanal Fisheries." *FAO Fisheries and Aquaculture Department*. Accessed December 30, 2012. http://www.fao.org/fishery/topic/14753/en.

FAO. 2012d. *The State of World Fisheries and Aquaculture*. Rome: Food and Agriculture Organization of the United Nations.

FAO. 2013. "Regional Fishery Bodies (RFB)." *FAO Fisheries and Aquaculture Department*. Accessed March 2, 2014. http://www.fao.org/fishery/rfb/en.

FAO. 2014. "FAO—EAFnet—EAF Projects." FAO Fisheries and Aquaculture Department. Accessed February 6, 2014. http://www.fao.org/fishery/eaf-net/topic/166265/en.

Felando, August, and Harold Medina. 2011. *The Tuna/Porpoise Controversy*. San Diego: Western Sky Press.

Fichou, Jean-Christohpe. 2004. Le Front Populaire et les Conserveurs de Sardines de Bretagne. *Annales de Bretagne et Des Pays de l'Quest* 1: 111–125.

Finley, Carmel. 2009. The Social Construction of Fishing, 1949. *Ecology and Society* 14 (1): art. 6.

Finley, Carmel. 2011. *All the Fish in the Sea: Maximum Sustainable Yield and the Failure of Fisheries Management*. Chicago: University of Chicago Press.

Finucane, Melissa L., Ali Alhakami, Paul Slovic, and Stephen M. Johnson. 2000. The Affect Heuristic in Judgments of Risks and Benefits. *Journal of Behavioral Decision Making* 13 (1): 1–17.

Fisheries of North America, The. 1966. Proceedings of the First North American Fisheries Conference. Washington, DC, April 30-May 5, 1965. Washington, DC: US Department of the Interior.

Fish Wise. 2014. "Fish Wise." Accessed February 15, 2014. http://www.fishwise.org.

Fitzmaurice, Andrew. 2014. *Sovereignty, Property and Empire, 1500–2000*. Cambridge: Cambridge University Press.

Folke, Carl, Thomas Hahn, Per Olsson, and Jon Norberg. 2005. Adaptive Governance of Social-Ecological Systems. *Annual Review of Environment and Resources* 30: 441–473.

Fraker, Mark A., and Bruce R. Mate. 1999. Seals, Sea Lions, and Salmon in the Pacific Northwest. In *Conservation and Management of Marine Mammals*, edited by John R. Twiss and Randall R. Reeves, 156–178. Washington, DC: Smithsonian Institution Press.

Fraser, Evan D. G. 2007. Travelling in Antique Lands: Using Past Famines to Develop an Adaptability/Resilience Framework to Identify Food Systems Vulnerable to Climate Change. *Climatic Change* 83 (4): 495–514.

Friedheim, Robert L., ed. 2001. *Toward a Sustainable Whaling Regime*. Seattle: University of Washington Press.

Fréon, Pierre, Marilú Bouchon, Christian Mullon, Christian García, and Miguel Ñiquen. 2008. Interdecadal Variability of Anchoveta Abundance and Overcapacity of the Fishery in Peru. *Progress in Oceanography* 79 (2–4): 401–412.

Frid, Chris L. J., Odette A. L. Paramor, and Catherine L. Scott. 2006. Ecosystem-Based Management of Fisheries: Is Science Limiting? *ICES Journal of Marine Science* 63: 1567–1572.

Frohman, Arthur H., John Mehos, Eric Turnill, and Wendell Earle. 1966. A Panel Considers: 'Sell up to Higher Profits.' In *Fisheries of North America,* 37–44. Washington, DC, April 30–May 5, 1965. Washington, DC: US Department of the Interior

Fryer, Charles E. 1884. The Salmon Fisheries. In *International Fisheries Exhibition,* vol. 2, 277–358. London: William Clowes and Sons.

Fulton, Thomas Wemyss. 1911. *The Sovereignty of the Sea: An Historical Account of the Claims of England to the Dominion of the British Seas, and of the Evolution of the Territorial Waters.* Edinburgh: William Blackwood and Sons.

Furuno. 2013. "Products—Depth and Fish Finders." Accessed February 20, 2013. http://www.furunousa.com/Products/Products.aspx?category=Products+%3a+Depth+%26+Fish+Finders.

Gadgil, Madhav, Per Olsson, Fikret Berkes, and Carl Folke. 2003. Exploring the Role of Local Ecological Knowledge in Ecosystem Management: Three Case Studies. In *Navigating Social-Ecological Systems: Building Resilience for Complexity and Change,* edited by Fikre Berkes, Johan Colding and Carl Folke, 189–209. Cambridge: Cambridge University Press.

Gallagher, Deborah Rigling, and Erika Weinthal. 2012. Business-State Relations and the Environment: The Evolving Role of Corporate Social Responsibility. In *Comparative Environmental Politics: Theory Practice and Prospects,* edited by Paul F. Steinberg and Stacy D. Van Deveer, 143–170. Cambridge, MA: MIT Press.

Garcia, Serge M., and Ignacio De Leiva Moreno. 2001. Global Overview of Marine Fisheries. In *Reykjavik Conference on Responsible Fisheries in the Marine Ecosystem, 1–4 October 2001.* Rome: Food and Agriculture Organization of the United Nations.

Garthwaite, G. A. 1997. Chilling and Freezing of Fish. In *Fish Processing Technology,* 2nd ed., edited by G. M. Hall, 93–118. London: Blackie Academic and Professional.

Gezelius, Stig S., and Maria Hauck. 2011. Toward a Theory of Compliance in State-Regulated Livelihoods: A Comparative Study of Compliance Motivations in Developed and Developing World Fisheries. *Law & Society Review* 45 (2): 435–470.

Gibbs, William Edward. 1922. *The Fishing Industry.* London: Isaac Pitman & Sons.

Giddens, Anthony. 1979. *Central Problems in Social Theory: Action, Structure, and Contradiction in Social Analysis.* Berkeley: University of California Press.

Gilbert, Charles H. 1912. *Age at Maturity of Pacific Coast Salmon of the Genus Oncorhynchus.* Washington, DC: Bureau of Fisheries.

Gilman, Eric L. 2011. Bycatch Governance and Best Practice Mitigation Technology in Global Tuna Fisheries. *Marine Policy* 35 (5): 590–609.

Gilovich, Thomas, Dale W. Griffen, and Daniel Kahneman, eds. 2013. *Heuristics and Biases: The Psychology of Intuitive Judgement*, 14th ed. Cambridge: Cambridge University Press.

Glantz, Michael H. 1979. Science, Politics and Economics of the Peruvian Anchoveta Fishery. *Marine Policy* 3 (3): 201–210.

Goldenberg, Suzanne. 2014. "Marine Mining: Underwater Gold Rush Sparks Fears of Ocean Catastrophe." *The Guardian*, March 1. Accessed March 3, 2014. http://www.theguardian.com/environment/2014/mar/02/underwater-gold-rush-marine-mining-fears-ocean-threat.

González-Laxe, Fernando. 2005. The Precautionary Principle in Fisheries Management. *Marine Policy* 29 (6): 495–505.

Goode, G. Brown. 1884. A Review of the Fishery Industries of the United States and the Work of the United States Fish Commission. In *International Fisheries Exhibition*, vol. 5, 1–82. London: William Clowes and Sons.

Gopinath, Munisamy, Daniel Pick, and Yonghai Li. 2003. Concentration and Innovation in the US Food Industries. *Journal of Agricultural & Food Industrial Organization* 1 (1): 1–21.

Gordon, David M. 2006. *Nachituti's Gift: Economy, Society, and Environment in Central Africa*. Madison: University of Wisconsin Press.

Gordon, H. Scott. 1954. The Economic Theory of a Common-Property Resource: The Fishery. *Journal of Political Economy* 62 (2): 124–142.

Goss, Jasper, David Burch, and Roy E. Rickson. 2000. Agri-Food Restructuring and Third World Transnationals: Thailand, the CP Group and the Global Shrimp Industry. *World Development* 28 (3): 513–530.

Gosse, Philip. 2007. *The History of Piracy*, 2nd ed. Mineola, NY: Dover.

Gould, Richard A. 2011. *Archaeology and The Social History of Ships*, 2nd ed. Cambridge, MA: Cambridge University Press.

Gould, W. T. S. 2009. *Population and Development*. New York: Routledge.

Graham, Michael. 1935. Modern Theory of Exploiting a Fishery, and Application to North Sea Trawling. *ICES Journal of Marine Science* 10 (3): 264–274.

Graham, Michael. 1943. *The Fish Gate*. London: Faber and Faber.

Green, H. J. 1884. The Herring Fisheries. In *International Fisheries Exhibition*, vol. 11, 127–168. London: William Clowes and Sons.

Greenpeace. 2014. "Oceans: Greenpeace International." http://www.greenpeace.org/international/en/campaigns/oceans.

Grieve, Chris, and Katherine Short. 2007. *Implementation of Ecosystem-Based Management in Marine Capture Fisheries*. Gland, Switzerland: Global Marine program of WWF International.

Grossman, Gene M., and Elhanan Helpman. 1994. Endogenous Innovation in the Theory of Growth. *Journal of Economic Perspectives* 8 (1): 23–44.

Grossman, Gene M., and Elhanan Helpman. 2001. *Special Interest Politics*. Cambridge, MA: MIT Press.

Gruszczynski, Lukasz. 2012. Trade, Investment and Risk: Re-Tuning Tuna? Appellate Body Report in US – Tuna II. *European Journal of Risk Regulation* 3: 430–437.

Guerra, A. 1972. Fisheries Management in Peru. *Marine Pollution Bulletin* 3 (3): 39–40.

Gulland, J. A. 1971. Science and Fishery Management. *ICES Journal of Marine Science* 33 (3): 471–477.

Gulland, J. A. 1974. *The Management of Marine Fisheries*. Seattle: University of Washington Press.

Gunderson, Lance H., and C. S. Holling, eds. 2002. *Panarchy: Understanding Transformations in Human and Natural Systems*. Washington: Island Press.

Gunderson, Lance H., C. S. Holling, and Garry D. Peterson. 2002. Surprises and Sustainability: Cycles of Renewal in the Everglades. In *Panarchy: Understanding Transformations in Human and Natural Systems*, edited by Lance H. Gunderson and C. S. Holling, 315–332. Washington, DC: Island Press.

Haakonsen, Jan M. 1992. Industrial vs Artisanal Fisheries in West Africa: The Lessons to Be Learnt. In *Fishing for Development: Small Scale Fisheries in Africa*, edited by Inge Tvedten and Björn Hersoug, 33–53. Uppsala, Sweden: Nordiska Afrikainstitutet.

Haas, Peter M. 1989. Do Regimes Matter? Epistemic Communities and Mediterranean Pollution Control. *International Organization* 43 (3): 377–403.

Haas, Peter M. 1992. Introduction: Epistemic Communities and International Policy Coordination. *International Organization* 46 (1): 1–35.

Haas, Peter M., Robert O. Keohane, and Marc A. Levy, eds. 1995. *Institutions for the Earth: Sources of Effective International Environmental Protection*, 3rd ed. Cambridge, MA: MIT Press.

Hall, G., and N. Ahmad. 1997. Surimi and Fish-Mince Products. In *Fish Processing Technology*, 2nd ed., edited by G. M. Hall, 74–92. New York: VHC Publishers.

Hall, Martin A., Dayton L. Alverson, and Kaija I. Metuzals. 2000. By-Catch: Problems and Solutions. *Marine Pollution Bulletin* 41 (1–6): 204–219.

Hamilton, Amanda, Antony Lewis, Mike A. McCoy, Elizabeth Havice, and Liam Campling. 2011. *Market and Industry Dynamics in the Global Tuna Supply Chain*. Honiara, Solomon Islands: Pacific Islands Forum Fisheries Agency.

Hamilton, Lawrence C., Steingrímur Jónsson, Helga Ögmundardóttir, and Igor M. Belkin. 2004. Sea Changes Ashore: The Ocean and Iceland's Herring Capital. *Human Dimensions of the Arctic System* 57 (4): 325–335.

Hanes, Phyllis. 1980. "Julia Child Finds New Faces at the Fish Market." *The Christian Science Monitor*, February 4. http://www.csmonitor.com/1980/0204/020450.html.

Hannesson, Rognvaldur. 2004. *The Privatization of the Oceans*. Cambridge, MA: MIT Press.

Hannesson, Rögnvaldur, Kjell G. Salvanes, and Dale Squires. 2010. Technological Change and the Tragedy of the Commons: The Lofoten Fishery over 130 Years. *Land Economics* 86 (4): 746–765.

Hardin, Garrett. 1968. Tragedy of the Commons. *Science* 162 (3859): 1243–1248.

Hardin, Russell. 1995. *One for All: The Logic of Group Conflict*. Princeton, NJ: Princeton University Press.

Harrison, John C., and Jason Schratweiser. 2008. The Role of Non-Governmental Organizations in Recreational Fisheries Management: Challenges, Responsibilities, and Possibilities. In *Global Challenges in Recreational Fisheries*, edited by Øystein Aas, 324–337. Oxford: Blackwell.

Hart, Paul J. B., and Tony J. Pitcher. 2001. Conflict, Consent, and Cooperation: An Evolutionary Perspective on Individual Human Behavior in Fisheries Management. In *Reinventing Fisheries Management*, edited by Tony J. Pitcher, Paul J. B. Hart, and Daniel Pauly, 215–226. Dordrecht: Kluwer.

Hatcher, Aaron. 2000. Subsidies for European Fishing Fleets: the European Community's Structural Policy for Fisheries 1971–1999. *Marine Policy* 24 (2): 129–140.

Hawkshaw, Robert Stephen, Sarah Hawkshaw, and U. Rashid Sumaila. 2012. The Tragedy of the "Tragedy of the Commons": Why Coining Too Good a Phrase Can Be Dangerous. *Sustainability* 4 (12): 3141–3150.

Hayden, Anne, and Philip Conkling. 2007. Toward Ecosystem-Based Management. *National Fisherman* 88 (2): 9.

Hennessey, T., and M. Healey. 2000. Ludwig's Ratchet and the Collapse of New England Groundfish Stocks. *Coastal Management* 28 (3): 187–213.

Hérubel, Marcel Adolphe. 1912. *Sea Fisheries: Their Treasures and Toilers*. London: T. F. Unwin.

Hervas, Susana, Kai Lorenzen, Michael A. Shane, and Mark A. Drawbridge. 2010. Quantiative Assessment of a White Seabass (*Atractoscion nobilis*) Stock Enhancement Program in California: Post-Release Dispersal, Growth and Survival. *Fisheries Research* 105: 237–243.

Hewison, Grant J. 1996. The Precautionary Approach to Fisheries Management: An Environmental Perspective. *International Journal of Marine and Coastal Law* 11 (3): 301–332.

Higgs, Robert. 1987. *Crisis and Leviathan: Critical Episodes in the Growth of an American Government.* New York: Oxford University Press.

Hilborn, Ray, Jean-Jacques Maguire, Ana M. Parma, and Andrew A. Rosenberg. 2001. The Precautionary Approach and Risk Management: Can They Increase the Probability of Successes in Fishery Management? *Canadian Journal of Fisheries and Aquatic Sciences* 58 (1): 99–107.

Hilborn, Ray, J. M. Lobo Orensanz, and Ana M. Parma. 2005. Institutions, Incentives and the Future of Fisheries. *Philosophical Transactions of the Royal Society of London. Series B, Biological Sciences* 360 (1453): 47–57.

Hilborn, Ray, and Carl J. Walters. 1992. *Quantitative Fisheries Stock Assessment: Choice, Dynamics and Uncertainty.* New York: Chapman and Hall.

Hirschman, Albert O. 1970. *Exit, Voice, and Loyalty: Responses to Decline in Firms, Organizations, and States.* Cambridge, MA: Harvard University Press.

Hjort, Johan. 1932. A Brief History of Whaling. *Polar Record* 1 (3): 28–32.

Hobart, W. L. 1996. *Baird's Legacy: The History and Accomplishments of NOAA's National Marine Fisheries Service, 1871–1996.* Washington, DC: National Marine Fisheries Service.

Holdsworth, Edmund W. H. 1874. *Deep-Sea Fishing and Fishing Boats.* London: Edward Stanford.

Holland, Daniel S., and Jon G. Sutinen. 1999. An Empirical Model of Fleet Dynamics in New England Trawl Fisheries. *Canadian Journal of Fisheries and Aquatic Sciences* 56 (2): 253–264.

Hollick, Ann L. 1981. *US Foreign Policy and the Law of the Sea.* Princeton: Princeton University Press.

Holling, C. S. 1973. Resilience and Stability of Ecological Systems. *Annual Review of Ecology and Systematics* 4: 1–23.

Holt, Sidney. 1978. Marine Fisheries. In *Ocean Yearbook 11*, edited by Elisabeth Mann Borgese and Norton Ginsburg, 38–83. Chicago: University of Chicago Press.

Holt, Sidney. 1985. Whale Mining, Whale Saving. *Marine Policy* 9 (3): 192–213.

Holt, Sidney J. 2004. Foreword to the 2004 edition. In *On the Dynamics of Exploited Fish Stocks*, 2nd edition, edited by Raymond J. H. Beverton and Sidney J. Holt, i–xxiii. Caldwell, NJ: Blackburn Press.

Homans, Frances R., and James E. Wilen. 2005. Markets and Rent Dissipation in Regulated Open Access Fisheries. *Journal of Environmental Economics and Management* 49 (2): 381–404.

Hondex. 2013. "Welcome to Hondex World." Accessed February 20, 2013. http://www.hondex.co.jp.

Horner, W. F. A. 1997. Canning Fish and Fish Products. In *Fish Processing Technology*, 2nd ed., edited by G. M. Hall, 119–159. London: Blackie Academic and Professional.

Horowitz, Roger, Jeffrey M. Pilcher, and Sydney Watts. 2004. Meat for the Multitudes: Market Culture in Paris, New York City, and Mexico City over the Long Nineteenth Century. *American Historical Review* 109 (4): 1055–1083.

Howarth, Richard B. 2009. Discounting, Uncertainty, and Revealed Time Preference. *Land Economics* 85 (1): 24–40.

Howell, David L. 1995. *Capitalism from Within: Economy, Society, and the State in a Japanese Fishery*. Berkeley: University of California Press.

Hubrecht, Ambrosius Arnold Willem. 1884. Oyster Culture and Oyster Fisheries in the Netherlands. In *International Fisheries Exhibition*, vol. 5., 83–112. London: William Clowes and Sons.

Hurd, Archibald, and Henry Castle. 1913. *German Sea-Power: Its Rise, Progress, and Economic Basis*. London: John Murray.

Huxley, Thomas. 1884. Inaugural Address by Professor Huxley. In *International Fisheries Exhibition*, vol. 4, 1–22. London: William Clowes and Sons.

IATTC. 1954. Sixth Meeting Inter-American Tropical Tuna Commission. August 11, 1954. San Jose, Costa Rica. La Jolla, CA: Inter-American Tropical Tuna Commission.

IATTC. 1955. Summary Minutes of the Seventh Meeting of the Inter-American Tropical Tuna Commission. July 14, Panama City, Panama. La Jolla, CA: Inter-American Tropical Tuna Commission.

IATTC. 1956. Summary Minutes of the Eighth Meeting of the Inter-American Tropical Tuna Commission. July 30, San Diego, California. La Jolla, CA: Inter-American Tropical Tuna Commission.

IATTC. 1957. Summary Minutes of the Ninth Meeting of the Inter-American Tropical Tuna Commission. March 12, San Jose, Costa Rica. La Jolla, CA: Inter-American Tropical Tuna Commission.

IATTC. 1958. Summary Minutes of the Tenth Meeting of the Inter-American Tropical Tuna Commission. February 11, Panama City, Panama. La Jolla, CA: Inter-American Tropical Tuna Commission.

IATTC. 1959. Summary Minutes of the Eleventh Meeting of the Inter-American Tropical Tuna Commission. February 5, San Pedro, California. La Jolla, CA: Inter-American Tropical Tuna Commission.

IATTC. 1960. Summary Minutes of the Twelfth Meeting of the Inter-American Tropical Tuna Commission. February 23–24, San José, Costa Rica. La Jolla, CA: Inter-American Tropical Tuna Commission.

IATTC. 1961. *Annual Report of the Inter-American Tropical Tuna Commission.* La Jolla, CA: Inter-American Tropical Tuna Commission.

IATTC. 1969. *Annual Report of the Inter-American Tropical Tuna Commission.* La Jolla, CA: Inter-American Tropical Tuna Commission.

IATTC. 1992. *IATTC Resolution to the La Jolla Agreement.* La Jolla, CA: Inter-American Tropical Tuna Commission.

IATTC. 2001. Dolphin-Safe Program Press Release. San Salvador, El Salvador La Jolla, CA: Inter-American Tropical Tuna Commission.

IATTC. 2013a. "AIDCP Dolphin Safe." Accessed February 9, 2014. https://www.iattc.org/DolphinSafeENG.htm.

IATTC. 2013b. "IATTC Resolutions and Recommendations 1998-2013." Accessed February 9, 2014. http://iattc.org/Resolutions ENG.htm.

IBISWorld. 2014. "Seafood Preparation in the US." Accessed October 25, 2014. http://clients 1.ibisworld.com/reports/us/industry/default.aspx?entid=257.

ICCAT. 1966. International Convention for the Conservation of Tropical Tunas. *Official Journal of the European Communities* 162: 35–42.

ICCAT. 2008. ICCAT 40th Anniversary Report 1966-2006. Madrid: International Commission for the Conservation of Atlantic Tunas

ICES. 2012. ICES Historical Catch Statistics 1903–1949. *International Council for the Exploration of the Sea.* Accessed February 5, 2013. http://www.ices.dk/fish/CATCHSTATISTICS.asp.

Imperial, Mark T., Tim Hennessey, and Donald Robadue, Jr. 1993. The Evolution of Adaptive Management for Estuarine Ecosystems: The National Estuary Program and Its Precursors. *Ocean and Coastal Management* 20 (2): 147–180.

Inglehart, Ronald. 1995. Public Support for Environmental Protection: Objective Problems and Subjective Values in 43 Societies. *PS: Political Science & Politics* 28 (1): 57–72.

Innis, Harold Adams. 1940. *The Cod Fisheries: The History of an International Economy*. New Haven, CT: Yale University Press.

International Fisheries Exhibition. 1884. Compiled Literature of the International Fisheries Exhibition, London 1883, in 12 volumes. London: William Clowes and Sons.

INTERPOL. 2013. "Project Scale Projects Environmental Crime Crime Areas Internet Home—INTERPOL." Accessed February 15, 2014. http://www.interpol.int/Crime-areas/Environmental-crime/Projects/Project-Scale.

Islam, Rafiqul M. 1991. The Proposed "Driftnet-Free Zone" in the South Pacific and the Law of the Sea Convention. *International and Comparative Law Quarterly* 40 (1): 184–198.

ITLOS. 2013. Tribunal Invites States Parties to the Convention, the Sub-Regional Fisheries Commission and Other Intergovernmental Organizations to Present Written Statements on IUU Fishing Activities by 29 November 2013. Press Release 194. Hamburg: International Tribunal for the Law of the Sea.

Jackson, Jeremy B. C. 2001. What Was Natural in the Coastal Oceans? *Proceedings of the National Academy of Sciences of the United States of America* 98 (10): 5411–5418.

Jacquet, Jennifer, John Hocevar, Sherman Lai, Patricia Majluf, Nathan Pelletier, Tony Pitcher, Enric Sala, Rashid Sumaila, and Daniel Pauly. 2009. Conserving Wild Fish in a Sea of Market-Based Efforts. *Oryx* 44 (1): 45–56.

Jakobsson, J., and G. Stefánsson. 1999. Management of Summer-Spawning Herring off Iceland. *ICES Journal of Marine Science* 56: 827–833.

Jasanoff, Sheila S. 1987. Contested Boundaries in Policy-Relevant Science. *Social Studies of Science* 17 (2): 195–230.

Jeffers, Jennifer. 2010. Climate Change and the Arctic: Adapting to Changes in Fisheries Stocks and Governance Regimes. *Ecology Law Quarterly* 37 (3): 917–978.

Jefferson, Thomas, and Lorenzo Sabine. (1791) 2011. *Cod and Whale Fisheries: Report of Hon. Thomas Jefferson, Secretary of State, on the Subject of Cod and Whale Fisheries, Made to the House of Representatives, February 1, 1791. Also, Report of Lorenzo Sabine, Esq., on the Principal Fisheries of the American Seas*. Charleston: Nabu Press.

Jenkins, James Travis. 1920. *The Sea Fisheries*. London: Constable and Company.

Jenkins, James Travis. 1921. *A History of the Whale Fisheries: From the Basque Fisheries of the Tenth Century to the Hunting of the Finner Whale at the Present Date*. London: H. F. & G. Whitherby.

Jenkins, Lekelia D. 2012. Reducing Sea Turtle Bycatch in Trawl Nets: A History of NMFS Turtle Excluder Device (TED) Research. *Marine Fisheries Review* 74 (2): 26–45.

Jensson, Paul. 1988. Daily Production Planning in Fish Processing Firms. *European Journal of Operational Research* 36: 410–415.

Jentoft, Svein. 2000. Legitimacy and Disappointment in Fisheries Management. *Marine Policy* 24 (2): 141–148.

Jentoft, Svein, Bonnie J. McCay, and Douglas C. Wilson. 1998. Social Theory and Fisheries Co-Management. *Marine Policy* 22 (4–5): 423–436.

Jentoft, Svein, and Ratana Chuenpagdee. 2009. Fisheries and Coastal Governance as a Wicked Problem. *Marine Policy* 33 (4): 553–560.

JFIC. 1999. "Satellite." *Japan Foundation Information Center.* Accessed http://www.jafic.or.jp/sat/index.html.

Johnston, Douglas M. 1965. *The International Law of Fisheries: A Framework for Policy-Oriented Inquiries*, 2nd ed. New Haven, CT: Yale University Press.

Joncas, Louis Z. 1884. The Fisheries of Canada. In *International Fisheries Exhibition*, vol. 5, 113–168. London: William Clowes and Sons.

Jones, Bryan D., and Frank R. Baumgartner. 2005. *The Politics of Attention: How Government Prioritizes Problems*. Chicago: University of Chicago Press.

Jones, Dixon D., and Carl J. Walters. 1976. Catastrophe Theory and Fisheries Regulation. *Journal of the Fisheries Research Board of Canada* 33 (12): 2829–2833.

Joseph, James. 2003. Managing Fishing Capacity of the World Tuna Fleet. In *FAO Fisheries Circular 982*. Rome: Food and Agriculture Organization of the United Nations.

Joyner, Christopher C., and Zachary Tyler. 2000. Marine Conservation versus International Free Trade: Reconciling Dolphins with Tuna and Sea Turtles with Shrimp. *Ocean Development and International Law* 31 (1–2): 127–150.

Juda, Lawrence. 1996. *International Law and Ocean Use Management*. London: Routledge.

Juda, Lawrence. 1999. Considerations in Developing a Functional Approach to the Governance of Large Marine Ecosystems. *Ocean Development and International Law* 30 (2): 89–125.

Lane, Daniel E. 1988. Investment Decision Making by Fishermen. *Canadian Journal of Fisheries and Aquatic Sciences* 45: 782–796.

Larkin, P. A. 1977. An Epitaph for the Concept of Maximum Sustainable Yield. *Transactions of the American Fisheries Society* 106 (1): 1–11.

Lasswell, Harold Dwight. (1936) 2011. *Politics: Who Gets What, When, and How*. Whitefish, MT: Literary Liscensing, LLC.

Lauer, Matthew, and Shankar Aswani. 2010. Indigenous Knowledge and Long-Term Ecological Change: Detection, Interpretation, and Responses to Changing Ecological Conditions in Pacific Island Communities. *Environmental Management* 45 (5): 985–997.

Lavigne, David M., Victor B. Scheffer, and Stephen R. Kellert. 1999. The Evolution of North American Attitudes toward Marine Mammals. In *Conservation and Management of Marine Mammals*, edited by John R. Twiss, Jr. and Randall R. Reeves, 10–47. Washington, DC: Smithsonian Institution Press.

Le Gal, Yves. 2009. "2009: The Concarneau Marine Biology Laboratory Celebrates Its 150th Anniversary." *Paris: La lettre du Collège de France*. Accessed May 19, 2013. http://lettre-cdf.revues.org/779.

Le Gallic, Bertrand. 2008. The Use of Trade Measures against Illicit Fishing: Economic and Legal Considerations. *Ecological Economics* 64 (4): 858–866.

Lejano, Raul P. 2006. *Frameworks for Policy Analysis: Merging Text and Context*. New York: Routledge.

Lehodey, P., J. Alheit, M. Barange, T. Baumgartner, G. Beaugrand, K. Drinkwater, J. M. Fromentin, 2006. Climate Variability, Fish, and Fisheries. *Journal of Climate* 19 (20): 5009–5030.

Lett, F., and A. C. Kohler. 1976. Recruitment: A Problem of Multispecies Interaction and Environmental Perturbation with Special Reference to Gulf of St. Lawrence Herring (Culpea *hurengus L.*). In *Proceedings of the 26th Annual Meeting of the International Commission for the Northwest Atlantic Fisheries*, no. 3763, June, Dartmouth, Canada.

Levinson, Marc. 2006. *The Box: How the Shipping Container Made the World Smaller and the World Economy Bigger*. Princeton, NJ: Princeton University Press.

Levy, David L., and Peter J. Newell, eds. 2005. *The Business of Global Environmental Governance*. Cambridge, MA: MIT Press.

Lindblom, Charles E. 1977. *Politics and Markets: The World's Political-Economic Systems*. New York: Basic Books.

Lindegren, Martin, Rabea Diekmann, and Christian Möllmann. 2010. Regime Shifts, Resilience and Recovery of a Cod Stock. *Marine Ecology Progress Series* 402: 239–253.

Liu, Jianguo, Thomas Dietz, Stephen R. Carpenter, Carl Folke, Marina Alberti, Charles L. Redman, Stephen H. Schneider, Elinor Ostrom, Alice N. Pell, Jane Lubchenco, William Taylor, Zhiyun Ouyang, Peter Deadman, Timothy Kratz, and William Provencher. 2007. Coupled Human and Natural Systems. *Ambio* 36 (8): 639–649.

Lowe, Justin. 1996. Earth Island Institute. *Environment: Science and Policy for Sustainable Development* 38 (9): 43–44.

Ludwig, Donald, Ray Hilborn, and Carl Waters. 1993. Uncertainty, Resource Exploitation, and Conservation: Lessons from History. *Science* 260 (5104): 17–19.

Lugten, Gail. 2010. *The Role of International Fishery Organizations and Other Bodies in the Conservation and Management of Living.* Aquatic Resources. FAO Fisheries and Aquaculture Circular no. 1054. Rome: Food and Agriculture Organization of the United Nations.

Lyman, Jonathan. 2008. Subsistence versus Sport: Cultural Conflict on the Frontiers of Fishing. In *Global Challenges in Recreational Fisheries,* edited by Øystein Aas, 292–302. Oxford: Blackwell.

Mabee, Bryan. 2009. Pirates, Privateers and the Political Economy of Private Violence. *Global Change, Peace & Security* 21 (2): 139–152.

MacNeil, M. Aaron, Nicholas A. J. Graham, Joshua E. Cinner, Nicholas K. Dulvy, Philip A. Loring, Simon Jennings, Nicholas V. C. Polunin, Aaron T. Fisk, and Tim R. McClanahan. 2010. Transitional States in Marine Fisheries: Adapting to Predicted Global Change. *Philosophical Transactions of the Royal Society of London. Series B, Biological Sciences* 365 (1558): 3753–3763.

Maitland, James Ramsey Gibson, and Francis Day. 1884. Fish Culture. In *International Fisheries Exhibition,* vol. 2, 1–118. London: William Clowes and Sons.

Makino, Mitsutaku. 2011. *Fisheries Management in Japan: Its Institutional Features and Case Studies.* New York: Springer.

Mancini, Agnese, Jesse Senko, Ricardo Borquez-Reyes, Juan Guzman Póo, Jeffrey A. Seminoff, and Volker Koch. 2011. To Poach or Not to Poach an Endangered Species: Elucidating the Economic and Social Drivers Behind Illegal Sea Turtle Hunting in Baja California Sur, Mexico. *Human Ecology* 39 (6): 743–756.

Mansfield, Becky. 2003. Spatializing Globalization: A 'Geography of Quality' in the Seafood Industry. *Economic Geography* 79 (1): 1–16.

Marasco, Richard J., Daniel Goodman, Churchill B. Grimes, Peter W. Lawson, Andre E. Punt, and Terrance J. Quinn, II. 2007. Ecosystem-Based Fisheries Management: Some Practical Suggestions. *Canadian Journal of Fisheries and Aquatic Sciences* 64 (6): 928–939.

March, James, and Herbert Simon. (1958) 1993. *Organizations,* 2nd ed. Cambridge, MA: Blackwell.

Marine Stewardship Council. 2014. "Certified Sustainable Seafood." Accessed February 15, 2014. http://www.msc.org.

Marx, Karl. (1887) 1977. *Capital,* vol. 1. New York: Vintage Books.

Mason, Fred. 2002. The Newfoundland Cod Stock Collapse: A Review and Analysis of Social Factors. *Electronic Green Journal* 1 (17): 1–21.

Mathiesen, Karl. 2013. "Russia Blocks Antarctica Marine Sanctuary Plan." *The Guardian*, July 16. Accessed March 4, 2014. http://www.theguardian.com/world/2013/jul/16/antarctica-marine-reserve-plan.

Matthíasson, Thórólfur. 2003. Closing the Open Sea: Development of Fishery Management in Four Icelandic Fisheries. *Natural Resources Forum* 27 (1): 1–18.

Maunder, Mark N. 2003. Paradigm Shifts in Fisheries Stock Assessment: From Integrated Analysis to Bayesian Analysis and Back Again. *Natural Resource Modeling* 16 (4): 465–475.

Maury, Matthew Fontaine. 1851. Whale Chart [preliminary Sketch]. Scale Not Given. *National Observer*.

May, R. M., ed. 1984. *Exploitation of Marine Communities: Report of the Dahlem Workshop on Exploitation of Marine Communities, Berlin 1984, April 1–6,* Life Sciences Research Report 32. New York: Springer-Verlag.

May, Robert M., John R. Beddington, Colin W. Clark, Sidney J. Holt, and Richard M. Laws. 1979. Management of Multispecies Fisheries. *Science* 205 (4403): 267–277.

McClanahan, Tim, and Juan Carlos Castilla. 2007. *Fisheries Management: Progress Toward Sustainability*. Oxford: Wiley-Blackwell.

McDonald, Mary G. 2000. Food Firms and Food Flows in Japan 1945–98. *World Development* 28 (3): 487–512.

McFarland, Raymond. 1911. *A History of the New England Fisheries with Maps*. New York: D. Appleton.

McGoodwin, James R. 1990. *Crisis in the World's Fisheries: People, Problems, and Policies*. Stanford, CA: Stanford University Press.

Meadowcroft, James. 2012. Greening the State? In *Comparative Environmental Politics: Theory Practice and Prospects*, edited by Paul F. Steinberg and Stacy D. Van Deveer, 63–88. Cambridge, MA: MIT Press.

Meadows, Donella, Jorgen Randers, and Dennis Meadows. 2004. *Limits to Growth: The 30-Year Update*. White River Junction, VT: Chelsea Green.

Meadows, Donella H., Dennis L. Meadows, Jorgen Randers, and William W. Behrens, III. 1974. *The Limits to Growth*. London: Pan Books.

Meliado, Fabrizio. 2012. Fisheries Management Standards in the WTO Fisheries Subsidies Talks: Learning How to Discipline Environmental PPMs? *Journal of World Trade* 46 (5): 1083–1146.

Michener, William K., Elizabeth R. Blood, Keith L. Bildstein, Mark M. Brinson, and Leonard R. Gardner. 1997. Climate Change, Hurricanes and Tropical Storms, and Rising Sea Level in Coastal Wetlands. *Ecological Applications* 7 (3): 770–801.

Milazzo, Matteo. 1998. *Subsidies in World Fisheries: A Reexamination.* World Bank Technical Paper no. 406, Fisheries Series. Washington, DC: World Bank.

Milliman, Scott R. 1986. Optimal Fishery Management in the Presence of Illegal Activity. *Journal of Environmental Economics and Management* 13 (4): 363–381.

Ministry of Economic Affairs. 1958. *Fisheries of Taiwan, Republic of China.* Taipei: Ministry of Economic Affairs, the Republic of China.

Ministry of the Environment. 1999. "Til laks åt alle kan ingen gjera? (lit: To Eat Salmon All Can Not Do?)" *Norwegian Ministry of the Environment.* Accessed July 12, 2013. http://www.regjeringen.no/nb/dep/md/dok/nou-er/1999/nou-1999-09/7.html?id=141607.

Miyake, Makoto Peter, Patrice Guillotreau, Chin-Hwa Sun, and Gakushi Ishimura. 2010. *Recent Developments in the Tuna Industry.* FAO Technical Paper 543. Rome: Food and Agriculture Organization of the United Nations.

Miyake, Makoto Peter, Naozumi Miyabe, and Hideki Nakano. 2004. *Historical Trends of Tuna Catches in the World.* FAO Technical Paper 467. Rome: Food and Agriculture Organization of the United Nations.

Molenaar, Erik J. 2012. Arctic Fisheries and International Law: Gaps and Options to Address Them. *Carbon and Climate Law Review* 1: 63–77.

Moloney, David G., and Peter H. Pearse. 1979. Quantitative Rights as an Instrument for Regulating Commercial Fisheries. *Journal of the Fisheries Research Board of Canada* 36 (7): 859–866.

Monterey Bay Aquarium. 2014. "Seafood Watch". Accessed February 15, 2014. http://www.seafoodwatch.org/cr/seafoodwatch.aspx.

Moon, Hyo-Bang, and Hee-Gu Choi. 2009. Human Exposure to PCDDs, PCDFs and Dioxin-like PCBs Associated with Seafood Consumption in Korea from 2005 to 2007. *Environment International* 35 (2): 279–284.

Moore, Jeffrey E., Bryan P. Wallace, Rebecca L. Lewison, Ramúnas Žydelis, Tara M. Cox, and Larry B. Crowder. 2009. A Review of Marine Mammal, Sea Turtle and Seabird Bycatch in USA Fisheries and the Role of Policy in Shaping Management. *Marine Policy* 33 (3): 435–451.

Moser, Susanne C. 2007. In the Long Shadows of Inaction: The Quiet Building of a Climate Protection Movement in the United States. *Global Environmental Politics* 7 (2): 124–144.

Moser, Susanne C., and John Tribbia. 2006. Vulnerability to Inundation and Climate Change Impacts in California: Coastal Managers' Attitudes and Perceptions. *Marine Technology Society Journal* 40 (4): 35–44.

Munro, Gordon R., and Trond Bjørndal. 1999. The Economics of Fisheries Management: A Survey. In *The International Yearbook of Environmental and Resource Economics 1998/1999: A Survey of Current Issues*, edited by Tom Tietenberg and Henk Folmer, 153–188. Cheltenham, England: Edward Elgar.

Murphy, Dale D. 2006. The Tuna-Dolphin Wars. *Journal of World Trade* 40 (4): 597–618.

Murphy, Garth I. 1972. Fisheries in Upwelling Regions: With Special Reference to Upwelling Waters. *Geoforum* 3 (3): 63–71.

Murray, Steven N., Richard E. Ambrose, James A. Bohnsack, Louis W. Botsford, Mark H. Carr, Gary E. Davis, Paul K. Dayton, 1999. No-Take Reserve Networks: Sustaining Fishery Populations and Marine Ecosystems. *Fisheries* 24 (11): 11–25.

Muscolino, Micah S. 2009. *Fishing Wars and Environmental Change in Late Imperial and Modern China*. Cambridge, MA: Harvard University Asian Center.

Myers, Nancy. 2002. The Precautionary Principle Puts Values First. *Bulletin of Science, Technology & Society* 22: 210–219.

NASA. 2013a. "NASA Earth Observatory: Home." *National Aeronautics and Space Administration*. Accessed February 20, 2013. http://earthobservatory.nasa.gov.

NASA. 2013b. "On Earth: Jet Propulsion Laboratory California Institute of Technology." *National Aeronautics and Space Administration*. Accessed February 20, 2013. http://onearth.jpl.nasa.gov.

Nash, Richard D. M., Mark Dickey-Collas, and Laurence T. Kell. 2009. Stock and Recruitment in North Sea Herring (*Clupea Harengus*); Compensation and Depensation in the Population Dynamics. *Fisheries Research* 95 (1): 88–97.

Naylor, Rosamond, and Marshall Burke. 2005. Aquaculture and Ocean Resources: Raising Tigers of the Sea. *Annual Review of Environment and Resources* 30 (1): 185–218.

Newell, Allen, and Herbert A. Simon. 1972. *Human Problem Solving*. Englewood Cliffs, NJ: Prentice-Hall.

Newell, Dianne. 1997. *Tangled Webs of History: Indians and the Law in Canada's Pacific Coast Fisheries*. Toronto: University of Toronto Press.

NFUSO. 2013. "NFUSO Homepage." *Nakano Engineering Co., Ltd.* Accessed February 20, 2013. http://www.nakanoeng.com/fuso-marine/index-t.htm.

O. J. B. 1937. International Agreement for the Regulation of Whaling. *Nature* 12 (3): 180–181.

Ocean Imaging. 2013. "Corporate Profile." Accessed February 11, 2013. http://www.oceani.com/Company Info.html.

O'Connor, Sue, Rintaro Ono, and Chris Clarkson. 2011. Pelagic Fishing at 42, 000 Years before the Present and the Maritime Skills of Modern Humans. *Science* 334 (6059): 1117–1121.

Oda, Shigeru. 1989. *International Control of Sea Resources*. Leiden, Netherlands: A. W. Stythoff.

OECD. 1965. *Financial Support to the Fishing Industry*. Paris: Organization for Economic Co-operation and Development.

OECD. 1970. *Fishery Subsidies and Economics*. Paris: Organization for Economic Co-operation and Development.

Officer, Lawrence H. 2013. "Dollar-Pound Exchange Rate From 1791." *Measuring Worth*. Accessed March 10, 2013. http://measuringworth.com/exchangepound.

Officer, Lawrence H. 2014. "Exchange Rates Between the United States Dollar and Forty-One Currencies." *Measuring Worth*. Accessed August 26, 2014. http://www.measuringworth.com/datasets/exchangeglobal/result.php?year_source=1960&year_result=2010&country E%5B%5D=Japan.

Officer, Lawrence H., and Samuel H. Williamson. 2014a. "Computing 'Real Value' Over Time With a Conversion Between U. K. Pounds and U. S. Dollars, 1774 to Present." *Measuring Worth*. Accessed August 21, 2014. http://www.measuringworth.com/exchange.

Officer, Lawrence H., and Samuel H. Williamson. 2014b. "Six Ways to Compute the Relative Value of a UK Pound Amount, 1270 to Present." *Measuring Worth*. Accessed August 21, 2014. http://www.measuringworth.com/ukcompare.

Okoshi, Narimori. 1884. A Sketch of the Fisheries of Japan. In *International Fisheries Exhibition*, vol. 5, 187-224. London: William Clowes and Sons.

Olson, Mancur. 1971. *The Logic of Collective Action: Public Goods and the Theory of Groups*. Cambridge, MA: Harvard University Press.

Olson, Mancur. 2000. *Power and Prosperity: Outgrowing Communist and Capitalist Dictatorships*. New York: Basic Books.

O'Neill, Kate. 2012. The Comparative Study of Environmental Movements. In *Comparative Environmental Politics: Theory Practice and Prospects*, edited by Paul F. Steinberg and Stacy D. Van Deveer, 115–142. Cambridge, MA: MIT Press.

Orbach, Michael K. 1977. *Hunters, Seamen, and Entrepreneurs: The Tuna Seinermen of San Diego*. Los Angeles, CA: University of California Press.

Österblom, Henrik, U. Rashid Sumaila, Örjan Bodin, Jonas Hentati Sundberg, and Anthony J. Press. 2010. Adapting to Regional Enforcement: Fishing down the Governance Index. *PLo S One* 5 (9): e12832.

Österblom, Henrik, and Ussif Rashid Sumaila. 2011. Toothfish Crises, Actor Diversity and the Emergence of Compliance Mechanisms in the Southern Ocean. *Global Environmental Change* 21 (3): 972–982.

Ostrom, Elinor. 1990. *Governing the Commons: The Evolution of Institutions for Collective Action*. Cambridge: Cambridge University Press.

Ostrom, Elinor. 2002. Book Review of "Panarchy: Understanding Transformations in Human and Natural Systems." *Ecological Economics* 49(4): 488–491.

Ostrom, Elinor. 2007. A Diagnostic Approach for Going beyond Panaceas. *Proceedings of the National Academy of Sciences of the United States of America* 104 (39): 15181–15187.

Ostrom, Elinor. 2009. A General Framework for Analyzing Sustainability of Social-Ecological Systems. *Science* 325 (5939): 419–422.

Ostrom, Elinor, Marco A. Janssen, and John M. Anderies. 2007. Going beyond Panaceas. *Proceedings of the National Academy of Sciences of the United States of America* 104 (39): 15176–15178.

Ostrom, Elinor, Roy Gardner, and James Walker, eds. 1994. *Rules, Games, and Common-Pool Resources*. Ann Arbor, MI: The University of Michigan Press.

Pala, Christopher. 2013. Something's Fishy. *Earth Island Journal* 28 (3): 48.

Palais, Hyman. 1959. England's First Attempt to Break the Commercial Monopoly of the Hanseatic League, 1377–1380. *American Historical Review* 64 (4): 852–865.

Pálsson, Gísli, and Agnar Helgason. 1995. Figuring Fish and Measuring Men: The Individual Transferable Quota System in the Icelandic Cod Fishery. *Ocean and Coastal Management* 28 (1): 117–146.

Parkes, Basil A. 1966. Great Britain: The Future of Fish Harvesting. In *Fisheries of North America,* 28–33. Washington, DC: US Department of the Interior.

Parkes, Graeme. 2000. Precautionary Fisheries Management: The CCAMLR Approach. *Marine Policy* 24: 83–91.

Parrish, James D. 1975. Marine Trophic Interactions by Dynamic Simulation of Fish Species. *Fish Bulletin* 73 (4): 695–716.

Pascoe, Sean, and Andy Revill. 2004. Costs and Benefits of Bycatch Reduction Devices in European Brown Shrimp Trawl Fisheries. *Environmental and Resource Economics* 27 (1): 43–64.

Patterson, K. R., J. Zuzunaga, and G. Cárdenas. 1992. Size of the South American Sardine (*Sardinops sagax*) Population in the Northern Part of the Peru Upwelling Ecosystem after Collapse of Anchoveta (*Engraulis ringens*) Stocks. *Canadian Journal of Fisheries and Aquatic Sciences* 49 (9): 1762–1769.

Paulik, Gerald J. 1971. Anchovies, Birds and Fishermen in the Peru Current. In *Environment, Resources, Pollution and Society*, edited by W. W. Murdock, 156–185. Stamford, CT: Sinauer.

Pauly, Daniel, Villy Christensen, Johanne Dalsgaard, Rainer Froese, and Francisco Torres. 1998. Fishing down Marine Food Webs. *Science* 279 (5352): 860–863.

Payne, John W., James R. Bettman, and Eric J. Johnson. 1993. *The Adaptive Decision Maker*. Cambridge: Cambridge University Press.

Pearl, Raymond, and Lowell J. Reed. 1920. On the Rate of Growth of the Population of the United States Since 1790 and Its Mathematical Representation. *Proceedings of the National Academy of Sciences of the United States of America* 6 (6): 275–288.

Pearson, John C., ed. 1972. *The Fish and Fisheries of Colonial North America: A Documentary History of the Fishery Resources of the United States and Canada, Part II: The New England States*. Washington, DC: US Department of Commerce.

Peel, Ellen, Russell Nelson, and C. Phillip Goodyear. 2003. Managing Atlantic Marlin as Bycatch under ICCAT. The Fork in the Road: Recovery or Collapse. *Marine & Freshwater Research* 54 (4): 575–584.

Perkins, Richard, and Eric Neumayer. 2007. Implementing Multilateral Environmental Agreements: An Analysis of EU Directives. *Global Environmental Politics* 7 (3): 13–41.

Perrin, William F. 2009. Early Days of the Tuna/Dolphin Problem. *Aquatic Mammals* 35 (2): 293–306.

PETA. 2014. Save The Sea Kittens. *People for the Ethical Treatment of Animals*. Accessed January 18, 2014. http://features.peta.org/PETASea Kittens.

Peterson, M. J. 1995. International Fisheries Management. In *Institutions for the Earth: Sources of Effective International Environmental Protection*. 3rd ed., edited by Peter M. Haas, Robert O. Koehane and Marc A. Levy, 249–308. Cambridge, MA: MIT Press.

Phelps, Samuel. 1818. *A Treatise on the Importance of Extending the British Fisheries*. London: W. Simpkin and R. Marshall.

Pietrowski, Brittney N., Reza Tahergorabi, Kristen E. Matak, Janet C. Tou, and Jacek Jaczynski. 2011. Chemical Properties of Surimi Seafood Nutrified with Ω-3 Rich Oils. *Food Chemistry* 129 (3): 912–919.

Pikitch, Ellen K. 1992. Objectives for Biologically and Technically Interrelated Fisheries. In *Fishery Science and Management: Objectives and Limitations*, edited by Warren S. Wooster, 107–136. London: Springer.

Pikitch, E. K., C. Santora, E. A. Babcock, A. Bakun, R. Bonfil, D. O. Conover, P. Dayton, 2004. Towards Ecosystem-Based Fisheries Management. *Science* 305 (5682): 346–347.

Platteau, Jean-Philippe. 1992. Small-Scale Fisheries and the Evolutionists Theory of Institutional Development. In *Fishing for Development: Small Scale Fisheries in Africa*, edited by Inge Tvedten and Bjørn Hersoug, 91–114. Uppsala, Sweden: Nordiska Afrikainstitutet.

Policansky, David. 2008. Trends and Developments in Catch and Release. In *Global Challenges in Recreational Fisheries*, edited by Øystein Aas 202–236. Oxford: Blackwell.

Potter, E. C. E., J. C. Maclean, R. J. Wyatt, and R. N. B. Campbell. 2003. Managing the Exploitation of Migratory Salmonids. *Fisheries Research* 62 (2): 127–142.

Prince, Eric D., Mauricio Ortiz, and Arietta Venizelos. 2002. A Comparison of Circle Hook and 'J' Hook Performance in Recreational Catch-and-Release Fisheries for Billfish. *American Fisheries Society Symposium* 14: 66–79.

Prochaska, Fred J. 1984. Principal Types of Uncertainty in Seafood Processing and Marketing. *Marine Resource Economics* 1 (1): 51–66.

Putnam, Robert D. 1988. Diplomacy and Domestic Politics: The Logic of Two-Level Games. *International Organization* 42 (3): 427–460.

Putnam, Robert D. 1993. *Making Democracy Work: Civic Traditions in Modern Italy*. Princeton, NJ: Princeton University Press.

Rabanal, Herminio R. 1988. *History of Aquaculture*. ASEAN/UNDP/FAO Regional Small-Scale Coastal Fisheries Development Project, RAS/84/016. Manilla: United Nations Development Program.

Radcliffe, William. 1921. *Fishing from the Earliest Times*. London: John Murray.

Ragen, Timothy J., Henry P. Huntington, and Grete K. Hovelsrud. 2008. Conservation of Arctic Marine Mammals Faced with Climate Change. *Ecological Applications* 18 (2): 166–174.

Ramsay, E. P. 1884. Notes on the Food Fisheries and Edible Mollusca of New South Whales. In *International Fisheries Exhibition*, vol. 5, 303–552. London: William Clowes and Sons.

Randhawa, Sabah U. 1995. Theory and Methodology: A Decision Aid for Coordinating Fishing and Fish Processing. *European Journal of Operational Research* 81: 62–75.

Raymarine. 2013. "Explanation of Raymarine Clear Pulse CHIRP Technology." Accessed February 20, 2013. http://www.raymarine.com/view/?id=3178.

Ricardo, David. (1821) 2004. *The Principles of Political Economy and Taxation*, 2nd ed. Mineola, NY: Dover.

Richards, Andrew H. 1994. Problems of Drift-Net Fisheries in the South Pacific. *Marine Pollution Bulletin* 29 (1–3): 106–111.

Riddle, Kevin W. 2006. Illegal, Unreported, and Unregulated Fishing: Is International Cooperation Contagious? *Ocean Development and International Law* 37 (3–4): 265–297.

Riddle, Mary Ellen. 2003. "Rising Fuel Prices Keep Some Fishing Vessels in Port." *Tribune Business News*, February 28.

Riffenburgh, Robert H. 1969. A Stochastic Model of Interpopulation Dynamics in Marine Ecology. *Journal of the Fisheries Research Board of Canada* 26 (11): 2843–2880.

Rijnsdorp, Adriaan D., Jan Jaap Poos, Floor J. Quirijns, Reinier Hille Ris Lambers, Jan W. De Wilde, and Willem M. Den Heijer. 2008. The Arms Race between Fishers. *Journal of Sea Research* 60 (1–2): 126–138.

Robinson, Robb. 1996. *Trawling: The Rise and Fall of the British Trawl Fishery*. Exeter, UK: University of Exeter Press.

Rogers, George W. 1979. Alaska's Limited Entry Program: Another View. *Journal of the Fisheries Research Board of Canada* 36 (7): 783–788.

Roland, Alex. 2007. Containers and Causality. *Technology and Culture* 48 (2): 386–392.

Royce, William F. 1987. *Fishery Development*. New York: Academic Press Inc.

Rudd, Murray A. 2004. An Institutional Framework for Designing and Monitoring Ecosystem-Based Fisheries Management Policy Experiments. *Ecological Economics* 48 (1): 109–124.

Russell, E. S. 1931. Some Theoretical Considerations on the 'Overfishing' Problem. *ICES Journal of Marine Science* 6 (1): 3–20.

Sabatier, Paul A. 1987. Knowledge, Policy-Oriented Learning, and Policy Change: An Advocacy Coalition Framework. *Knowledge* 8 (4): 649–692.

Sakaguchi, Isao. 2013. The Roles of Activist NGOs in the Development and Transformation of the IWC Regime: The Interaction of Norms and Power. *Journal of Environmental Studies and Sciences* 3 (2): 194–208.

Sampson, Henry. 1875. *A History of Advertising from the Earliest Times*. London: Chatto and Windus.

Sandweiss, Daniel H., Kirk A. Maasch, Fei Chai, C. Fred T. Andrus, and Elizabeth J. Reitz. 2004. Geoarchaeological Evidence for Multidecadal Natural Climatic Variability and Ancient Peruvian Fisheries. *Quaternary Research* 61 (3): 330–334.

Schaeffer, Milner B. 1957. Some Considerations of Population Dynamics and Economics in Relation to the Management of Commercial Fisheries. *Journal of the Fisheries Research Board of Canada* 14 (5): 669–681.

Scheiber, Harry N. 1989. Origins of the Abstention Doctrine in Ocean Law: Japanese-US Relations and the Pacific Fisheries 1937–1958. *Ecology Law Quarterly* 16 (23): 23–99.

Scheiber, Harry N., Kathryn J. Mengerink, and Song Yann-huei. 2008. Ocean Tuna Fisheries, East Asian Rivalries, and International Regulation: Japanese Policies and the Overcapacity/IUU Fishing Conundrum. *University of Hawaii Law Review* 30: 97–165.

Schindler, Julia. 2012. Rethinking the Tragedy of the Commons: The Integration of Socio-Psychological Dispositions. *Journal of Artificial Societies and Social Simulation* 15 (1): 1–15.

Schlag, Anne Katrin, and Kaja Ystgaard. 2013. Europeans and Aquaculture: Perceived Differences between Wild and Farmed Fish. *British Food Journal* 115 (2): 209–222.

Schlager, Edella. 1994. Fisher's Institutional Responses to Common-Pool Resource Dilemmas. In *Rules, Games, and Common-Pool Resources*, edited by Elinor Ostrom, Roy Gardner and James Walker, 247–265. Ann Arbor, MI: University of Michigan Press.

Schreiber, Milena Arias, and Andrew Halliday. 2013. Uncommon among the Commons? Disentangling the Sustainability of the Peruvian Anchovy Fishery. *Ecology and Society* 18 (2): 53–64.

Schrijver, Nico. 1997. *Sovereignty over Natural Resources: Balancing Rights and Duties*. Cambridge: Cambridge University Press.

Schumpeter, Joseph A. (1942) 1976. *Capitalism, Socialism and Democracy*, 3rd ed. New York: Harper Torchbooks.

Schwartz, Barry, Andrew Ward, John Monterosso, Sonja Lyubomirsky, Katherine White, and Darrin R. Lehman. 2002. Maximizing versus Satisficing: Happiness Is a Matter of Choice. *Journal of Personality and Social Psychology* 83 (5): 1178–1197.

Sea Star. 2013. "Sea Star Fisheries Information Service." Accessed February 21, 2013. http://www.kaisho-japan.com.

Sea View Fishing. 2013. "Sea View Fishing Services: Find Fish From Space." Accessed February 11, 2013. http://www.seaviewfishing.com.

Sen, Amartya. 1999. *Development as Freedom*. New York: Knopf.

Sen, D. P. 2005. *Advances in Fish Processing Technology*. New Delhi: Allied.

Seppelt, Ralf, Brian Fath, Benjamin Burkhard, Judy L. Fisher, Adrienne Grêt-Regamey, Sven Lautenbach, Petina Pert, Stefan Hotes, Joachim Spangenberg, Peter H. Verburg, and Alexander P. E. Van Oudenhoven. 2012. Form Follows Function? Proposing a Blueprint for Ecosystem Service Assessments Based on Reviews and Case Studies. *Ecological Indicators* 21: 145–154.

Sethi, S. N., J. K. Sundaray, A. Panigrahi, and Subhash Chand. 2011. Prediction and Management of Natural Disasters through Indigenous Technical Knowledge, with Special Reference to Fisheries. *Indian Journal of Traditional Knowledge* 10 (1): 167–172.

Shackeroff, Janna M., Lisa M. Campbell, and Larry B. Crowder. 2011. Social-Ecological Guilds: Putting People into Marine Historical Ecology. *Ecology and Society* 16 (1): 52.

Shaw-Lefevre, George. 1884. Principles of Fishery Legislation. In *International Fisheries Exhibition*, vol. 4, 83–114. London: William Clowes and Sons.

Shea, Amerose. 1884. Newfoundland: Its Fisheries and General Resources. In *International Fisheries Exhibition*, vol. 5, 225–252. London: William Clowes and Sons.

Sihley, Angus. 2008. Hayek, Novak & the Limits of Laissez Faire: The Cult of Capitalism. *Commonweal* (April 25): 18–21.

Simon, Herbert A. 1955. A Behavioral Model of Rational Choice. *Quarterly Journal of Economics* 69 (1): 99–118.

Simon, Herbert A. 1995. Rationality in Political Behavior. *Political Psychology* 16 (1): 45–61.

Simon, Herbert A. 1997. *An Empirically Based Microeconomics*. Cambridge, UK: Cambridge University Press.

Simrad. 2013. "Fish Finding Products—Simrad." *Simrad: Technology for Sustainable Fisheries*. Accessed February 20, 2013. http://www.simrad.com/www/01/NOKBG0237.nsf/All Web/8100201BB4906A03C12570D1004641EC?Open Document.

Skewgar, Elizabeth P., Dee Boersma, Graham Harris, and Guillermo Caille. 2005. Anchovy Fishery Threat to Patagonian Ecosystem. *Science* 315 (5808): 45.

Skinner, E. Benjamin. 2012. Fishing as Slaves on the High Seas. *Bloomberg Businessweek*, February 20.

Skud, B. E. 1973. Management of the Pacific Halibut Fishery. *Journal of the Fisheries Research Board of Canada* 30 (12): 2393–2398.

Slocombe, D. Scott. 1998. Lessons from Experience with Ecosystem-Based Management. *Landscape and Urban Planning* 40 (1–3): 31–39.

Smith, A. (1776) 1976. *An Inquiry into the Nature and Causes of the Wealth of Nations*. Edited by Edwin Cannan. Chicago: University of Chicago Press.

Smith, Martin D., and James E. Wilen. 2005. Heterogeneous and Correlated Risk Preferences in Commercial Fishermen: The Perfect Storm Dilemma. *Journal of Risk and Uncertainty* 31 (1): 53–71.

Smith, Tim D. 1994. *Scaling Fisheries: The Science of Measuring the Effects of Fishing, 1855–1955*. Cambridge: Cambridge University Press.

Smitt, F. A. 1884. The Swedish Fisheries. In *International Fisheries Exhibition*, vol. 5, 253–270. London: William Clowes and Sons.

Solá, Francisco-Garcia. 1884. The Fisheries of Spain. In *International Fisheries Exhibition*, vol. 5, 353–364. London: William Clowes and Sons.

Solomon, Susan Dahe Qin, Martin Manning, Melinda Marquis, Kristen Averyt, Melida M. B. Tignor, Henry LeRoy Miller, Jr., and, Zhenlin Chen, eds. 2007. *Climate Change 2007: The Physical Science Basis. Contribution of Working Group I to the Fourth Assessment Report of the Intergovernmental Panel on Climate Change*. Cambridge: Cambridge University Press.

Space Fish. 2013. "Space Fish LLP." Accessed February 21, 2013. http://www.spacefish.co.jp/index.html.

Sparholt, Henrik, and Robin M. Cook. 2010. Sustainable Exploitation of Temperate Fish Stocks. *Biology Letters* 6 (1): 124–127.

Sprout, Harold, and Margaret Sprout. 1979. *The Ecological Perspective on Human Affairs: With Special Reference to International Politics*. Westport, CT: Greenwood.

Squires, Dale. 1987. Fishing Effort: Its Testing, Specification, and Internal Structure in Fisheries Economics and Management. *Journal of Environmental Economics and Management* 14: 268–282.

Squires, Dale. 1988. Production Technology, Costs, and Multiproduct Industry Structure: An Application of the Long-Run Profit Function to the New England Fishing Industry. *Canadian Journal of Economics* 21 (2): 359–378.

Squires, Dale, Yongil Jeon, R. Quentin Grafton, and James Kirkley. 2010. Controlling Excess Capacity in Common-Pool Resource Industries: The Transition from Input to Output Controls. *Australian Journal of Agricultural and Resource Economics* 54 (3): 361–377.

Squires, Dale, and Niels Vestergaard. 2009. Technical Change and the Commons. *Review of Economics and Statistics* 95 (5): 1769–1787.

Starbuck, Alexander. 1878. *History of the American Whale Fishery from Its Earliest Inception to the Year 1876*. New York: Castle Books.

Stergiou, Konstantinos I., and Athanassios C. Tsikliras. 2011. Fishing Down, Fishing through and Fishing up: Fundamental Process versus Technical Details. *Marine Ecology Progress Series* 441: 295–301.

Stoett, Peter. 2011. Irreconcilable Differences: The International Whaling Commission and Cetacean Futures. *Review of Policy Research* 28 (6): 631–634.

Stokke, Olav Schram. 2009. Trade Measures and the Combat of IUU Fishing: Institutional Interplay and Effective Governance in the Northeast Atlantic. *Marine Policy* 33 (2): 339–349.

Su, Yi-Cheng, and Chengchu Liu. 2007. Vibrio Parahaemolyticus: A Concern of Seafood Safety. *Food Microbiology* 24 (6): 549–558.

Suisankyoku, Noshomusho. 1915. *Japan Special Catalogue: Fisheries*. Tokyo: Imperial Fisheries Bureau.

Sumaila, U. R., J. Alder, and H. Keith. 2006. Global Scope and Economics of Illegal Fishing. *Marine Policy* 30 (6): 696–703.

Sumaila, U., A. S. Khan, A. J. Dyck, R. Watson, G. Munro, P. Tydemers, and D. Pauly. 2010a. *Subsidizing Global Fisheries: A Summary of a New Scientific Analysis*. Washington. DC: The PEW Environmental Group.

Sumaila, U. R. Rashid, Ahmed S. Khan, Andrew J. Dyck, Reg R. Watson, Gordon G. Munro, Peter P. Tydemers, and Daniel D. Pauly. 2010b. A Bottom-up Re-Estimation of Global Fisheries Subsidies. *Journal of Bioeconomics* 12 (3): 201–225.

Sumaila, U. Rashid, and Daniel Pauly. 2011. The "March of Folly" in Global Fisheries. In *Shifting Baselines: The Past and the Future of Ocean Fisheries*, edited by Jeremy B. C. Jackson, Karen E. Alexander and Enric Sala, 21–32. Washington, DC: Island.

Sunstein, Cass R. 2006. The Availability Heuristic, Intuitive Cost-Benefit Analysis, and Climate Change. *Climatic Change* 77 (1–2): 195–210.

Sutinen, Jon G. 1993. Recreational and Commercial Fisheries Allocation with Costly Enforcement. *American Journal of Agricultural Economics* 75 (5): 1183–1187.

Sverrisson, Árni. 2002. Small Boats and Large Ships: Social Continuity and Technological Change in Icelandic Fisheries, 1800–1960. *Technology and Culture* 43 (2): 227–253.

Sydnes, Are K. 2001. New Regional Fisheries Management Regimes: Establishing the South East Atlantic Fisheries Organisation. *Marine Policy* 25 (5): 353–364.

Sydnes, Are K. 2002. Regional Fishery Organisations in Developing Regions: Adapting to Changes in International Fisheries Law. *Marine Policy* 26 (5): 373–381.

Takei, Yoshinobu. 2013. *Filling Regulatory Gaps in High Seas Fisheries: Discrete High Seas Fish Stocks, Deep-Sea Fisheries, and Vulnerable Marine Ecosystems*. Leiden, Netherlands: Martinus Nijhoff.

Talfourd, F. J Chater. 1884. The Relations of the State with Fishermen and Fisheries Including All Matters Dealing with Their Protection and Regulation. In *International Fisheries Exhibition*, vol. 9., 216–295. London: William Clowes and Sons.

Taylor, Graham D. 1995. The Collapse of the Northern Cod Fishery: A Historical Perspective. *Dalhousie Law Journal* 18 (1): 13–22.

Taylor, Steve L. 1986. Histamine Food Poisoning: Toxicology and Clinical Aspects. *Critical Reviews in Toxicology* 17 (2): 91–128.

Taylor, Steve L., and Julie A. Nordlee. 1993. Chemical Additives in Seafood Products. *Clinical Reviews in Allergy* 11 (2): 261–291.

Teuteberg, Hans Jurgen. 2009. The Industrialization of Seafood: German Deep-Sea Fishing and the Sale and Preservation of Fish, 1885–1930. In *Exploring the Food Chain: Food Production and Food Processing in Western Europe 1850–1990,* edited by Yves Segers, Jan Bieleman and Erik Buyst, 191–212. Turnhout, Belgium: Brepols.

The Tuna Scare. 1963. *Time Magazine* April 26: 101.

Thompson, William F. 1919. *The Scientific Investigation of Marine Fisheries, As Related to the Work of the Fish and Game Commission in Southern California.* Fish Bulletin 2. Sacramento, CA: California State Printing Office.

Thompson, William F., and F. Heward Bell. 1934. *Biological Statistics of the Pacific Halibut Fishery.* Report of the International Fisheries Commission 8. Seattle: International Fisheries Commission.

Thompson, William F., and Norman L. Freeman. 1930. *History of the Pacific Halibut Fishery.* Report of the International Fisheries Commission 5. Vancouver: Wrigley.

Thomson, Janice E. 1994. *Mercenaries, Pirates, and Sovereigns: State Building and Extraterritorial Violence in Early Modern Europe.* Princeton, NJ: Princeton University Press.

Tisdell, Clem, and Clevo Wilson. 2002. Ecotourism for the Survival of Sea Turtles and Other Wildlife. *Biodiversity and Conservation* 11 (9): 1521–1538.

Torell, Magnus. 2009. Some Institutional Implications of an Ecosystems Approach to Capture Fisheries Management. *Aquatic Ecosystem Health & Management* 12 (4): 440–443.

Torres-Irineo, Edgar, Daniel Gaertner, Alicia Delgado de Molina, and Javier Ariz. 2011. Effects of Time-Area Closure on Tropical Tuna Purse-Seine Fleet Dynamics through Some Fishery Indicators. *Aquatic Living Resources* 24 (4): 337–350.

Townsend, Ralph E. 1985. On "Capital-Stuffing" in Regulated Fisheries. *Land Economics* 61 (2): 195–197.

TRAFFIC. 2014. "Latest News from TRAFFIC." Accessed March 2, 2014. http://www.traffic.org/.

Troell, M., C. Halling, A. Neori, T. Chopin, A. H. Buschmann, N. Kautsky, and C. Yarish. 2003. Integrated Mariculture: Asking the Right Questions. *Aquaculture* 226: 69–90.

Tuna Back in Favor. 1963. *Time Magazine* November 29: 98.

Twigger, Robert. 1999. *Inflation: The Value of the Pound 1750–1998*. Research Paper 99/20. London: House of Commons Library.

UCS. 2014. "Independent Science, Practical Solutions." *Union of Concerned Scientists* Accessed March 2, 2014. http://www.ucsusa.org.

Unger, Richard W. 1978. The Netherlands Herring Fishery in the Late Middle Ages: The False Legend of Willem Beukels of Biervliet. *Viator* 9 (1): 335–356.

United Nations. 2010. Assumptions Underlying the 2010 Revision Fertility Assumptions. Rome: United Nations.

United Nations. 2013a. "Law of the Sea." Bulletin 27. Accessed March 2, 2013. http://www.un.org/Depts/los/convention_agreements/convention_overview_fish_stocks.htm.

United Nations. 2013b. "World Population Database." Accessed May 14, 2013. http://esa.un.org/wpp/unpp/panel_population.htm.

United Nations. 2014. "Chronological Lists of Ratifications of the UN Law of the Sea and Amendments." Accessed March 2, 2013. http://www.un.org/depts/los/reference_files/chronological_lists_of_ratifications.htm#Agreement for the implementation of the provisions of the Convention relating to the conservation and management of straddling fish stocks and highly migratory fish stocks.

UNEP. 1992. "Rio Declaration on Environment and Development." *United Nations Environment Programme*. Accessed January 28, 2014. http://www.unep.org/Documents. Multilingual/Default.asp?documentid=78&articleid=1163.

US Census Bureau. 2013a. "Historical Estimates of World Population." Accessed December 28, 2012. http://www.census.gov/population/international/data/worldpop/table_history.php.

US Census Bureau. 2013b. "International Data Base." Accessed May 14, 2013. http://www.census.gov/population/international/data/idb/information Gateway.php.

US Coast Guard. 2013. "Coast Guard History." *US Department of Homeland Security*. Accessed July 30, 2013. http://www.uscg.mil/history/faqs/when.asp.

US Congress. 1930. Second Deficiency Appropriation Bill for 1930: Hearings Before the Subcommittee of the Committee on Appropriations, United States Senate, 71st Congress, Second Session, on H. R. 12902 A Bill Making Appropriations to Supply Deficiencies in Certain Appropriations.

US Congress. 1932. Northern Pacific Halibut Fishery: Hearings Before The Committee on Merchant Marine, Radio, and Fisheries House of Representatives, 72nd Congress, First Session on H. R. 8084, A Bill for the Protection of the Northern Pacific Halibut Fishery.

US Congress. 1940. Replacement of Fishing Vessels Requisitioned by the Government: Hearings Before the Committee on Merchant Marine and Fisheries, House of Representatives, 76th Congress, Third Session, on H. R 10501, A Bill to Amend Section 509, As Amended, of the Merchant Marine Act, 1936.

US Congress. 1944. Extension of Lend-Lease Act: Hearings Before the Committee on Foreign Affairs, House of Representatives, 78th Congress, Second Session, on H. R. 4254, a Bill to Extend for One Year the Provisions of an Act to Promote the Defense of the United States.

US Congress. 1945. Lend-Lease: Hearings Before the Committee on Foreign Relations, United States Senate, 79th Congress, First Session, on H. R. 2013 An Act to Extend for One Year The Provisions of an Act to Promote the Defense of the United States.

US Congress. 1947a. Fish and Shellfish Problems: Hearings Before the Committee on Merchant Marine and Fisheries, House of Representatives, 80th Congress, First Session.

US Congress. 1947b. Operation of the Vessel "Pacific Explorer": Hearings Before the Subcommittee on Salt Water Fish and Shellfish Problems of the Committee on Merchant Marine and Fisheries, House of Representatives, 80th Congress, First Session.

US Congress. 1954. To Protect Rights of United States Vessels on the High Seas and in Territorial Waters of Foreign Countries: Hearing Before the Subcommittee on Interstate and Foreign Commerce, United States Senate, 83rd Congress, Second Session, on S. 3594, A Bill.

US Congress. 1958. New England Fisheries Subsidies: Hearings Before the Committee on Merchant Marine and Fisheries, House of Representatives, 85th Congress, Second Session, on H. R. 10529, to Provide a 5-Year Program of Assistance to Enable Depressed Segments of the Fishing Industry in the United States to Regain a Favorable Economic Status, and for Other Purposes.

US Congress. 1962. Conservation of Tropical Tuna: Hearings Before the Merchant Marine and Fisheries Subcommittee of the Committee on Commerce, United States Senate, 87th Congress, Second Session, on S. 2568, A Bill to Amend the Act of September 7, 1950, to Extend the Regulatory Authority of the Federal and State Agencies Concerned under the Terms of the Convention for the Establishment of an Inter-American Tropical Tuna Commission.

US Congress. 1963. Fishing Vessel Subsidies Part 2: Hearing before the Subcommittee on Fisheries and Wildlife Conservation of the Committee on Merchant Marine and Fisheries, House of Representatives, 88th Congress, First Session, on H. R. 2172, H. R. 2643, S. 1006, Bills to Amend the Act of June 12, 1960, for the Correction of Inequities in the Construction of Fishing Vessels, and for Other Purposes.

US Congress. 1967. Fisheries Convention, Maritime Conventions, Customs Conventions: Hearing Held before the Ad Hoc Subcommittee on Customs, Conventions and Maritime Matters of the Committee on Foreign Relations, vol. 1, United States Senate, 9th Congress, First Session.

US Congress. 1992. New England Groundfish Restoration. Hearing before the National Ocean Policy Study of the Committee on Commerce, Science, and Transportation, United States Senate, 102nd Congress, Second Session.

US Senate. 1840. Reports of the Majority and Minority of the Select Committee on the Origin and Character of Fishing Bounties and Allowances. S. Doc. 26–368. Washington DC: Blair and Rivers Printers.

Vedung, Evert, Ray C. Rist, and Marie-Louise Bemelmans-Videc, eds. 1998. *Carrots, Sticks and Sermons: Policy Instruments and Their Evaluation*. New Brunswick, NJ: *Trans-Action*.

Volpe, John P. 2005. Dollars without Sense: The Bait for Big-Money Tuna Ranching around the World. *Bioscience* 55 (4): 301–302.

Vintage ad. 1946. "Chicken of the Sea & White Star Tuna." *Vintage Ad Broswer*. Accessed May 3, 2013. http://www.vintageadbrowser.com/search?q=tuna.

Wallem, Fredrik M. 1884. Notes on the Fish Supply of Norway. In *International Fisheries Exhibition*, vol. 5, 271–302. London: William Clowes and Sons.

Walpole, Spencer. 1884a. The British Fish Trade. In *International Fisheries Exhibition*, vol. 1, 1–72. London: William Clowes and Sons.

Walpole, Spencer. 1884b. Fish Transport and Fish Markets. In *International Fisheries Exhbition*, vol. 4, 115–148. London: William Clowes and Sons.

Walsh, Virginia M. 2004. *Global Institutions and Social Knowledge: Generating Research at the Scripps Institution and the Inter-American Tropical Tuna Commission, 1900s–1990s*. Cambridge, MA: MIT Press.

Walters, Carl J. 1975. Optimal Harvest Strategies for Salmon in Relation to Environmental Variability and Uncertain Production Parameters. *Journal of the Fisheries Research Board of Canada* 32 (10): 1777–1784.

Walters, Carl J. 1984. Managing Fisheries under Biological Uncertainty. In *Exploitation in Marine Communities*, edited by R. M. May, 263–274. Berlin: Springer.

Walters, Carl J. 1986. *Adaptive Management of Renewable Resources*. Caldwell, NJ: Blackburn Press.

Walters, Carl J., and Ray Hilborn. 1976. Adaptive Control of Fishing Systems. *Journal of the Fisheries Research Board of Canada* 33 (1): 145–149.

Walters, Carl J., and Ray Hilborn. 1978. Ecological Optimization and Adaptive Management. *Annual Review of Ecology and Systematics* 8: 157–188.

Ward, J. M., J. E. Kirkley, R. Metzner, and S. Pascoe. 2004. *Measuring and Assessing Capacity in Fisheries: Basic Concepts and Management Options*. FAO Technical Paper 433. Rome: Food and Agriculture Organization of the United Nations.

Ward, Peter, and Sheree Hindmarsh. 2007. An Overview of Historical Changes in the Fishing Gear and Practices of Pelagic Longliners, with Particular Reference to Japan's Pacific Fleet. *Reviews in Fish Biology and Fisheries* 17 (4): 501–516.

Warner, William W. 1997. The Fish Killers. In *How Deep Is the Ocean? Historical Essays on Canada's Atlantic Fishery*, edited by James E. Candow and Carol Corbin, 223–242. Sydney: Cape Breton University.

Weber, J. Todd, Richard G. Hibbs, Jr., Ahmed Darwish, Mishu Ban, Andrew L. Corwin, Magda Rakha, Charles L. Hatheway, Said El Sharkawy, Sobhi Abd El-Rahim, Mohammed Fathi Shiba Al-Hamd, James E. Sarn, Paul A. Blake, and Robert V. Tauxe. 1993. A Massive Outbreak of Type E Botulism Associated with Traditional Salted Fish in Cairo. *Journal of Infectious Diseases* 167 (2): 451–454.

Webster, D. G. 2015. The Action Cycle/Structural Context Framework: A Fisheries Application. *Ecology* and *Society* 20 (1): art. 33.

Webster, D. G. 2009. *Adaptive Governance: The Dynamics of Atlantic Fisheries Management*. Cambridge, MA: MIT Press.

Webster, D. G. 2011. The Irony and the Exclusivity of Atlantic Bluefin Tuna Management. *Marine Policy* 35 (2): 249–251.

Webster, D. G. 2013. International Fisheries: Gauging the Potential for Multispecies Management. *Journal of Environmental Studies and Sciences* 3 (2): 169–183.

Wendt, Alexander. 1992. Anarchy Is What States Make of It: The Social Construction of Power Politics. *International Organization* 46 (2): 391–425.

White, Crow, Bruce E. Kendall, Steven Gaines, David A. Siegel, and Christopher Costello. 2008. Marine Reserve Effects on Fishery Profit. *Ecology Letters* 11 (4): 370–379.

Wick, Carl Irving. 1946. *Ocean Harvest: The Story of Commercial Fishing in Pacific Coast Waters*. Seattle: Superior.

Wilen, James E. 1988. Limited Entry Licensing: A Retrospective Assessment. *Marine Resource Economics* 5 (4): 313–324.

Wilen, James E., and Frances R. Homans. 1998. What Do Regulators Do? Dynamic Behavior of Resource Managers in the North Pacific Halibut Fishery 1935–1978. *Ecological Economics* 24 (2–3): 289–298.

Williamson, Samuel H. 2014. "Seven Ways to Compute the Relative Value of the US Dollar Amount, 1774 to Present." *Measuring Worth.* http://www.measuringworth.com/uscompare.

Wilson, J. 1982. The Economical Management of Multispecies Fisheries. *Land Economics Land Economics* 58 (4): 417–434.

Wilson, James A., Liying Yan, and Carl Wilson. 2007. The precursors of governance in the Maine lobster fishery. *Proceedings of the National Academy of Sciences of the United States of America* 104 (39): 15212–15217.

Windsor, M. L., and P. Hutchinson. 1990. International Management of Atlantic Salmon—the Role of NASCO. *Fisheries Research* 10 (1–2): 5–14.

Winston, Andrew S. 2014. *The Big Pivot: Radically Practical Strategies for a Hotter, Scarcer, and More Open World.* Cambridge, MA: Harvard Business Review Press.

Wirsenius, Stefan, Christian Azar, and Göran Berndes. 2010. How much land is needed for global food production under scenarios of dietary changes and livestock productivity increases in 2030? *Agricultural Systems* 9: 621–638.

Wong, Chu-Kwan, Patricia Hung, Kellie L. H. Lee, and Kai-Man Kam. 2005. Study of an Outbreak of Ciguatera Fish Poisoning in Hong Kong. *Toxicon* 46 (5): 563–571.

World Bank. 1996. *Guidelines for Integrated Coastal Zone Management.* Washington, DC: World Bank.

World Bank. 2012. "World Development Indicators Databank." Accessed December 28, 2012. *World Bank.* http://databank.worldbank.org/ddp/home.do?Step=3&id=4.

World Bank and FAO. 2009. *The Sunken Billions: The Economic Justification of Fisheries Reform.* Joint report from the World Bank and the Food and Agriculture Organization of the United Nations. Washington, DC: World Bank.

World Bank Group. 2014. *Global Economic Propsects: Shifting Priorities, Building for the Future,* vol. 9. Washington, D. C: The World Bank.

Wright, Andrew, and David J. Doulman. 1991. Drift-Net Fishing in the South Pacific: from Controversy to Management. *Marine Policy* 15 (5): 303–329.

Wright, Brian G. 2007. Environmental NGOs and the Dolphin-Tuna Case. *Environmental Politics* 9 (4): 82–103.

WTO. 1994. Agreement on Subsidies and Countervailing Measures ('SCM Agreement'). *World Trade Organization.* Accessed June 23, 2013. http://www.wto.org/english/tratop_e/scm_e/subs_e.htm.

WWF. 2001. "Fishery Subsidies by Country: Hard Facts, Hidden Problems—A Review of Current Data on Fishing Subsidies." *World Wildlife Fund.* Accessed March 2, 2014. http://www.nationmaster.com/graph/eco_fis_sub-economy-fishing-subsidies.

Ye, Yimin, Kevern Cochrane, and Yongsong Qiu. 2011. Using Ecological Indicators in the Context of an Ecosystem Approach to Fisheries for Data-Limited Fisheries. *Fisheries Research* 112 (3): 108–116.

Ye, Yimin, Kevern Cochrane, Gabriella Bianchi, Rolf Willmann, Jacek Majkowski, and Merete Tandstad. 2013. Rebuilding Global Fisheries: The World Summit Goal, Costs and Benefits. *Fish and Fisheries* 14 (2): 174–185.

Yen, Wei-Ching W. 1910. The Fisheries of China. *Bulletin of the Bureau of Fisheries* 38 (664): 369–373.

Young, Oran R. 1994. *International Governance: Protecting the Environment in a Stateless Society*. Ithaca, NY: Cornell University Press.

Young, Oran R., ed. 1999. *The Effectiveness of International Environmental Regimes: Causal Connections and Behavioral Mechanisms*. Cambridge, MA: MIT Press.

Young, Oran R. 2001. The Behavioral Effects of Environmental Regimes: Collective-Action vs Social-Practice Models. *International Environmental Agreement: Politics, Law and Economics* 1 (1): 9–29.

Young, Oran R. 2002. *The Institutional Dimensions of Environmental Change: Fit, Interplay, and Scale. Organization*. Cambridge, MA: MIT Press.

Young, Oran R. 2010. *Institutional Dynamics: Emergent Patterns in International Environmental Governance*. Cambridge, MA: MIT Press.

Young, Oran R., Frans Berkhout, Gilberto C. Gallopin, Marco A. Janssen, Elinor Ostrom, and Sandervan der Leeuw. 2006. The Globalization of Socio-Ecological Systems: An Agenda for Scientific Research. *Global Environmental Change* 16 (3): 304–316.

Young, Oran R., Eric F. Lambin, Frank Alcock, Helmut Haberl, Sylvia I. Karlsson, William J. Mcconnell, Tun Myint, Claudia Pahl-Wostl, Colin Polsky, P. S. Ramakrishnan, Heike Schroeder, Marie Scouvart, and Peter H. Verburg. 2006. A Portfolio Approach to Analyzing Complex Human-Environment Interactions: Institutions and Land Change. *Ecology and Society* 11 (2): 31.

Zahran, Sammy, Samuel D. Brody, Himanshu Grover, and Arnold Vedlitz. 2006. Climate Change Vulnerability and Policy Support. *Society & Natural Resources* 19 (9): 771–789.

Zhang, Yuying, Yunkai Li, and Yong Chen. 2012. Modeling the Dynamics of Ecosystem for the American Lobster in the Gulf of Maine. *Aquatic Ecology* 46 (4): 451–464.

Zimmern, Helen. 1889. *The Hansa Towns*, 3rd ed. London: T. Fisher Unwin.

Index

Politics, Science, and the Environment

Peter M. Haas and Sheila Jasanoff, editors

Peter Dauvergne, *Shadows in the Forest: Japan and the Politics of Timber in Southeast Asia*

Peter Cebon, Urs Dahinden, Huw Davies, Dieter M. Imboden, and Carlo C. Jaeger, eds., *Views from the Alps: Regional Perspectives on Climate Change*

Clark C. Gibson, Margaret A. McKean, and Elinor Ostrom, eds., *People and Forests: Communities, Institutions, and Governance*

The Social Learning Group, *Learning to Manage Global Environmental Risks. Volume 1: A Comparative History of Social Responses to Climate Change, Ozone Depletion, and Acid Rain. Volume 2: A Functional Analysis of Social Responses to Climate Change, Ozone Depletion, and Acid Rain*

Clark Miller and Paul N. Edwards, eds., *Changing the Atmosphere: Expert Knowledge and Environmental Governance*

Craig W. Thomas, *Bureaucratic Landscapes: Interagency Cooperation and the Preservation of Biodiversity*

Nives Dolsak and Elinor Ostrom, eds., *The Commons in the New Millennium: Challenges and Adaptation*

Kenneth E. Wilkening, *Acid Rain Science and Politics in Japan: A History of Knowledge and Action Toward Sustainability*

Virginia M. Walsh, *Global Institutions and Social Knowledge: Generating Research at the Scripps Institution and the Inter-American Tropical Tuna Commission, 1900s–1990s*

Sheila Jasanoff and Marybeth Long Martello, eds., *Earthly Politics: Local and Global in Environmental Governance*

Christopher Ansell and David Vogel, eds., *What's the Beef? The Contested Governance of European Food Safety*

Charlotte Epstein, *The Power of Words in International Relations: Birth of an Anti-Whaling Discourse*

Ann Campbell Keller, *Science in Environmental Politics: The Politics of Objective Advice*

Henrik Selin, *Global Governance of Hazardous Chemicals: Challenges of Multilevel Management*

Rolf Lidskog and Göran Sundqvist, eds., *Governing the Air: The Dynamics of Science, Policy, and-Citizen Interaction*

Andrew S. Mathews, *Instituting Nature: Authority, Expertise, and Power in Mexican Forests*

Eric Brousseau, Tom Dedeurwaerdere, and Bernd Siebenhüner, eds.,

Reflexive Governance and Global Public Goods

D. G. Webster, *Beyond the Tragedy in Global Fisheries*

Printed in the United States
by Baker & Taylor Publisher Services